3D AutoCAD 2006

One Step at a Time

Timothy Sean Sykes

Forager Publications
Spring, Texas

ISBN 0-9765888-3-8

Forager Publications

2043 Cherry Laurel

Spring, TX 77386

www.foragerpub.com

AUTHOR'S NOTES

Where to Find the Required Files (and Review Questions) for Using This Text:

All of the files required to complete the lessons in this text can be downloaded free of charge from this site:

http://www.uneedcad.com/2006/Files/

Select on the file called *PartI.zip* and save it to your computer. (Windows XP will open a zipped file. If you need a zip utility, however, I suggest the free WinZip download at: http://www.winzip.com/. The evaluation version will do to get your files. You'll have to follow WinZip procedures to unzip the files.)

Once you've downloaded the zip file, follow these instructions to unzip it with Windows XP.

1. Double-click on the file. Windows XP will display a window with the Step folder listed.
2. Right click on the Step folder and select **Copy** from the menu that appears.
3. Using Windows Explorer, navigate to the C-drive.
4. Select **Paste** from the Edit pull down menu.

To get the review questions, go to the website indicated at the end of each lesson. A PDF file will open in your Internet browser. These files are fully printable, or you can save them to your disk if you wish. The last page of each file contains the answers to the questions in that file.

Contacting the Author/Publisher:

Although we tried awfully hard to avoid errors, typos, and the occasional boo boos, I admit to complete fallibility. Should you find it necessary to let me know of my blunders, or to ask just about anything about the text – or even to make suggestions as to how to better the next edition, please feel free to contact me at: comments@foragerpub.com. I can't promise a fast response (although I'm usually pretty good about answering my email), but I can promise to read (almost) everything that comes my way.

Frequently Asked Questions (and responses to less-frequently-asked but more annoying questions):

I went to the web site but got a Page cannot be found *error. What gives?*

This is the most common complaint I get. The fix is simple: make sure you capitalize the "Files" in the address. The web is case-sensitive.

Contents

Chapter 1: "Z" Basics

1.1 The UCS Icon and the Right-Hand Rule 4
  1.1.1 The UCS Icon 4
  1.1.2 The Right-Hand Rule 8
1.2 Maneuvering Through Z-Space with the *VPoint* Command 8
  1.2.1 Using Coordinates to Assign a Vpoint 8
  1.2.2 Using the Compass to Assign a Vpoint 12
  1.2.3 Setting Viewpoints Using a Dialog Box 14
  1.2.4 The Camera Approach 17
1.3 Drawing with the Z-Axis 19
  1.3.1 Three-Dimensional Coordinate Entry 19
  1.3.2 Using the Thickness and Elevation System Variables 22
1.4 Three-Dimensional Viewing Made Easy 26
  1.4.1 The *Hide* Command 26
  1.4.2 The *Shademode* Command 28
1.5 Extra Steps 33
1.6 What Have We Learned? 33
1.7 Exercises 34

Chapter 2: More of Z Basics

2.1 WCS vs. UCS 43
2.2 The UCS Dialog Box (Manager) 51
2.3 Using Working Planes 53
2.4 Advanced Viewing Techniques 59
  2.4.1 3DOrbit 59
  2.4.2 A Continuous Three-Dimensional Orbit – *3DCOrbit* 68
2.5 Extra Steps 69
2.6 What Have We Learned? 69
2.7 Exercises 70

Chapter 3: Wireframes and Surface Models

3.1 *3DPoly* vs. *PLine* 78
3.2 Drawing in Three Directions at Once – Point Projection 78
3.3 Adding Surfaces – Regions, Solids, and 3D Faces 81
  3.3.1 Creating 3D Faces 81
  3.3.2 Invisible Edges in 3D Faces – SPLFrame and the *Edge* Command 84
  3.3.3 Solids and Regions 86
  3.3.4 Which Method Should You Use? 92
3.4 Extra Steps 93
3.5 What Have We Learned? 95
3.6 Exercises 95

Chapter 4: Predefined Surface Models

4.1 What Are Predefined Surface Models? 102
4.2 Drawing Predefined Surface Models 103
  4.2.1 Box 103
  4.2.2 Wedge 105
  4.2.3 Pyramid 106
  4.2.4 Cone 110
  4.2.5 Sphere 112
  4.2.6 Domes and Dishes 113
  4.2.7 Torus 115
4.3 Understanding the Limitations of Predefined Surface Models 116
4.4 Extra Steps 117
4.5 What Have We Learned? 117
4.6 Exercises 118

Chapter 5: Complex Surface Models

5.1 Controlling the Number of Surfaces – Surftab1 and Surftab2 125
5.2 Different Approaches for Different Goals 126
  5.2.1 Follow the Path – The *Tabsurf* Com'd 126
  5.2.2 Add a Surface Between Objects – The *Rulesurf* Command 129
  5.2.3 Creating Circular Surfaces – The *Revsurf* Command 131
  5.2.4 Using Edges to Define a Surface Plane – The *Edgesurf* Command 133
5.3 More Complex Surfaces 135
  5.3.1 Creating Meshes with the *3DMesh* Command 136
  5.3.2 Creating Meshes with the *PFace* Command 138
5.4 Extra Steps 142
5.5 What Have We Learned? 143
5.6 Exercises 144

Chapter 6: Z-Space Editing

6.1 Three-Dimensional Uses for Familiar (Two-Dimensional) Tools 151
  6.1.1 Trimming and Extending in Z-Space 151
  6.1.2 Aligning Three-Dimensional Objects 157
  6.1.3 Three-Dimensional Object Properties 158
  6.1.4 Modifying a 3D Mesh 161
6.2 Editing Tools Designed for Z-Space 170
  6.2.1 Rotating About an Axis – The *Rotate3d* Command 170
  6.2.2 Mirroring Three-Dimensional Objects – The *Mirror3d* Command 173

6.2.3 Arrayed Copies in Three Dimensions – The *3DArray* Command 177
6.3 Extra Steps 181
6.4 What Have We Learned? 181
6.5 Exercises 182

Chapter 7: Solid Modeling Building Blocks

7.1 What Are Solid Modeling Bldg Blocks? 187
7.2 Extruding 2D Regions and Solids 187
7.3 Drawing the Solid Modeling Bldg Blocks 191
7.3.1 Box 191
7.3.2 Wedge 193
7.3.3 Cones and Cylinders 195
7.3.4 Sphere 198
7.3.5 Torus 199
7.4 Creating More Complex Solids Using the *Revolve* Command 201
7.5 Extra Steps 204
7.6 What Have We Learned? 205
7.7 Exercises 205

Chapter 8: Composite Solids

8.1 Solid Construction Tools 212
8.1.1 Union 213
8.1.2 Subtract 214
8.1.3 Intersect 216
8.1.4 Slice 217
8.1.5 Interfere 222
8.2 Using Some Old Friends on Solids – *Fillet* and *Chamfer* 224
8.3 Creating Cross Sections the Easy Way – The *Section* Command 227
8.4 Extra Steps 229
8.5 What Have We Learned? 229
8.6 Exercises 230

Chapter 9: Editing 3D Solids

9.1 A Single Command, But It Does So Much - *SolidEdit* 238
9.2 Changing Faces – The Face Category 238
9.2.1 Changing the Thickness of a 3D Solid Face – the Extrude Option 239
9.2.2 Moving a Face on a 3D Solid 241
9.2.3 Rotating Faces on a 3D Solid 243
9.2.4 Offsetting Faces on a 3D Solid 245
9.2.5 Tapering Faces on a 3D Solid 246
9.2.6 Deleting 3D Solid Faces 248
9.2.7 Copying 3D Solid Faces as Regions or Bodies 249
9.2.8 Changing the Color of a Single Face 251
9.3 Modifying Edges – The Edge Category 252
9.4 Changing the Whole 3D Solid – The Body Category 254
9.4.1 Imprinting an Image onto a 3D Solid 255
9.4.2 Separating 3D Solids with the seParate Solids Routines 257
9.4.3 Clean 258
9.4.4 Shell 259
9.4.5 Checking to Be Certain You Have an ACIS Solid 260
9.5 Extra Steps 261
9.6 What Have We Learned? 261
9.7 Exercises 262

Chapter 10: Three-Dimensional Blocks and Three-Dimensional Plotting Tools

10.1 Using Blocks in Z-Space 270
10.1.1 Three-Dim. Blocks and the UCS 270
10.1.2 Inserting Three-Dimensional Blocks 271
10.1.3 Making Good Use of Attributes 275
10.2 Plotting a 3D Solid 277
10.2.1 Setting Up the Plot – the *Solview* Command 277
10.2.2 Creating the Plot Images – The *Soldraw* and *Solprof* Commands 283
10.3 Extra Steps 287
10.4 What Have We Learned? 287
10.5 Exercises 288

Chapter 11: Is It Real or Is It Rendered?

11.1 What Is Rendering and Why Is It So Challenging? 297
11.2 Beyond Shademode – The *Render* Com'd 297
11.3 Adding Materials to Make Your Solids Look Real 308
11.4 Special Effects – Adding Other Graphic Images to Your Drawing 316
11.5 Lights 321
11.6 Creating a Scene 329
11.7 Extra Steps 332
11.8 What Have We Learned? 332
11.9 Exercises 333

Appendices

Appendix – A: Drawing Scales 338
Appendix – B: Project Drawings 339
Appendix – D: Additional Projects 342

Lesson 1

Following this lesson, you will:

- ✓ *Know how to maneuver in three-dimensional space*
  - o *Understand the **VPoint** & **VPoint** commands*
  - o *Understand the Right-Hand Rule*
  - o *Know how to use the **Plan** command*
- ✓ *Know how to draw simple three-dimensional objects*
  - o *Know how to use **Elevation** and **Thickness** in your drawings*
- ✓ *Know some of the basic tricks used to view a three-dimensional drawing*
- ✓ *Know how to use the **Hide** and **Shademode** commands*

"Z" Basics

Thanks, Mr. Woopie! You're the greatest.

Tennessee Tuxedo

Does anyone not remember Mr. Woopie's coveted 3D BB? How many times did he pull that tiny block from his closet, stretch it into a full-size blackboard, and help Tennessee Tuxedo devise yet another wonderful scheme?

Who would have thought – way back in those simple cartoon days – that one day you'd be learning to use your very own 3D BB?!

You're about to go where no board draftsman has gone before. You are about to enter Z-Space. (Hear that cool science fiction music playing in the background? Keep your eyes open for Dr. Who.) Z-Space is that area defined by the Z-axis. Remember the X- and Y-axes? They travel left to right and top to bottom on a sheet of paper (or drafting board). The Z-axis rises and falls into the space above and below the paper. It takes all three axes to create a three-dimensional object. (AutoCAD hasn't quite reached into that fourth dimension of which Einstein spoke ... or those other dimensions Stephen Hawking explains so well – at least to other PHDs. But three dimensions are quite enough to keep me confused.)

1.1 The UCS Icon and the Right-Hand Rule

Before entering Z-Space (the third dimension), you must learn how to keep your bearings. That is, you must learn to tell up from down, top from bottom, and left from right, regardless of your orientation within the drawing (feel like an astronaut?). It's not as easy as it sounds. Place your hand over Figure 1.002 and look carefully at Figure 1.001. Are you looking at the top or bottom of the object? Now look at Figure 1.002. How did you do?

I've removed the hidden lines in Figure 1.002 (more on how I did that in Section 1.4.1). However, it isn't always practical to do that, so AutoCAD has provided a method to determine your orientation at any time and at any place in the drawing. We'll use the UCS icon (Figure 1.003).

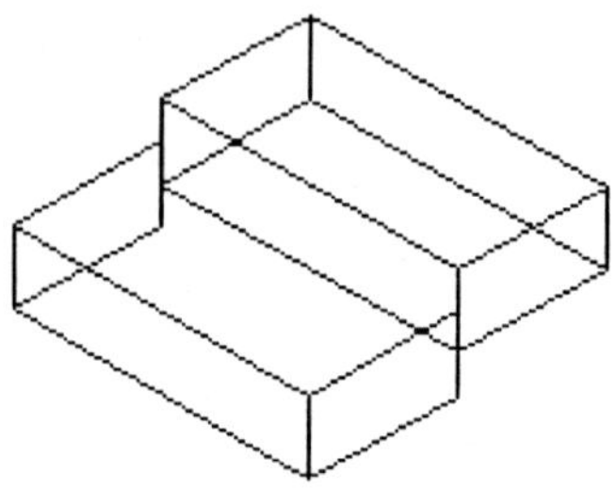

Figure 1.001

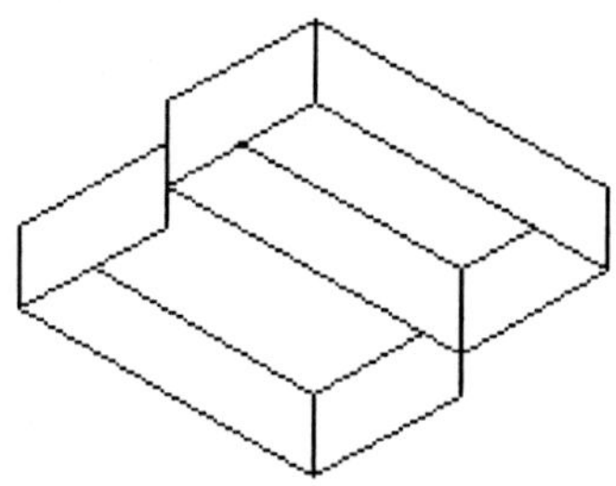

Figure 1.002

1.1.1 The UCS Icon

In our basic text, I told you how to turn off the UCS icon. In two-dimensional space, it has little use. However, to operate in three-dimensional space, you must activate the UCS icon or risk being forever lost. (The icon acts like a lighthouse on a dark and stormy night – giving the mariner a reference point from which to navigate.) Use the ***UCSIcon*** command to control the UCS icon. It works like this:

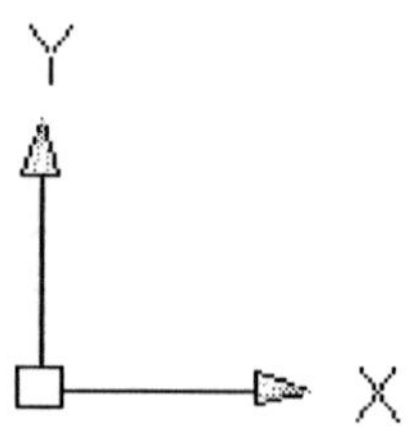

Figure 1.003:
UCS Icon in 2D Space

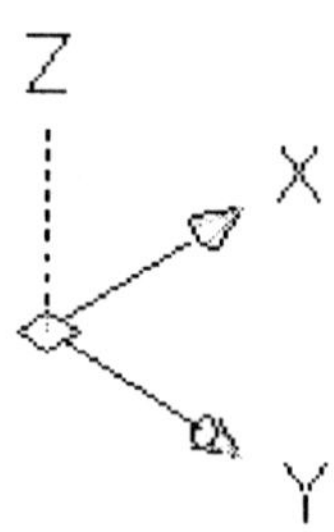

Figure 1.004:
UCS Icon in 3D Space

Command: *ucsicon*

Enter an option [ON/OFF/All/Noorigin/ ORigin/Properties] <ON>:

The simple command provides options that will prove indispensable in a three-dimensional drawing:

- The **ON/OFF** options are self-explanatory. They turn the icon **On** or **Off**. They work only in the current viewport in Model Space or on the Paper Space icon in Paper Space.
- The **All** option prompts as follows:

  Enter an option [ON/OFF/Noorigin/ORigin/Properties] <OFF>:

  These options perform the same function as the first tier of options, but operate on *all* the viewports.
- The UCS icon remains in the lower-left part of the graphics screen by default. Normal display manipulations (zooms and pans) won't affect it.

  With the **ORigin** option, however, you can attach the UCS icon to the 0,0,0 coordinate where it will remain regardless of the display (as long as the 0,0,0 coordinate is on the screen). This will provide a fixed point in space (a lighthouse) from which you can navigate. If the 0,0,0 coordinate leaves the graphics area through zooming or panning, the UCS icon returns to the lower-left corner of the screen.
- **Noorigin** disables the **ORigin** setting.
- The **Properties** option calls the UCS icon dialog box shown in Figure 1.005. Here you can set up the icon's physical properties to suit your preferences.
  - In the **UCS icon style** frame, you can tell AutoCAD to use a two-dimensional (**2D**) or three-dimensional (**3D**) icon. AutoCAD will use the 3D icon by default. (We used the 2D icon in earlier releases of AutoCAD. It's considerably more complicated to understand, so we'll welcome the 3D icon and ignore its honorable ancestor.)

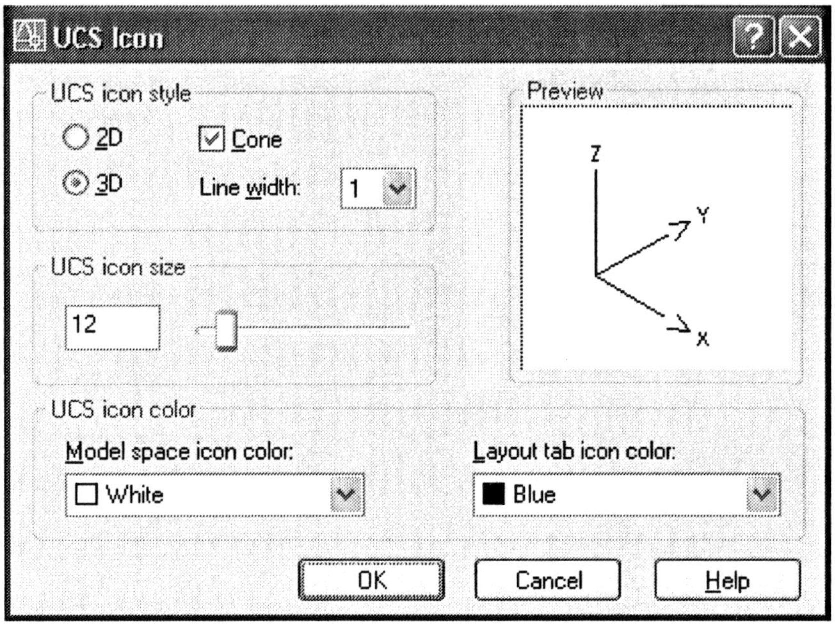

Figure 1.005

    You can use conical pointers at the ends of the icon arrows (the default setting) or remove them in favor of simple arrowheads by removing the check next to **Cone**. You can also control the **Line width** of the icon using the selection box in the lower right corner of the frame.
  - The **UCS icon size** frame presents two tools – a text box and a slider bar – which allow you to control the size of the icon. The default works well, but some operators prefer something a bit smaller and less obtrusive. Three-dimensional novices might prefer something larger until they become accustomed to Z-Space.

- The UCS icon color frame allows you more control over the color of the icon. Most people leave this at its default, although an occasional change in color does tend to relieve the monotony.

Let's experiment a bit with the UCS icon.

> You can access the various options of the ***UCSIcon*** command using the View pull-down menu. Follow this path:
>
> *View – Display – UCS Icon – [option]*

Do This: 1.1.1.1	**Manipulating the UCS Icon**

I. Open the *ucs practice.dwg* file in the C:\Steps3D\Lesson01 folder. The drawing looks like Figure 1.006. (Don't let the funny angle of the crosshairs or UCS icon bother you – you're viewing the model from an angle in Z-Space. More on this setup in Section 1.2.)

II. Follow these steps.

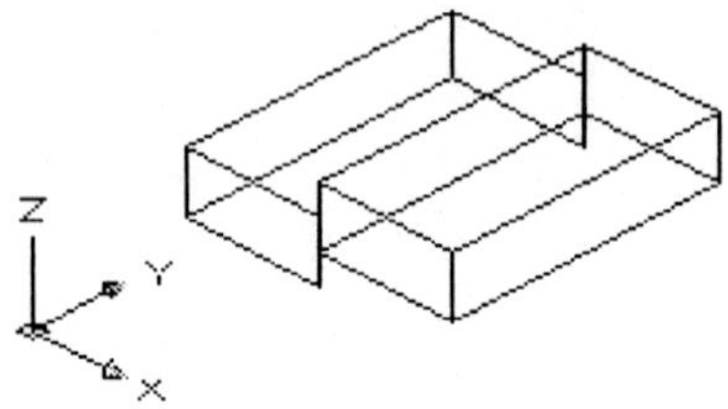

Figure 1.006

TOOLS	**COMMAND SEQUENCE**	**STEPS**
No Button Available	**Command:** ***ucsicon***	1. Enter the ***UCSIcon*** command. [Note: There are two UCS toolbars but neither will help in this exercise. We'll look at those tools in our next lesson.]
	**Enter an option [ON/OFF/All/ Noorigin/ ORigin/Properties] <ON>:** ***OR***	2. Use the **ORigin** option (enter ***or*** at the prompt or select **ORigin** from the cursor menu). Notice that the icon moves to the **0,0,0** coordinate.
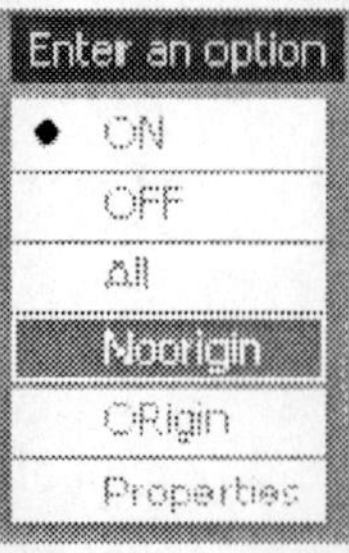	**Command: ucsicon** **Enter an option [ON/OFF/All/ Noorigin/ ORigin/Properties] <ON>:** ***n***	3. Now disable the **ORigin** option as indicated. The icon returns to the lower-left quadrant of the screen.
 Named Views Button	**Command:** ***v***	4. Use the ***View*** command to change the view to **negative z space** (already created for you). The view changes to that shown in Figure 1.007. Notice the UCS icon. The Z-axis appears dashed. This indicates that you're seeing the model from underneath.

TOOLS	COMMAND SEQUENCE	STEPS

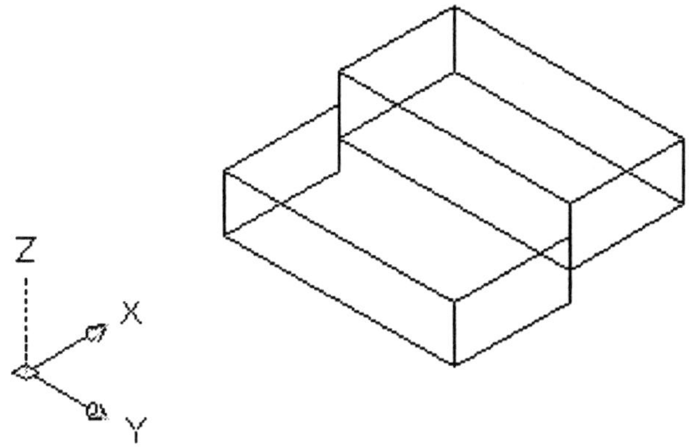

Figure 1.007

TOOLS	COMMAND SEQUENCE	STEPS
		5. Use the ***Vports*** command to set up **Four: Equal** viewports. Notice that each viewport has a UCS icon (Figure 1.008).

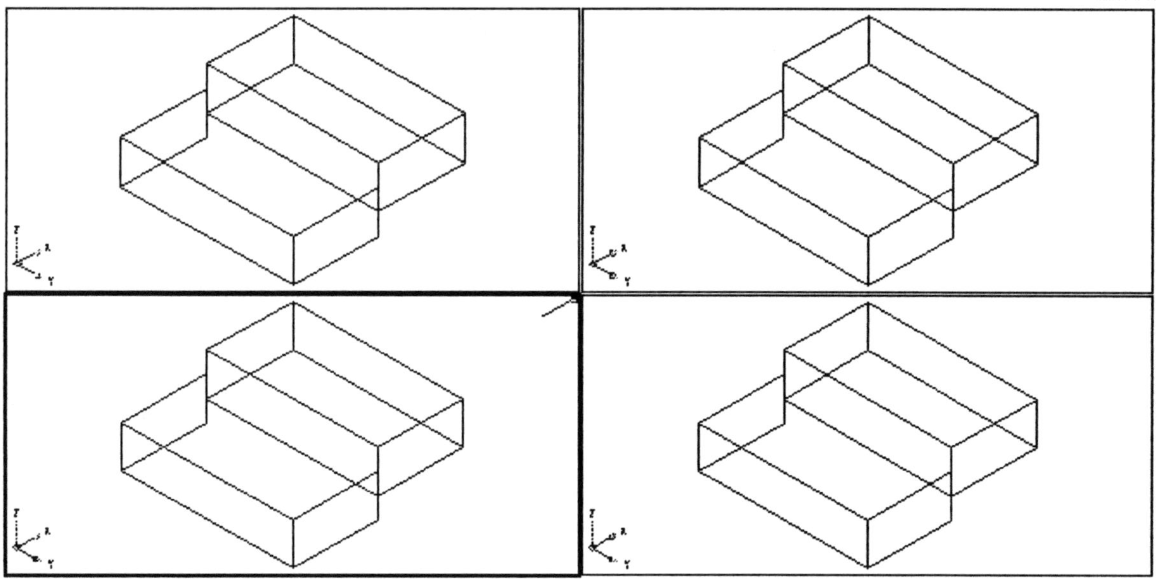

Figure 1.008

TOOLS	COMMAND SEQUENCE	STEPS
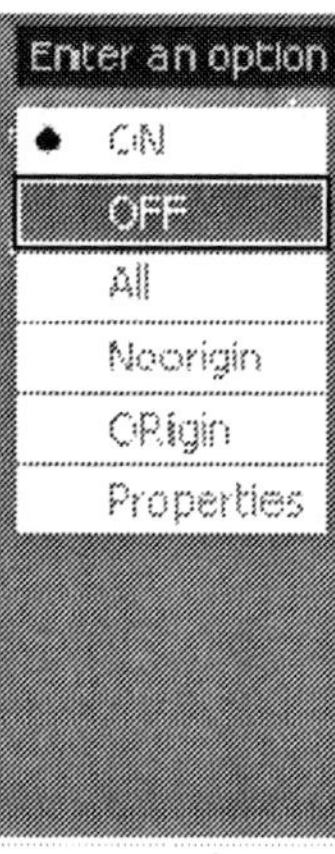	**Command: *ucsicon*** **Enter an option [ON/OFF/All/ Noorigin/ ORigin/Properties] <ON>: *OFF*** **Command: *ucsicon*** **Enter an option [ON/OFF/All/Noorigin/ ORigin/Properties] <OFF>: *a*** **Enter an option [ON/OFF/Noorigin/ORigin/ Properties] <OFF>: *off***	6. Use the ***UCSIcon*** command to turn **OFF** the icon. Notice that only the icon in the currently active viewport (lower right) disappears. Now use the ***UCSIcon*** command to turn **OFF** the icon in **All** the viewports.

Tools	Command Sequence	Steps
	**Command:** ***qsave***	7. Save and exit the drawing.

These simple tricks to manipulate and read the UCS icon will become second nature with experience. But there is another tool you can use to ease the learning curve – the Right-Hand Rule.

1.1.2 The Right-Hand Rule

The Right-Hand Rule provides another means of navigating Z-space. It works very much like the UCS icon but is a bit more "handy." It works like this (refer to Figure 1.009):

1. Make a fist with your right hand.
2. Extend the index finger upward (no no – the *index* finger!).
3. Extend the thumb at a right angle to the index finger.
4. Extend the middle finger at a right angle to the index finger (pointing outward).

How's that for feeling really awkward?

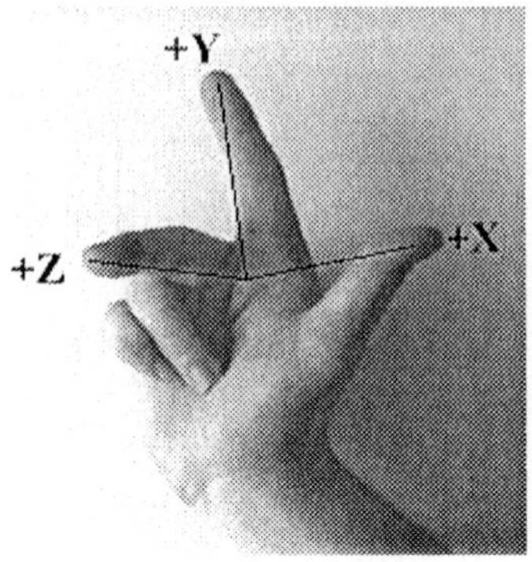

Figure 1.009

Each finger serves a purpose; each indicates the positive direction of one of the XYZ axes as indicated in the picture. See how it correlates with the UCS icon? Using the right hand in this fashion, you'll always be able to determine the third axis if you know the other two. Simply orient your right hand with the appropriate fingers pointing along the known axes and the location of the unknown axis becomes clear!

1.2 Maneuvering Through Z-Space with the *VPoint* Command

Okay. Now you know how to determine your own orientation in Z-Space, but how do you reorient the model itself? In other words, how do you turn the model this way and that in order to work on all sides of it?

Not to worry; AutoCAD has provided a simple but powerful tool to help you – the ***VPoint*** command.

It might help your understanding if I begin by telling you that the model will not actually move or rotate. Using the ***VPoint*** command, the model holds still while you change position.

Let's use an airplane as an example. The airplane is a three-dimensional object (it has length, width, and height or thickness). To draw the top of the airplane, we'll climb into a helicopter and hover above it for a better view. To draw the front, we must fly to the front of the airplane. Side and bottom views will also require us to fly to a better vantage point.

AutoCAD's helicopter is the ***VPoint*** command (perhaps better defined as *Vantage* Point).

There are three approaches to using the ***VPoint*** command: coordinate input, compass, and dialog box. Let's look at each.

1.2.1 Using Coordinates to Assign a Viewpoint

The ***VPoint*** command prompt looks like this:

Command: *vpoint* (or *–vp*)

Current view direction: VIEWDIR=1.0000,-1.0000,-1.0000

Specify a view point or [Rotate] <display compass and tripod>: ***[enter a coordinate]***

- The default response to this prompt requires you to enter the coordinates from which you wish to view the model. Does that sound simple? Only if you know the coordinates, you say? Well, it's actually easier than that! AutoCAD won't read the XYZ coordinates entered as actual coordinates but rather as a ratio. That is (keeping it simple), a coordinate of 1,-1,1 tells AutoCAD that you wish to stand – in relation to the model – a step to the right (+1 on the X-axis), back a step (-1 on the Y-axis), and up a step (+1 on the Z-axis).

> The default viewpoint for new or two-dimensional drawings is **0,0,1**. This means that you view the drawing from above (+1 on the Z-axis). Draftsmen generally refer to this as the *Plan* view.

Remembering the simple ratio approach, what do you think a coordinate of 4,-2,1 will mean?[*]

> When the *absolute* value of all three axis entries is the same (as in 1,1,1 or 1,-1,1), you have an isometric view of the drawing. (*Iso* means one – you're using the same absolute *value* on each axis).
>
> When only two of the absolute values are the same (as in 2,2,1 or –2,2,1), you have a *dimetric* view. (*Di* means two.)
>
> What do you suppose you have when all three of the absolute values are different? Of course, it's a *trimetric* view.

Why the ratio approach? When you change your vantage or viewpoint (your *VPoint*), AutoCAD will automatically zoom to the drawing's extents (move as close as possible while showing you everything on the model)!

> You already know that a plan view is achieved when the VPoint coordinates are 0,0,1. These coordinates allow you to view the model from a step up (+Z). What do you suppose happens when you set the coordinates to 1,0,0 or 0,-1,0?
>
> These coordinates provide elevations of the model – 1,0,0 provides a right elevation (a step to the right or +X) and 0,-1,0 provides a front elevation (a step back or –Y). Which coordinates would provide a back or left elevation?[**]
>
> AutoCAD lets you know when you're in an elevation view by changing the appropriate axis on the UCS icon to a small circle (Figure 1.010).

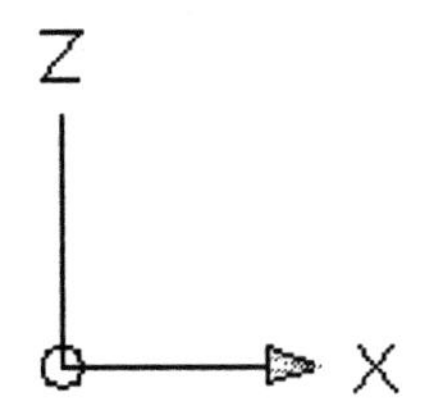

Figure 1.010

- The **Rotate** option of the ***VPoint*** command allows you to specify your vantage point by the angle at which you wish to see the model. The prompts look like this:

Command: ***vpoint***

Current view direction: VIEWDIR=0.0000,-1.0000,0.0000

Specify a view point or [Rotate] <display compass and tripod>: ***r [tell AutoCAD you wish to use the Rotate option]***

---

[*] It means: I want to stand twice as far to the right as I'm standing back, and twice as far back as I'm standing above the model.

[**] Back elevation coordinates = 0,1,0; left elevation coordinates = -1,0,0

Enter angle in XY plane from X axis: ***[indicate the two-dimensional angle (the angle along the ground) at which you wish to see the model]***

Enter angle from XY plane: ***[indicate the three-dimensional angle (the upward angle – like the angle of the sun in the sky) at which you wish to see the model]***

This will become clearer in our next exercise.

We'll look at the compass and tripod in Section 1.2.2, but first let's try the coordinate approach to the ***VPoint*** command.

> You can access several preset viewpoints using buttons on the View toolbar or by selecting the view from the View pull-down menu. Follow this path:
>
> *View – 3D Views – [selection]*

Do This: 1.2.1.1	**The Coordinate Approach to Setting the Viewpoint**

I. Open the *VPoint practice.dwg* file in the C:\Steps3D\Lesson01 folder. The drawing looks like Figure 1.011. It's currently in the plan view

II. Follow these steps.

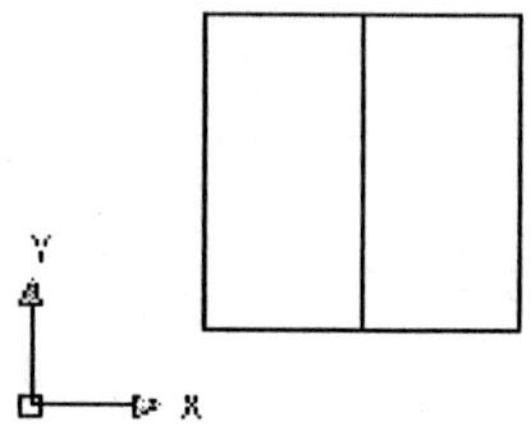

Figure 1.011

TOOLS	COMMAND SEQUENCE	STEPS
No Button Available	**Command:** ***-vp***	1. Enter the ***VPoint*** command.
 SE Isometric View Button	**Command:** ***-vp*** **Current view direction: VIEWDIR=0.0000,0.0000,1.0000** **Specify a view point or [Rotate] <display compass and tripod>:** ***1,-1,1***	2. AutoCAD tells you the current coordinate setting and prompts you to either **Specify a view point** or to **Rotate**. Let's use the first option. Tell AutoCAD to move to the right, back, and up as indicated (the SE Isometric View). Your drawing looks like Figure 1.012. (You can achieve the same results as Steps 1 and 2 using the **SE Isometric View** button on the View toolbar.)

TOOLS	COMMAND SEQUENCE	STEPS
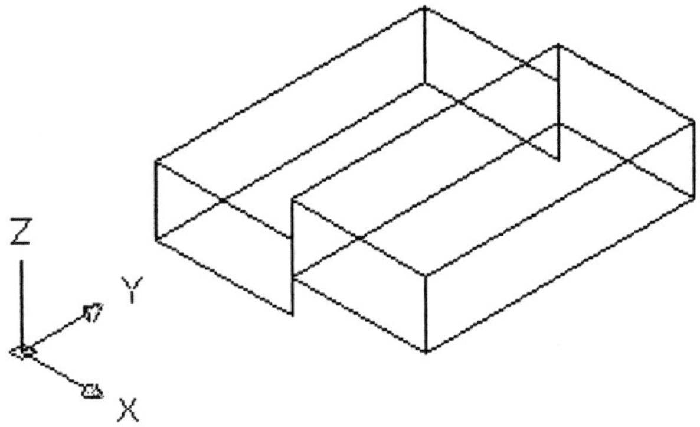 Figure 1.012		
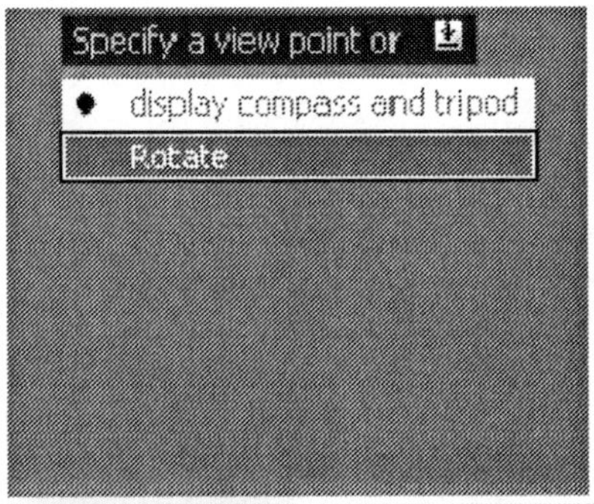	**Command: *[enter]*** **Current view direction: VIEWDIR=1.0000,-.0000,1.0000** **Specify a view point or [Rotate] <display compass and tripod>: *r***	3. Let's try the **Rotate** option. Repeat the ***VPoint*** command and select the **Rotate** option.
	Enter angle in XY plane from X axis <315>: *225*	4. AutoCAD asks for a two-dimensional angle. We'll stand in the –X,-Y quadrant of the XY plane. Enter the angle indicated.
	Enter angle from XY plane <35>: *45*	5. Now AutoCAD asks for a three-dimensional angle (the angle up or down in Z-space). We'll stand above the model. Enter the angle indicated. Your drawing now looks like Figure 1.2.013.
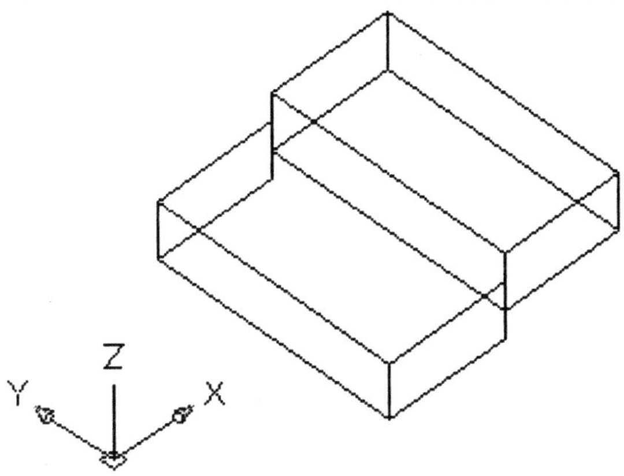 Figure 1.013		

Tools	Command Sequence	Steps
	Command: ***[enter]*** **Current view direction: VIEWDIR=-0.8660,-0.8660,1.2247** **Specify a view point or [Rotate] <display compass and tripod>:** ***0,-1,0***	6. Repeat the ***VPoint*** command. This time, enter coordinates for a front elevation as indicated. Your drawing looks like Figure 1.014. Notice the UCS icon. (*Caution:* Do *not* use the **Front View** button on the View toolbar as it will also change the UCS.)

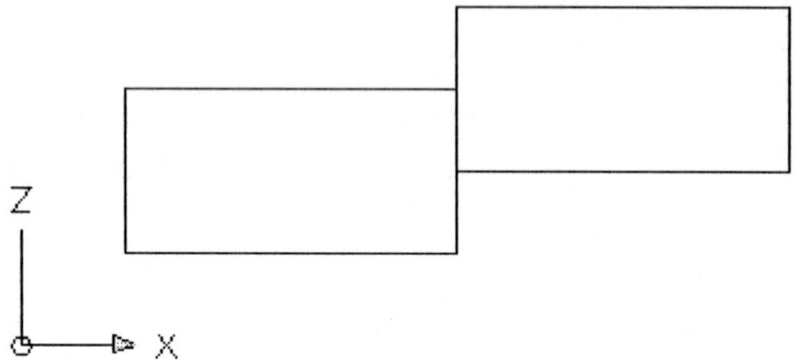

Figure 1.014

Tools	Command Sequence	Steps
		7. Experiment with different coordinate and angular entries. Use different combinations of positive and negative numbers, and watch the UCS icon (and use the Right-Hand Rule) to keep track of your orientation.
	**Command:** ***qsave***	8. Save the drawing but don't exit.

1.2.2 Using the Compass to Assign a Viewpoint

Often the precise coordinate or angle from which you view the model won't be all that important. You'll be in a hurry, know the general area in which you wish to stand, and want to go there quickly. For these times, AutoCAD provides the **compass and tripod** option of the ***VPoint*** command.

Access the compass and tripod by hitting enter at the **Specify a view point or [Rotate]<display compass and tripod>** prompt. AutoCAD presents the tripod and compass screen (Figure 1.015).

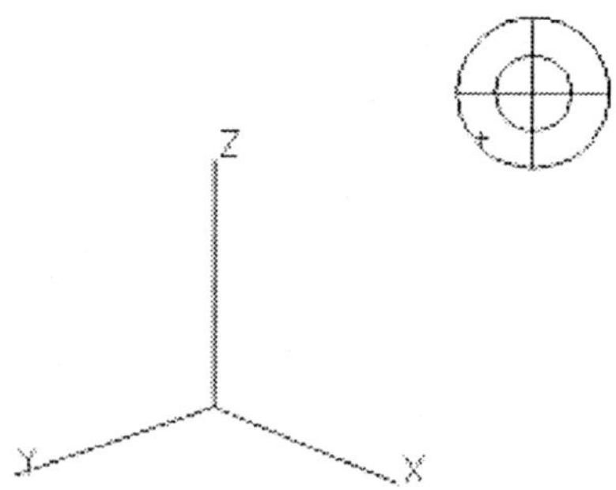

Figure 1.015

- Although it's fast, the tripod is difficult to use. Imagine your drawing resting at the vertex (center) of the tripod. Use the mouse to maneuver the XYZ-axes as desired. When you're happy with the orientation, left-click to return to the graphics screen.
- The compass in the upper-right quadrant of the screen is much easier to use than the tripod. This ingenious tool represents the positive and negative regions of a model. Refer to Figure 1.016.

Imagine that your model exists at the core of a globe. Where you stand on the globe determines your view of the model. The VPoint compass is a two-dimensional representation of that globe. Use it to show AutoCAD where you wish to stand.

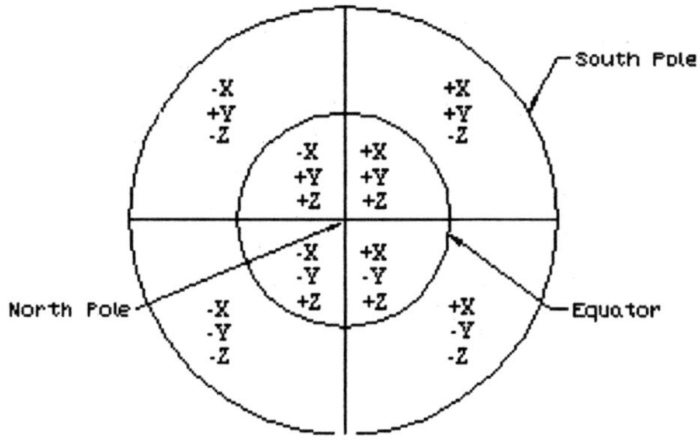

Figure 1.016

The inside of the compass's inner circle represents the upper (or northern) hemisphere of the globe. Since this is the upper part of the globe, we see the model from the top – so the area inside the inner circle is on the +Z-axis. The four quadrants of the inner circle represent the positive and negative X- and Y-axes as shown.

Between the inner and outer circles of the compass is the lower (or southern) hemisphere of the globe. Since this is the lower part of the globe, we see the model from the bottom – so this area is on the –Z-axis. The four outer quadrants represent the positive and negative X- and Y-axes as shown.

The compass may seem confusing at first, but (except for the toolbar) it's considerably faster than any other method of VPoint selection once you get used to it.

You can create a series of viewpoints by saving each, using the same ***View*** command that was so beneficial in two-dimensional AutoCAD.

Let's try the compass approach to the ***VPoint*** command.

Do This: 1.2.2.1	**Using the VPoint Compass**

I. Be sure you're still in the *vpoint practice.dwg* file in the C:\Steps3D\Lesson01 folder. If not, please open it now.

II. Follow these steps.

TOOLS	COMMAND SEQUENCE	STEPS
	Command: *-vp*	1. Begin the ***VPoint*** command.
	Current view direction: VIEWDIR=0.0000,-1.0000,0.0000 **Specify a view point or [Rotate] <display compass and tripod>: *[enter]***	2. Hit ***enter*** at the first prompt to access the tripod and compass.

Tools	Command Sequence	Steps
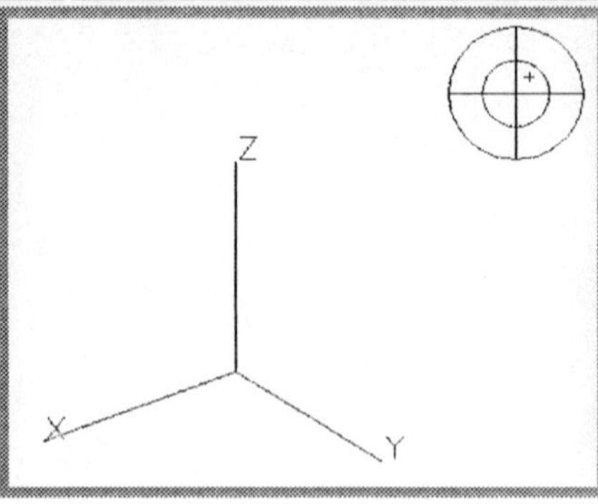		3. Place the compass cursor in the +X+Y+Z coordinates area as indicated, and then pick once with the left mouse button to return to the graphics screen. Your drawing looks something like Figure 1.017.

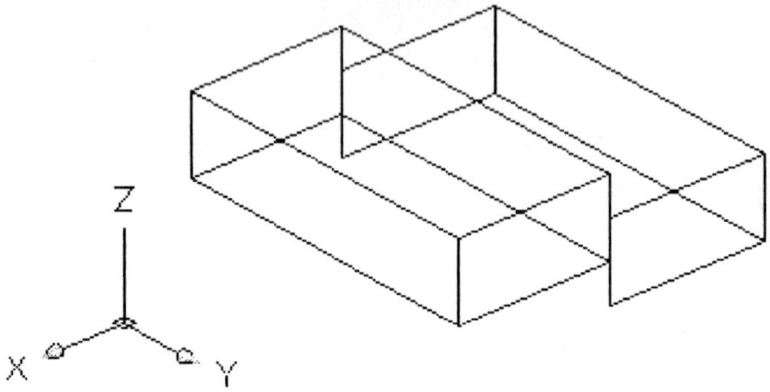

Figure 1.017

Tools	Command Sequence	Steps
	Command: ***-vp***	4. Experiment using different quadrants (and locations within quadrants) of the compass. Try using the tripod instead of the compass. Which is easier?
	**Command:** ***qsave***	5. Save the drawing but don't exit.

1.2.3 Setting Viewpoints Using a Dialog Box

Setting viewpoints using a dialog box is very similar to setting viewpoints using the **Rotate** option of the ***VPoint*** command.

Access the Viewpoint Presets dialog box (Figure 1.018) with the ***VPoint*** command or the ***VP*** hotkey. [You can also access the dialog box using the View pull down menu. Follow this path: *View – 3dViews – Viewpoint Presets.*]

The first things you'll notice on the dialog box are the two large drawings in the center. The first (on the left) looks something like a compass, and the second looks like half a compass. Use the compass on the left to set the two-dimensional angle (*in* the XY-plane); use the compass on the right to set the three-dimensional angle (up or down *from* the XY plane).

You can set the angles in two ways – by keyboard entry in the text boxes below the compasses or by mouse selection on the compasses themselves.

At the top of the dialog box, AutoCAD asks you to identify the angle in **Absolute to WCS** or **Relative to UCS** terms. We'll learn more about the WCS (World Coordinate System) and UCS (User Coordinate System) in Lesson 2. For now, leave the viewing angles set to **Absolute to WCS**.

The long button across the bottom of the dialog box is our first "Hail Mary" button for this text. If you read the basic text, you know to use a Hail Mary button in an emergency. AutoCAD designed this button to return the drawing to the plan view from anywhere. Set the **Absolute to WCS** option at the top of the dialog box and then pick the **Set to Plan View** button to return to a "normal" two-dimensional view of your model. This is quite handy when you become lost in Z-space.

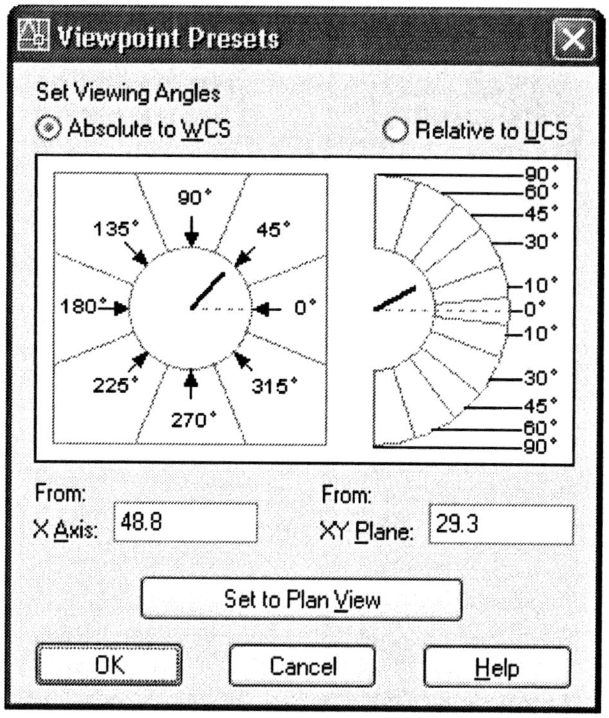

Figure 1.018

The keyboard equivalent of the **Set to Plan View** "Hail Mary" button is the ***Plan*** command. It looks like this:

Command: ***plan***

Enter an option [Current ucs/Ucs/World] <Current>:

Again, we'll look at the WCS and UCS in Lesson 2. Enter ***w*** for the **World** option and hit ***enter*** to do what the **Set to Plan View** button does – return to a normal two-dimensional view of your model.

Let's try the dialog box.

Do This: 1.2.3.1	Using the Viewpoint Presets Dialog Box

I. Be sure you're still in the *vpoint practice.dwg* file in the C:\Steps3D\Lesson01 folder. If not, please open it now.

II. Follow these steps.

TOOLS	COMMAND SEQUENCE	STEPS
No Button Available	**Command:** ***vp***	1. Enter the ***VPoint*** command. AutoCAD presents the Viewpoint Presets dialog box (Figure 1.018).

Tools	**Command Sequence**	**Steps**
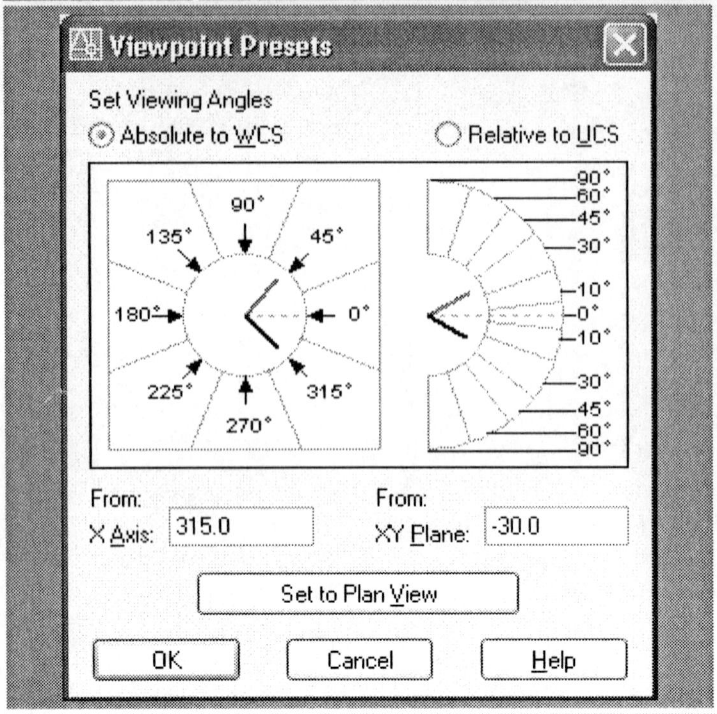		2. Set the viewpoint to **315°** on the left compass. Set the viewpoint to **-30°** on the right compass. (You can do this by typing the numbers into their respective text boxes or by picking on the numbers on the compasses themselves.) Pick the **OK** button to complete the command. Your drawing looks like Figure 1.019. Notice the UCS icon.

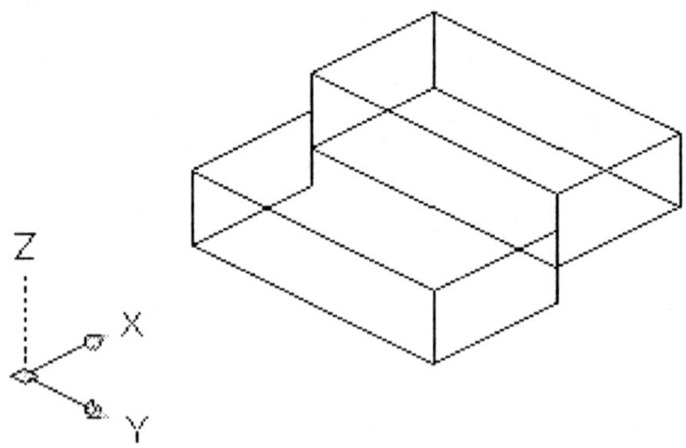

Figure 1.019

	Command: ***[enter]***	3. Repeat the ***VPoint*** command.
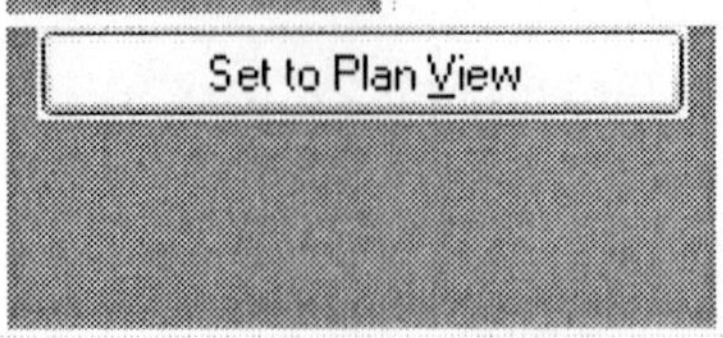		4. Pick the **Set to Plan View** button and complete the command. Your drawing looks like Figure 1.020.

TOOLS	COMMAND SEQUENCE	STEPS
	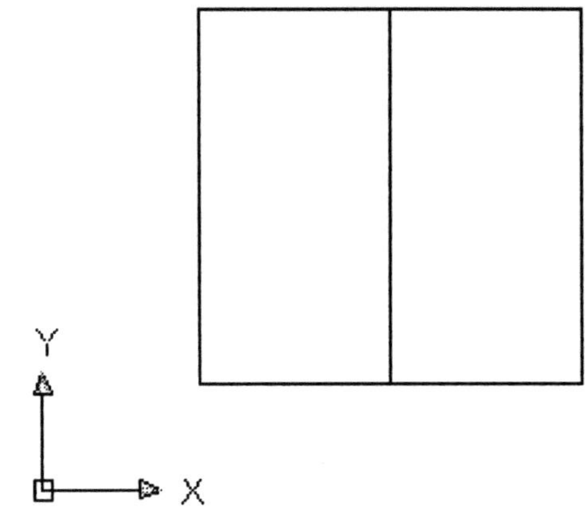 Figure 1.020	
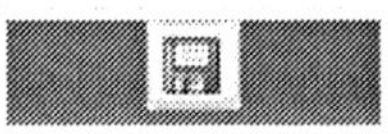	**Command:** ***qsave***	5. Save the drawing.

1.2.4	The Camera Approach

At first glance, the camera approach seems like the best approach to adjusting how you view objects in a three-dimensional drawing. It works by locating the camera (your vantage point) and then identifying the position of its target. Unfortunately, the view you get often – well, always – includes the camera's location.

AutoCAD uses a fairly straightforward command sequence for its camera:

> **Command:** ***camera***
>
> **Specify new camera position <0,0,0>:** ***[locate the camera – your vantage point]***
>
> **Specify new camera target <0,0,0>:** ***[locate the camera's target – what you want to see]***

Let me show you.

Do This: 1.2.4.1	Change the Camera Position and Target

III. Open the *CameraPositionExercise* drawing file in the C:\Steps3D\Lesson01 folder. The drawing looks like Figure 1.020. (Don't worry about the shading and UCS icon, you'll learn how to do all this in this text.)

IV. Follow these steps.

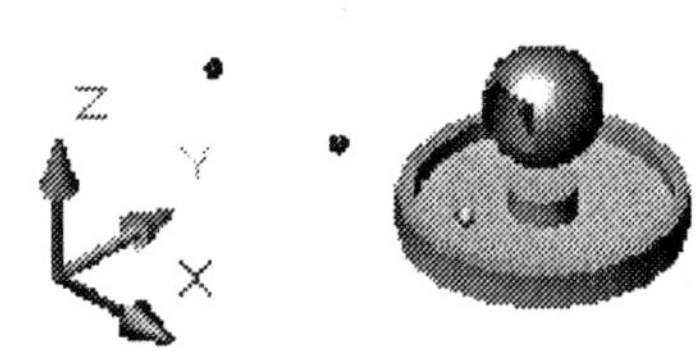

Figure 1.020

TOOLS	COMMAND SEQUENCE	STEPS
 	**Command:** ***camera***	1. Enter the ***Camera*** command. Alternately, you can pick the **Camera** button on the Views toolbar.

TOOLS	COMMAND SEQUENCE	STEPS
	Specify new camera position <37.4069,-38.9906,35.1925>: _cen of	2. Select the center of the red sphere as your vantage point. (Use the center OSNAP.)
	Specify new camera target <0.2102,-1.7939,-2.0042>: _ins of	3. Select the insertion point of the larger sphere block as your target. AutoCAD adjusts the view (Figure 1.021). Notice that the red sphere appears in the view.

Figure 1.021

		4. Repeat Steps 1 through 3 using the other spheres as camera positions.
		5. Repeat Steps 1 through 3 using the insertion point of the block as your camera position and the smaller spheres as your target.
	Command: *quit*	6. Exit the drawing without saving.

You'll create the nifty pyramid-in-a-sphere in later lessons.

We've spent a great deal of time in this lesson on a single concept – viewpoints. I can't overemphasize the importance of being comfortable with viewpoints.

Consider the astronaut floating in space. A basic understanding of geography will tell him where he is in terms of continent, nation, or even city (XY-space). But if he doesn't understand his altitude (his Z-space position), he can't know if he's coming or going. He might wind up on the wrong planet altogether!

If necessary, repeat this lesson to this point to get comfortable with viewpoints and the VPoint command (the astronaut's navigator). Experiment with viewpoints in different viewports (each viewport can have its own viewpoint). Then proceed to the next section where we'll take our first tender steps in creating three-dimensional objects.

Return Policy:

If for any reason you are not happy with this purchase, please return within 30 days of receipt for a refund. Please include this packing slip along with your return. We recommend that you reuse the original packing materials to ensure that items are not damaged in shipment. Please allow a minimum of 4 weeks for your refund to be processed. You will receive an email confirmation from amazon.co.uk when your refund is issued.

Address returns to:

Alibris Book Returns
Unit 2, Maida Vale Business Centre
Maida Vale Road
Cheltenham, Glos., Gl53 7ER
UNITED KINGDOM

Reason for Return:

☐ Item not as described ☐ Item different than expected ☐ Incorrect item received ☐ Other: ____________________

Problems? Questions? Suggestions?

If you have any questions regarding this order, please contact Alibris Customer Service at: amazon_uk@alibris.com.

Alibris Order # 23930333-1

Amazon.co.uk Order # 736-1601869-5892107

PO Line ID:

alibris™

Books You Thought You'd Never Find

www.alibris.com

Ship To: KEN FISHER
C/O Vibro_Meter UK, The Laurel
Jays Close, Viables Ind. Est., Basingstoke,, Hampshire, , RG22 4BS
UNITED KINGDOM (GBR)

Phone:

Contact: KEN FISHER

Order #	Order Date	Title and Description	Quantity	Ship Date
23930333-1	17/03/08	**3d Autocad 2006: One Step at a Time** Author: Sykes, Timothy Sean ISBN: 0976588838 Alibris ID: 1516446225 Condition: Unread, not previously owned. May show signs of wear including remainder marks or stickers on book or cover. Notes: Trade paperback (US). Glued binding. 344 p. Contains: Illustrations.	1	20/03/08

ROUTE LEG	APPRVD DT	APPRVD BY
IPAROW		

Thank you for buying from Alibris!

3

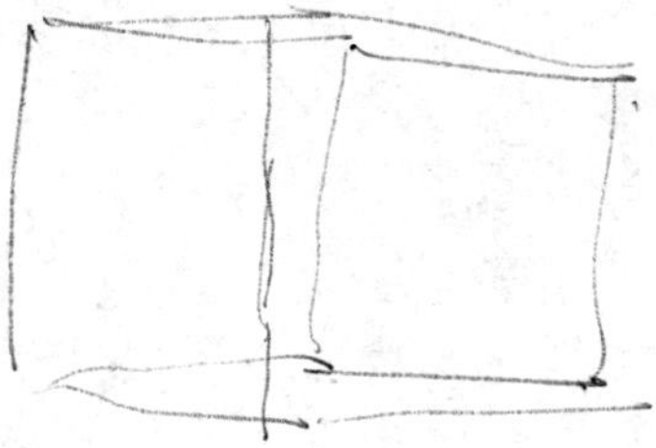

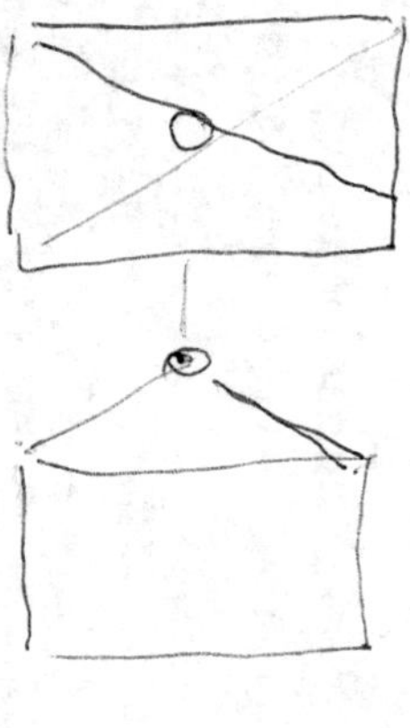

0, 0, 1

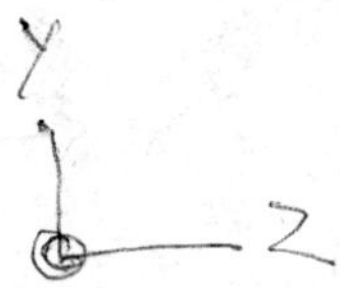

0, -Y, 0

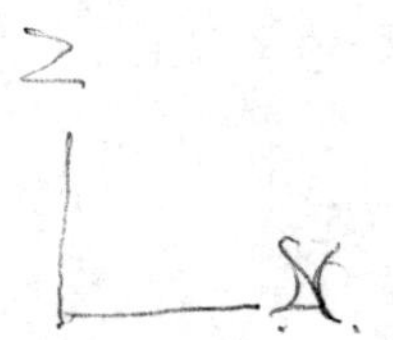

1.3 Drawing with the Z-Axis

I wish I could begin this section by saying that three-dimensional drafting is no different from two-dimensional drafting. However, that simply isn't the case. (I could lie if you'd prefer.) You've already seen that you'll have to master more navigation tools. Additionally, you must consider a couple things that have no meaning in a two-dimensional drawing – the Z-coordinate (or *elevation*) and the *thickness* of the object you're drawing.

Let's take a look at these.

1.3.1 Three-Dimensional Coordinate Entry

We must begin our study of three-dimensional drafting as we began our basic text – with a look at the Cartesian Coordinate System. This time, we need to understand how it works with the Z-axis. Consider the following table.

SYSTEM	ABSOLUTE	RELATIVE	POLAR/SPHERICAL/CYLINDRICAL
2D Entry	X,Y	@X,Y	@Dist<Angle
3D Entry	X,Y,Z	@X,Y,Z	@truedist<Xyangle<Zangle or @dist<2Dangle,Z-dist

The Absolute and Relative systems are easy enough to understand. Simply add the location on the Z-axis to the X and Y locations to have an XYZ-coordinate. But Polar Coordinate entry might need some explaining. (Refer to Figures 1.023 and 1.024.)

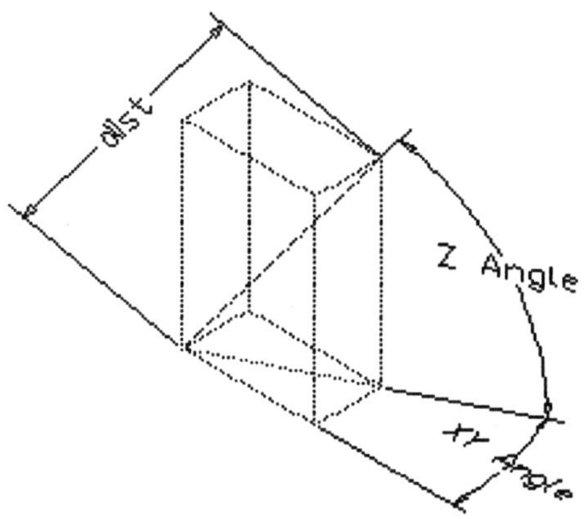

Figure 1.023: Spherical Coordinate Entry

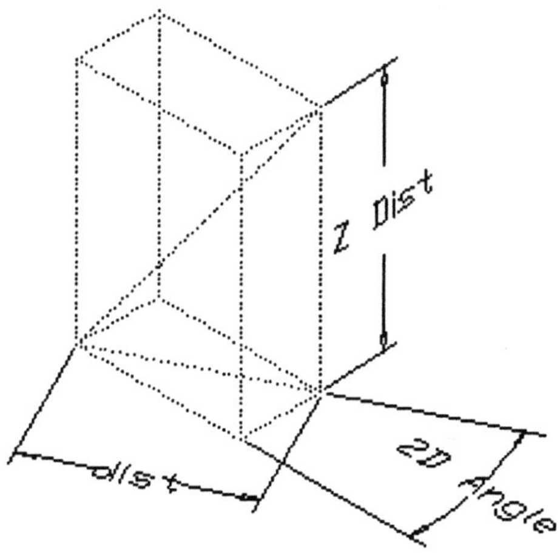

Figure 1.024: Cylindrical Coordinate Entry

The first formula for 3D entry of polar coordinates looks like this:

@truedist<XYangle<Zangle

(Read this line as: *at a true distance of___ ... at an XY angle of___ ... and a Z angle of___.*)

We call this type of coordinate entry *Spherical* (Figure 1.023). Begin Spherical coordinate entry with the true three-dimensional length of the line (or other object). Follow with the same two-dimensional (XY) angle you've always used for Polar Coordinates. Follow that with the three-dimensional angle (above or below the XY-plane).

The other formula for 3D entry of polar coordinates looks like this:

@dist<XYangle,Z-dist

(Read this line as: *at a distance of___ ... and an XY angle of___ ... at a Z distance of___.*)

We call this type of coordinate entry *Cylindrical* (Figure 1.024). Like the spherical coordinate entry method, begin cylindrical coordinate entry with the distance of the line *but this time, use only the XY distance*. Follow with the two-dimensional angle, but conclude with the distance *along the Z-axis* (perpendicular to the XY-plane).

We'll use these methods to draw the stick figure of a house in our next exercise.

Do This: 1.3.1.1	Three-Dimensional Coordinate Entry

I. Open the *1-3D-1.dwg* file in the C:\Steps3D\Lesson01 folder. This drawing has been set up with a viewpoint of 1,-2,1, and layers have been created for you.

II. Be sure that layer **obj1** is current. (Toggle dynamic input on or off as needed.)

III. Follow these steps.

TOOLS	COMMAND SEQUENCE	STEPS
	**Command:** ***l*** **Specify first point:** ***1,1*** **Specify next point or [Undo]:** ***4,1***	1. We'll begin using absolute coordinates. Draw a line as indicated. Don't exit the command. Notice that a Z-coordinate entry isn't required if the value is **0**.
	Specify next point or [Undo]: ***4,1,2***	2. Now draw a line straight up into Z-Space.
	Specify next point or [Close/ Undo]: ***1,1,2*** **Specify next point or [Close/ Undo]:** ***c***	3. Continue the line to complete the first side of the house.
	**Command:** ***[enter]*** **Specify first point:** ***1,3*** **Specify next point or [Undo]:** ***@3,0*** **Specify next point or [Undo]:** ***@0,0,2*** **Specify next point or [Close/ Undo]:** ***@-3,0,0*** **Specify next point or [Close/ Undo]:** ***c***	4. We'll use relative coordinates as indicated to draw the other side.
	**Command:** ***[enter]***	5. Using the Endpoint OSNAP, connect the two walls as shown in Figure 1.025.

TOOLS	COMMAND SEQUENCE	STEPS

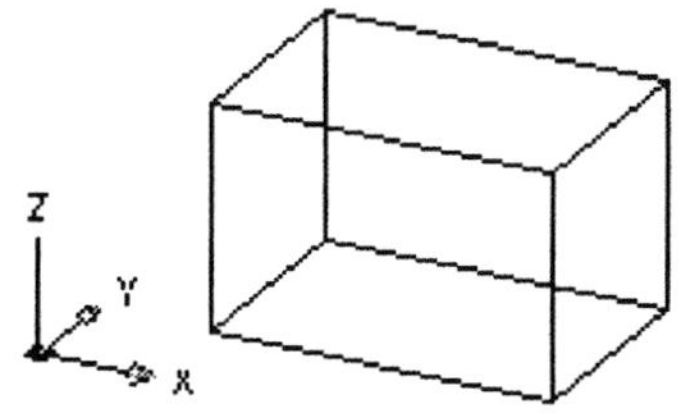

Figure 1.025

TOOLS	COMMAND SEQUENCE	STEPS
	Command: ***saveas***	6. Remember to save occasionally. Save the drawing as My Stick House in the C:\Steps3D\Lesson01 folder.
	**Command:** ***l*** **Specify first point:** ***mid***	7. Now use spherical coordinates as indicated to locate the peak of the roof. Begin at the midpoint of the upper line marking the west side of the house.
	Specify next point or [Undo]: ***@1.5<0<60*** **Specify next point or [Undo]:** ***[enter]***	8. We'll want a **1.5"** line at **0°** on the XY-plane and **60°** upward.
	**Command:** ***[enter]*** **Specify first point:** ***mid*** **Specify next point or [Undo]:** ***@1.5<180<60*** **Specify next point or [Undo]:** ***[enter]***	9. Repeat Steps 7 and 8 to locate the roof peak on the opposite wall. Your drawing looks like Figure 1.026.

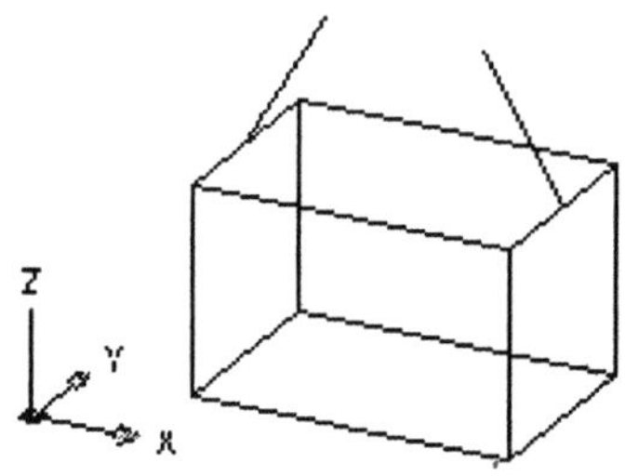

Figure 1.026

TOOLS	COMMAND SEQUENCE	STEPS
		10. Using the Endpoint OSNAP, draw the roof as shown in Figure 1.027. Erase the roof peak locators.

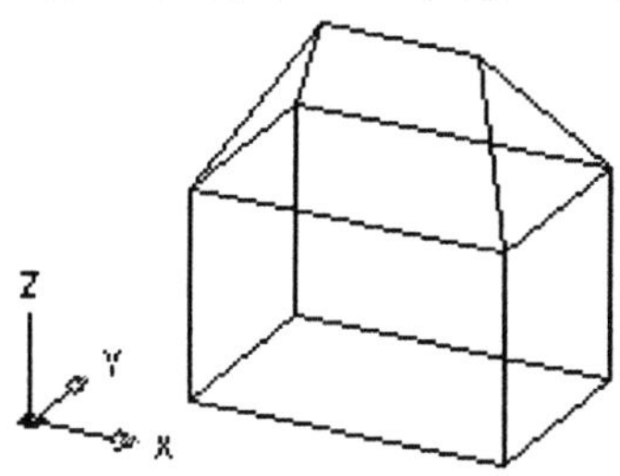

Figure 1.027

TOOLS	COMMAND SEQUENCE	STEPS
	Command: ***qsave***	11. Save the drawing but don't exit.
	Command: ***vp***	12. Take a few moments and examine the model using different viewpoints. Return to a location of 1,-2,1 when you've finished.

Congratulations! You've created your first three-dimensional drawing.

You may have noticed some three-dimensional idiosyncrasies. If not, let me list a few.

- Ortho works only on the XY-plane.
- Object and Polar Tracking work only on the XY-plane.
- OSNAPs work on most objects regardless of their XYZ-coordinates.
- It's impossible to locate a point in a three-dimensional drawing by arbitrarily picking a point on the screen. Remember that your screen is two-dimensional. An arbitrary point lacks a definition on one of the X-, Y-, or Z-axes, so there's no guarantee where it may actually be. *Always identify a three-dimensional point using a coordinate entry method or an OSNAP!* (Now you see why we put so much emphasis on coordinates and OSNAPs in the basic book!)
- Use three-dimensional coordinates just as you did two-dimensional coordinates when modifying the drawing (copying, moving, etc).
- Only very rarely will you want to create a three-dimensional object using a simple ***Line*** command. (After all, how often does your boss request stick figure drawings?)

Let's look next at adding a new dimension to our objects. We'll call our new dimension *thickness* and do marvelous things with it.

1.3.2 Using the Thickness and Elevation System Variables

Thickness is that property of an object that takes it from a stick figure to a true three-dimensional object. Until now, all the objects you've drawn have been stick figures. That is, they've all existed on a single two-dimensional plane. Even the lines we used to create our stick house in the last exercise existed in two-dimensional planes (although they crossed three-dimensional space).

> Point filters are also very useful in Z-Space. Refer to Section 3.5.1 in the basic *One Step at a Time* text (Part I) for details on the use of point filters.

Clear as milk? Let me help.

Consider each line created in our stick house. (Consider each line individually – not as it relates to other lines or coordinates.) Describe each in terms of length, width, and height. In each case, you can describe the line using two of the three terms mentioned. In no case can you use all three terms. Consider the first line you drew. It has a length of 3 units and a width (or lineweight) of 0.01" (AutoCAD's default). But it has no height – no *thickness*.

Drawing with thickness is as easy as setting the system variable, like this:

Command: ***thickness*** **(or** ***th*****)**

Enter new value for THICKNESS <0.0000>: ***.25***

All objects drawn while the **Thickness** system variable is set to **.25** will have a thickness of .25. This brings up another very important point: *Always remember to reset the* ***Thickness*** *system variable to* ***0*** *when you've finished drawing an object.*

Let's redraw our walls using **Thickness**.

> You can access the ***Thickness*** command from the Format pull-down menu. Follow this path:
> *Format – Thickness*

Do This: 1.3.2.1	Drawing with Thickness

I. Be sure you're still in the *My Stick House.dwg* file in the C:\Steps3D\Lesson01 folder. If not, please open it now.

II. Follow these steps.

TOOLS	COMMAND SEQUENCE	STEPS
No Button Available	**Command:** ***th*** **Enter new value for THICKNESS <0.0000>:** ***2***	1. Set the **Thickness** system variable to 2.
	Command: ***l*** **Specify first point:** ***6,1*** **Specify next point or [Undo]:** ***@3<0*** **Specify next point or [Undo]:** ***@2<90*** **Specify next point or [Close/Undo]:** ***@3<180*** **Specify next point or [Close/Undo]:** ***c***	2. Draw the walls as indicated.
	Command: ***th*** **Enter new value for THICKNESS <2.0000>:** ***0***	3. Reset the **Thickness** to ***0***.

Notice that we used simple two-dimensional coordinate entry. The **Thickness** system variable took care of the Z requirements.

Another very useful tool to employ in three-dimensional drafting is the **Elevation** system variable. When we drew our original house, we had to enter a location on the Z-axis as part of the coordinate whenever the value of Z wasn't zero (whenever it was above or below zero-Z). The **Elevation** system variable is designed to minimize the need for entering that third number.

The command looks like this:

Command: ***elevation***

Enter new value for ELEVATION <0.0000>: ***[enter the desired elevation]***

Once set, AutoCAD draws all objects at the identified elevation.

> Use the ***Elev*** command to set both the **Elevation** and **Thickness** system variables at once. The command looks like this:
>
> **Command:** ***elev***
>
> **Specify new default elevation <0.0000>:** ***[enter the elevation]***
>
> **Specify new default thickness <0.0000>:** ***[enter the thickness]***

Let's try using elevation.

Do This: 1.3.2.2	**Drawing with Thickness and Elevation**

I. Be sure you're still in the *My Stick House.dwg* file in the C:\Steps3D\Lesson01 folder. If not, please open it now.

II. Follow these steps.

TOOLS	COMMAND SEQUENCE	STEPS
No Button Available	**Command:** ***elev*** **Specify new default elevation <0.0000>:** ***2*** **Specify new default thickness <0.0000>:** ***1.299***	1. Set the **Elevation** to ***2*** and the **Thickness** to ***1.299***.
	Command: ***l*** **Specify first point:** ***6.75,2*** **Specify next point or [Undo]:** ***8.25,2*** **Specify next point or [Undo]:** ***[enter]***	2. Draw a line as indicated. Notice that, although no Z-axis location is given, AutoCAD assumes an elevation of 2.
	Command: ***th*** **Enter new value for THICKNESS <1.2990>:** ***0***	3. Reset the **Thickness** to ***0***.
	Command: ***l***	4. Draw the roof as shown in Figure 1.028.

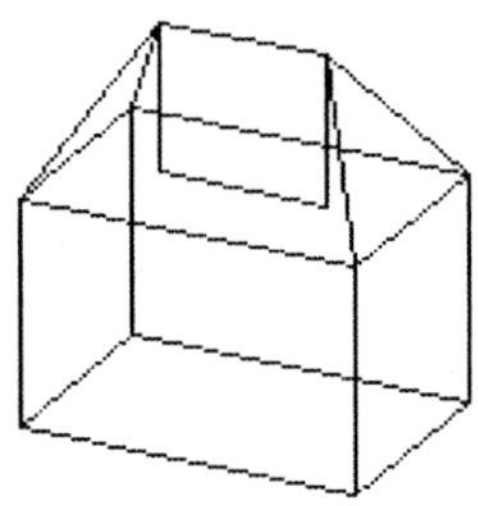

Figure 1.028

	Command: ***props***	5. And now let's use the Properties Palette to modify the roof. Select the roof locator line and then enter the ***Properties*** command.

Tools	Command Sequence	Steps

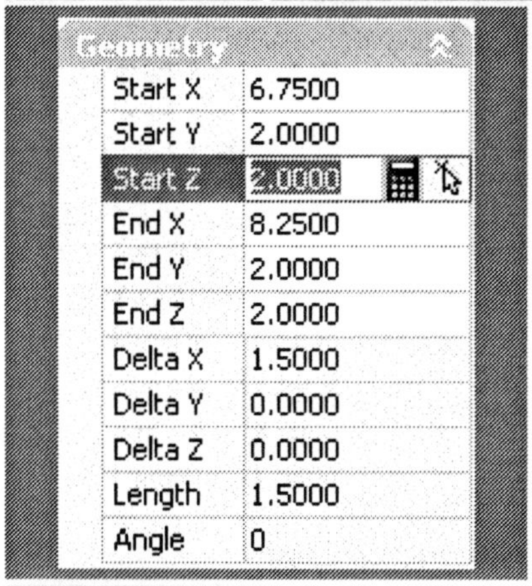

6. In the Geometry section of the Properties Palette, pick on **Start Z**. Notice that a **Pick** button appears to the right.

7. Pick on the **Pick** button. AutoCAD highlights that location and allows you to redefine it by picking another point on the screen. Using the Endpoint OSNAP, pick the upper point of that same line (the western point of the roof peak).

8. Repeat Steps 6 and 7 to relocate the **End Z** point. Your drawing looks like Figure 1.029.

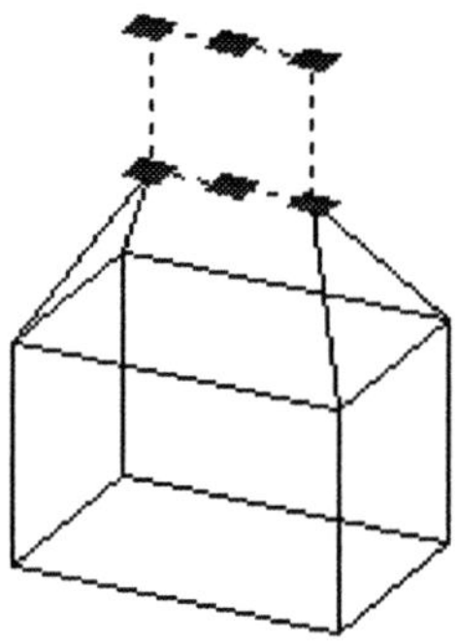

Figure 1.029

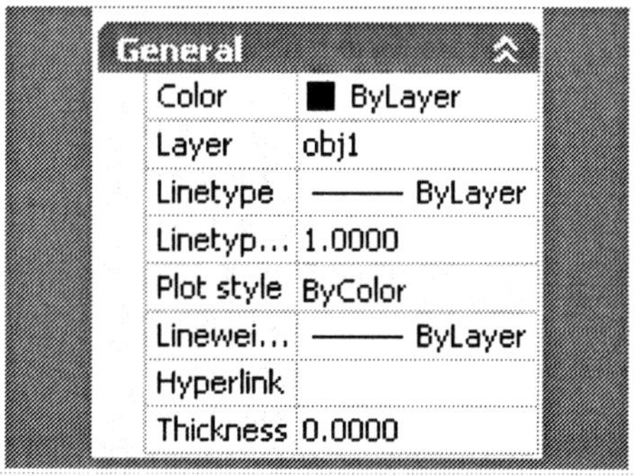

9. Now change the thickness of the line to ***0***. The **Thickness** property is located in the General section of the Properties Palette.

Clear the grips.

Your drawing now looks like Figure 1.030.

TOOLS	COMMAND SEQUENCE	STEPS

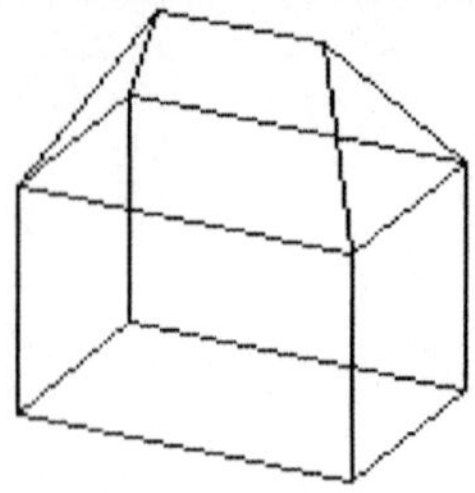

Figure 1.030

	**Command:** ***qsave***	10. Save the drawing.

> Of course, it would've been a lot easier to simply erase the roof peak line and redraw it without thickness. But I wanted to give you some experience working with the Properties Palette and 3-dimensional objects.

Do the two houses look the same? They should. But don't let that fool you. There are subtle (and remarkable) differences. We'll look at those in a few moments. First, let me list some things to remember about drawing with thickness and elevation (don't you just love my lists?).

- Always remember to reset **Thickness** and **Elevation** to zero when you've finished drawing an object. Otherwise, you may have to redo the next object when it's drawn with incorrect properties.
- Use the Properties Palette to change the thickness and elevation of drawn objects.
- You can enter positive or negative values for **Elevation** and **Thickness**.
- Closed objects behave differently from opened objects when drawn with thickness. We'll see examples of this in our next section.

1.4 Three-Dimensional Viewing Made Easy

Have you tried to view your houses from different viewpoints yet? Is it difficult to tell up from down in your drawing? Look back at Figures 1.001 and 1.002; remember how we needed the UCS icon to tell where we were?

Our next two commands will help us to see our drawings more clearly. The first – ***Hide*** – is simple and fast. The second – ***Shademode*** – is slower but more colorful.

1.4.1 The *Hide* Command

The ***Hide*** command removes "hidden" lines. That is, it removes lines that you won't normally see because they're behind other objects in the drawing. (It doesn't remove lines drawn with the **Hidden** line type.)

It looks like this:

Command: ***hide***

Regenerating model.

It's just that simple. But the results are like a light in a dark tunnel. Give it a try in the *My Stick House* drawing. Your drawing will look like Figures 1.031 and 1.032. Notice the difference between the two houses. The first (western) is unaffected by the command. You drew this house with lines only. (Like the little piggy's house that was made of sticks. There's nothing to hide behind.) The second (eastern) house, however, looks quite different. This house has solid walls (like the little piggy's brick house) – albeit with a stick roof (that ole wolf wouldn't think to climb, would he?). You used lines with thickness – or three-dimensional objects – to draw the walls.

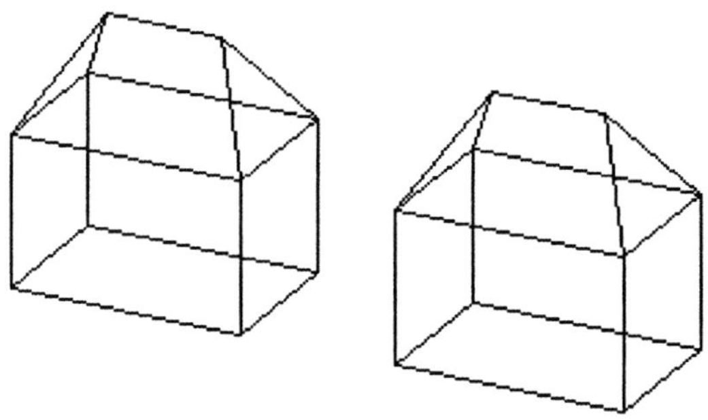

Figure 1.031: Before the ***Hide*** Command

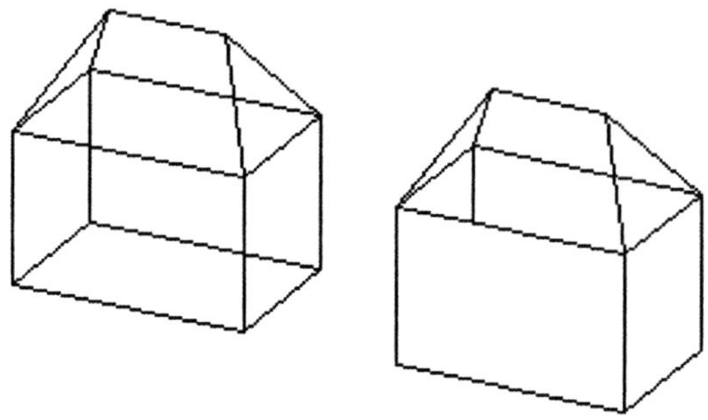

Figure 1.032: After the ***Hide*** Command

You can change the viewpoint (and use modifying tools such as ***Move*** and ***Copy***) while hidden lines are removed, but the commands will cause hidden lines to be visible again. Repeat the ***Hide*** command after the modifying procedure or change in view to again remove hidden lines. Try using the ***Hide*** command from different viewpoints. (To restore all hidden lines to view, regenerate the drawing).

This might be a good time to examine how thickness affects closed objects.

Do This: 1.4.1.1	**Thickness and Closed Objects**

I. Open the *Closed Objects.dwg* file in the C:\Steps3D\Lesson01 folder. The drawing looks like Figure 1.033. (Each object has been drawn with a thickness of 1.)

II. Follow these steps.

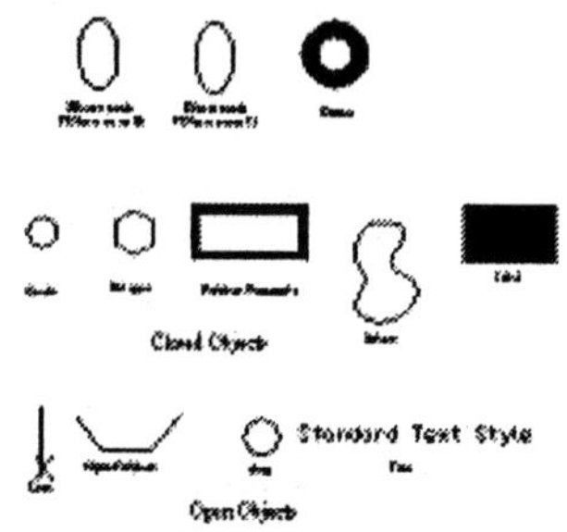

Figure 1.033

TOOLS	COMMAND SEQUENCE	STEPS
	**Command: *-vp***	1. Set the viewpoint to ***1,-1,1***.
No Button Available	**Command: *hide***	2. Enter the ***Hide*** command. Your drawing looks like Figure 1.4.034.

Tools	Command Sequence	Steps

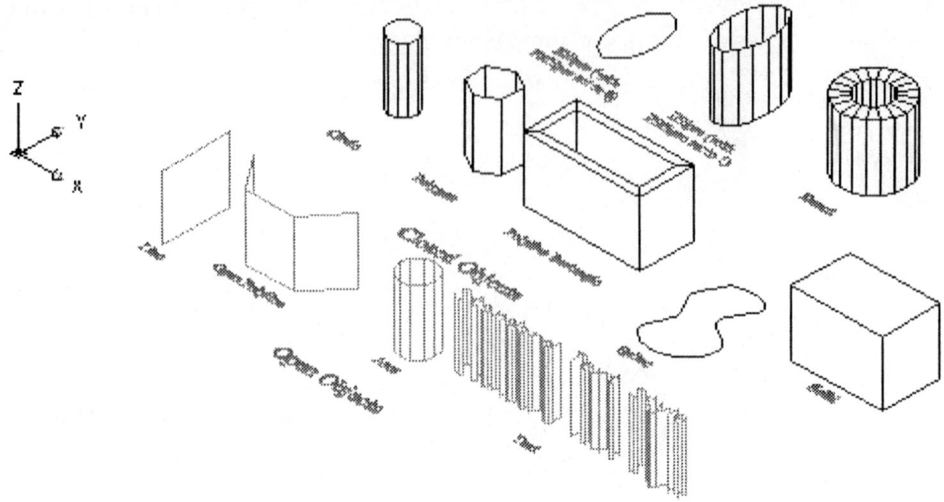

Figure 1.034

Let's consider each item.

- Closed objects:
  - Circles become cylinders with solid cores.
  - Polygons and closed polylines contain length, width, and height as shown, but the area enclosed by the polyline is unaffected by thickness.
  - Splines can't contain thickness.
  - Solids become solid blocks.
  - Ellipses drawn with the **Pellipse** system variable set to **0** can't contain thickness.
  - Ellipses drawn with the **Pellipse** system variable set to **1** behave like polylines.
  - Donuts behave like polylines.
- Open Objects:
  - Lines contain length, width, and height as shown.
  - Open polylines contain length, width, and height as shown.
  - Arcs contain length, width, and height. Closed arcs (shown here) look like circles but have no solid core. (Use these to create "holes.")
  - AutoCAD standard text contains length, width, and height as shown. Notice, however, that the callouts (drawn using a true-type font) can't contain thickness.

1.4.2 The *Shademode* Command

The ***Hide*** command serves to remove hidden lines and make it easier for you to tell up from down in a drawing. The **Shademode** system variable, however, actually fills in a solid object or three-dimensional face and gives it more of a *real* appearance. Compare Figure 1.035 with Figure 1.036. Shading gives the model a fuller, more complete look.

Notice that the UCS icon has changed. This colorful tool is the Shademode icon. It provides directional information just as the UCS icon does, but its presence lets you know that you're in one of the **Shademode** options.

But the comparison with the ***Hide*** command doesn't end here. Whereas modification, regeneration, or changes in view will make hidden lines visible again, **Shademode** causes no such trauma to your model. Once active, **Shademode** will remain active until reset to the **2D wireframe** option.

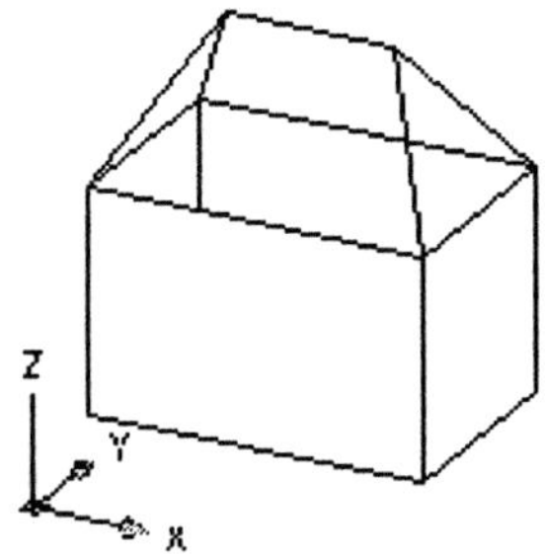

Figure 1.035: After the ***Hide*** Command

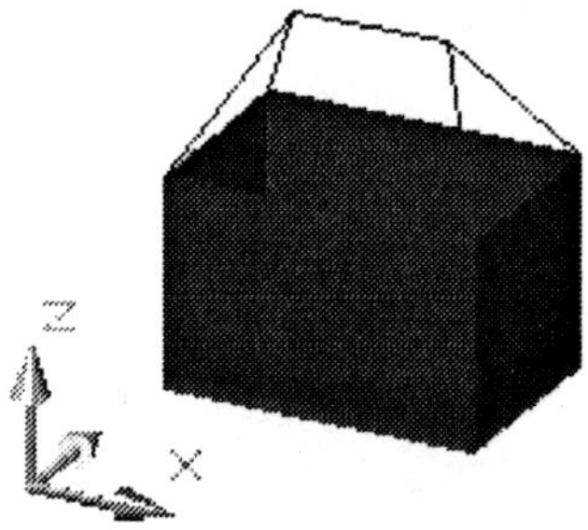

Figure 1.036: Shaded House

> Some surfaces on a shaded model appear to have shadows. This helps distinguish one surface from another. The light source for this shadow comes from behind your left shoulder. We'll discuss assigning other light sources in Lesson 11.

The ***Shademode*** command sequence looks like this:

Command: ***shademode***

Current mode: 2D wireframe

Enter option [2D wireframe/3D wireframe/Hidden/Flat/Gouraud/fLat+edges/ gOuraud+edges] <2D wireframe>: ***[enter your choice of shading modes]***

Let's look at each option.

- Remember the **2D wireframe** option! It's your ticket out of the land of Shademode. Use it to clear any shading in the model and replace the Shademode icon with the UCS icon.
- At first, the **3D wireframe** option appears the same as the previous option. But the **3D wireframe** option won't display raster images, linetypes, and lineweights (the **2D wireframe** option will).
- The **Hidden** option is similar to the ***Hide*** command (it removes hidden lines). However, the lines will remain hidden even during regeneration or modification. As with the **3D wireframe** option, raster images, linetypes, and lineweights will disappear when you activate the **Hidden** option.
- The **Flat** option shades the faces but doesn't show (or highlight) the edges. The appearance is like that resulting from a flat paint. Materials assigned to the object will show (more on materials in Lesson 11).
- **Gouraud** presents the best image of all. This option shades faces and smoothes edges for a more realistic appearance. Materials assigned to the object will show.
- The other two options – **flat+edges** and **gOuraud+edges** – work the same as the **Flat** and **Gouraud** options except that edges will be highlighted.

Let's experiment.

> You can also find the **Shademode** options by cursor or dynamic input menu once you've entered the command. Alternately, you can select the option from the View pull down menu. Follow this path:
>
> *View – Shade – [option]*

Do This: 1.4.2.1	Shading Objects

I. Reopen the *My Stick House.dwg* file in the C:\Steps3D\Lesson01 folder.

II. Zoom in around the house on the right (the one with the walls drawn with thickness).

III. Follow these steps.

TOOLS	COMMAND SEQUENCE	STEPS
	**Command:** ***props***	1. Change the linetype of the rooflines to **Dashed**. (You'll need to load the **Dashed** linetype first.) Your drawing looks like Figure 1.037.

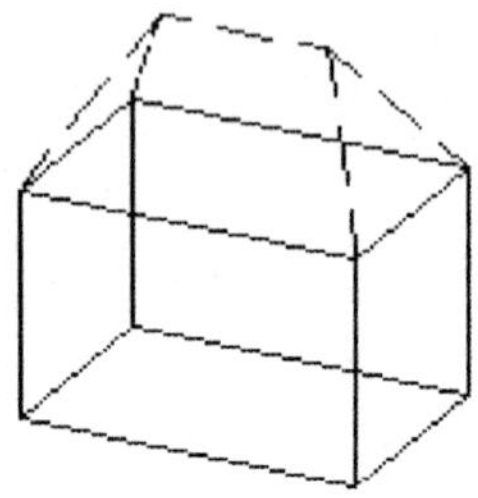

Figure 1.037

TOOLS	COMMAND SEQUENCE	STEPS
No Button Available	**Command:** ***shademode***	2. Enter the ***Shademode*** command. There's no button or hotkey; however, there's a Shade toolbar that provides buttons for each of the **Shademode** options. (Note: It isn't necessary to enter the ***Shademode*** command if you're going to use the buttons on the Shade toolbar.)
 3D Wireframe Button (Shade Toolbar)	**Current mode: 2D wireframe** **Enter option [2D wireframe/3D wireframe/Hidden/Flat/Gouraud/fLat+edges/gOuraud+edges] <2D wireframe>:** ***3***	3. Tell AutoCAD you wish to use the 3D wireframe option. The UCS icon changes to the Shademode icon and the hidden lines become solid (Figure 1.038).

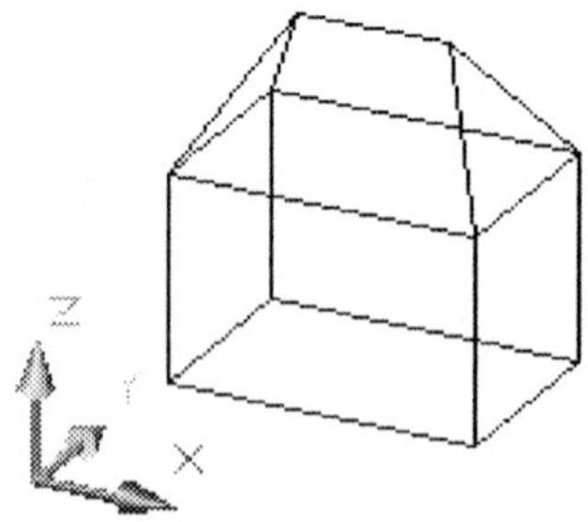

Figure 1.038

TOOLS	COMMAND SEQUENCE	STEPS
Hidden Button	**Command:** ***[enter]*** **Current mode: 3D wireframe** **Enter option [2D wireframe/3D wireframe/Hidden/Flat/Gouraud/ fLat+edges/gOuraud+edges] <3D wireframe>:** ***h***	4. Try the **Hidden** option. AutoCAD removes hidden lines (just as the ***Hide*** command did).
Flat Shade Button	**Command:** ***[enter]*** **Current mode: Hidden** **Enter option [2D wireframe/3D wireframe/Hidden/Flat/Gouraud/ fLat+edges/gOuraud+edges] <Hidden>:** ***f***	5. Now **Flat** shade the house. Your drawing looks like Figure 1.039.
	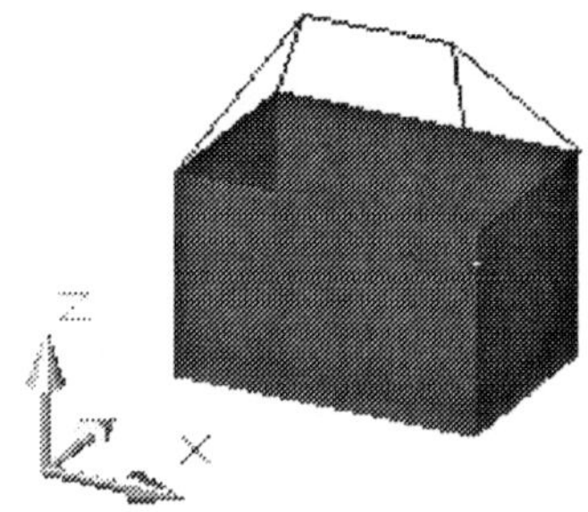 Figure 1.039	
Flat + Edges Button	**Command:** ***[enter]*** **Current mode: Flat** **Enter option [2D wireframe/3D wireframe/Hidden/Flat/Gouraud/ fLat+edges/gOuraud+edges] <Flat>:** ***l***	6. Compare the **fLat+edges** option (Figure 1.040) with the **Flat** option you used in Step 5.
	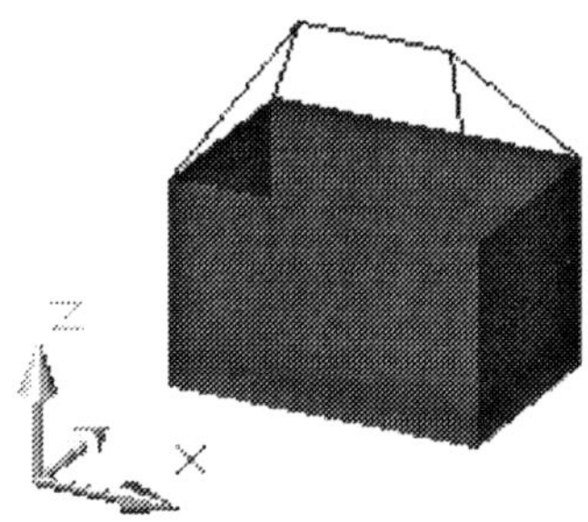 Figure 1.040	
		7. Return to the *Closed Objects.dwg* file.

TOOLS	COMMAND SEQUENCE	STEPS
Gouraud Button	**Command:** ***shademode*** **Current mode: 2D wireframe** **Enter option [2D wireframe/3D wireframe/Hidden/Flat/Gouraud/ fLat+edges/gOuraud+edges] <2D wireframe>:** ***g***	8. Now try the **Gouraud** option. Your drawing looks like Figure 1.041.

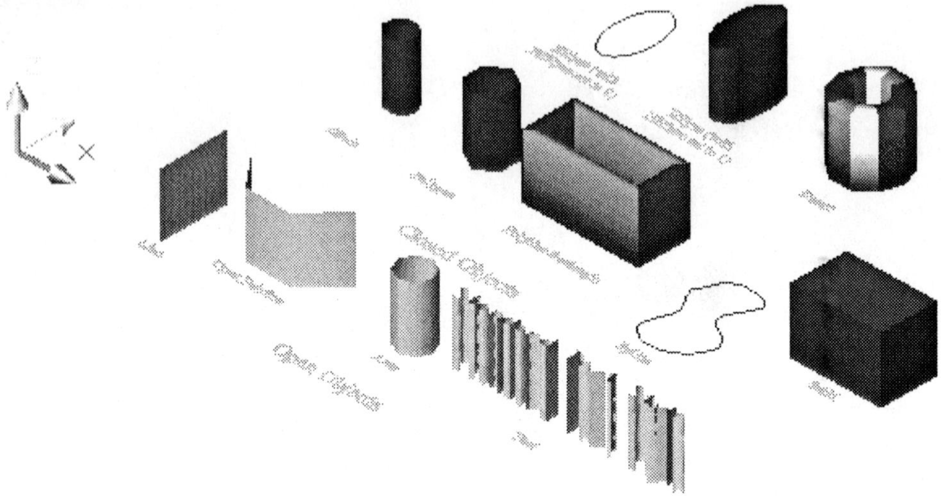

Figure 1.041

TOOLS	COMMAND SEQUENCE	STEPS
Gouraud + Edges Button	**Command:** ***[enter]*** **Current mode: Gouraud** **Enter option [2D wireframe/3D wireframe/Hidden/Flat/Gouraud/ fLat+edges/gOuraud+edges] <Gouraud>:** ***o***	9. Now compare the **gOuraud+edges** option (Figure 1.042) to the **Gouraud** option you used in Step 8.

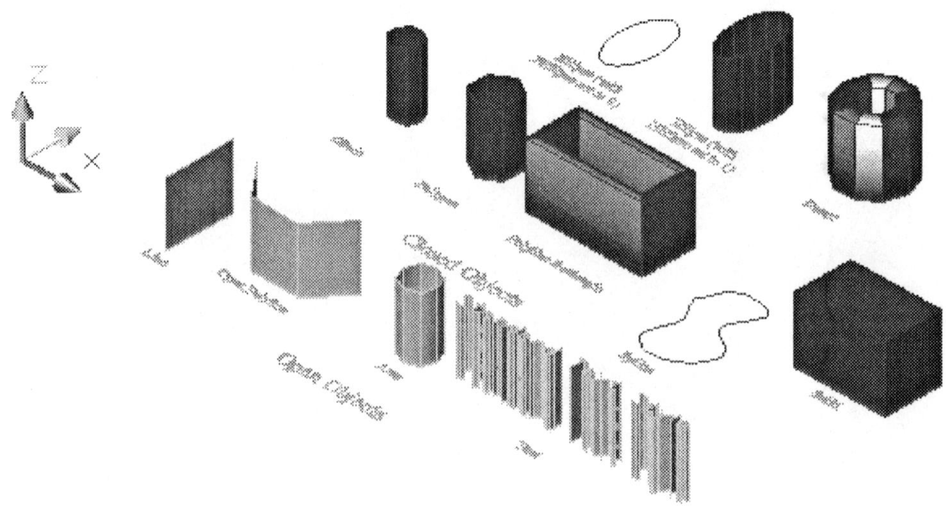

Figure 1.042

TOOLS	COMMAND SEQUENCE	STEPS
2D Wireframe Button	**Command:** ***[enter]*** **Current mode: Gouraud+Edges** **Enter option [2D wireframe/3D wireframe/Hidden/Flat/Gouraud/ fLat+edges/gOuraud+edges] <Gouraud+Edges>:** ***2***	10. Restore the drawing to a 2D wireframe and the Shademode icon to the UCS icon.
	Command: ***quit***	11. Exit both drawing]s without saving.

1.5 Extra Steps

- Repeat all of Exercise 1.4.2.1 using the *Closed Objects.dwg* file in the C:\Steps3D\Lesson01 folder to see how shading affects the different objects. View each Shademode option from at least five different viewpoints.

  Open some of the three-dimensional sample files that ship with AutoCAD (look in the \Sample subfolder of the AutoCAD folder). Practice your viewpoint manipulation and shading options. I recommend the following files: *Welding Fixture Model.dwg* or *Hotel Model.dwg*.
- Print/Plot a drawing or two using Paper Space and shading.

1.6 What Have We Learned?

Items covered in this lesson include:

- *The UCS icon and the Right-Hand Rule*
- *Adjusting your view of the model*
  - *Isometric view*
  - *Dimetric view*
  - *Trimetric view*
  - *Plan view*
  - *Coordinate approach*
  - *Compass and tripod approaches*
  - *Dialog box approach*
- *Coordinates in a three-dimensional drawing*
  - *Cartesian Coordinate entry*
  - *Spherical and Cylindrical Coordinate entry*
- *Commands*
  - ***Ucsicon***
  - ***VPoint***
  - ***VPoint***
  - ***Thickness***
  - ***Elevation***
  - ***Elev***
  - ***Hide***
  - ***Shademode***
  - ***Shade***

What a list! Stop and catch your breath!

We covered a tremendous amount of new material in this lesson. But it's all fundamental to three-dimensional drafting. After some practice, you'll find this material as easy as two-dimensional work.

I must caution you, however, about just how *fundamental* this material is. You must achieve at least a small degree of expertise with these methods and those in Lesson 2 to be able to function effectively in Z-space. But that's the benefit of a good textbook! You can repeat the lesson(s) until you're comfortable.

We've seen how to maneuver around a three-dimensional model, how to improve our view for better navigation (and aesthetics), and some fundamentals of creating three-dimensional objects. But there's much more to consider! How do you draw a three-dimensional object at an angle – such as creating a solid roof instead of the stick figure outline on your stick house? Is there an easier way to rotate your viewpoint in Z-space for a better view?

We'll continue our study of Z-Basics in Lesson 2 where we'll answer some of these questions (and present some others). But first, we should practice what we've learned so far.

1.7 Exercises

1. through 8. Create the "w" drawings in Appendix B (Refer to the "su" drawings for a clearer image.) Follow these guidelines.
    1.1. Don't try to create the dimensions yet.
    1.2. Start each drawing from scratch and use the default settings.
    1.3. Adjust the viewpoint as needed to help your drawing.
    1.4. Save each drawing as *My [title].dwg* (as in *MyB-1w.dwg*) in the C:\Steps3D\Lesson01 folder.

9. Create the drawing in Figure 1.043 according to the following parameters:
    9.1. Start the drawing from scratch and use the default settings.
    9.2. Set the viewpoint to 8.5,11,5.75.
    9.3. Don't attempt to draw the dimensions yet.
    9.4. Use a thickness setting of 0.
    9.5. The depth of the piece is ½".
    9.6. Save the drawing as *My Twisted Y.dwg* in the C:\Steps3D\Lesson01 folder.

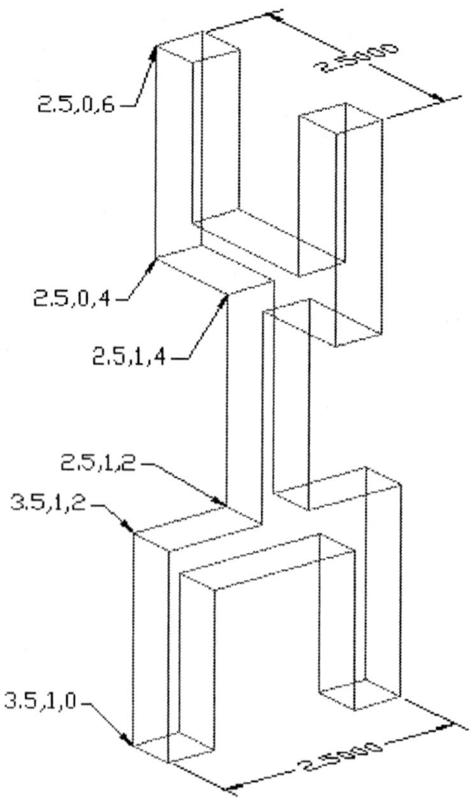

Figure 1.043

10. Create the drawing in Figure 1.044 according to the following parameters:
    10.1. Start the drawing from scratch and use the default settings.
    10.2. Don't attempt to draw the dimensions yet.
    10.3. Text is in Paper Space and is 3/16" and 1/8".
    10.4. Title block text is ¼", 3/16", and 1/8".
    10.5. The viewpoint for each viewport is indicated.
    10.6. Use lines with thickness.
    10.7. Use the Hide command to achieve the diametric view.
    10.8. Save the drawing as *My Block.dwg* in the C:\Steps3D\Lesson01 folder.

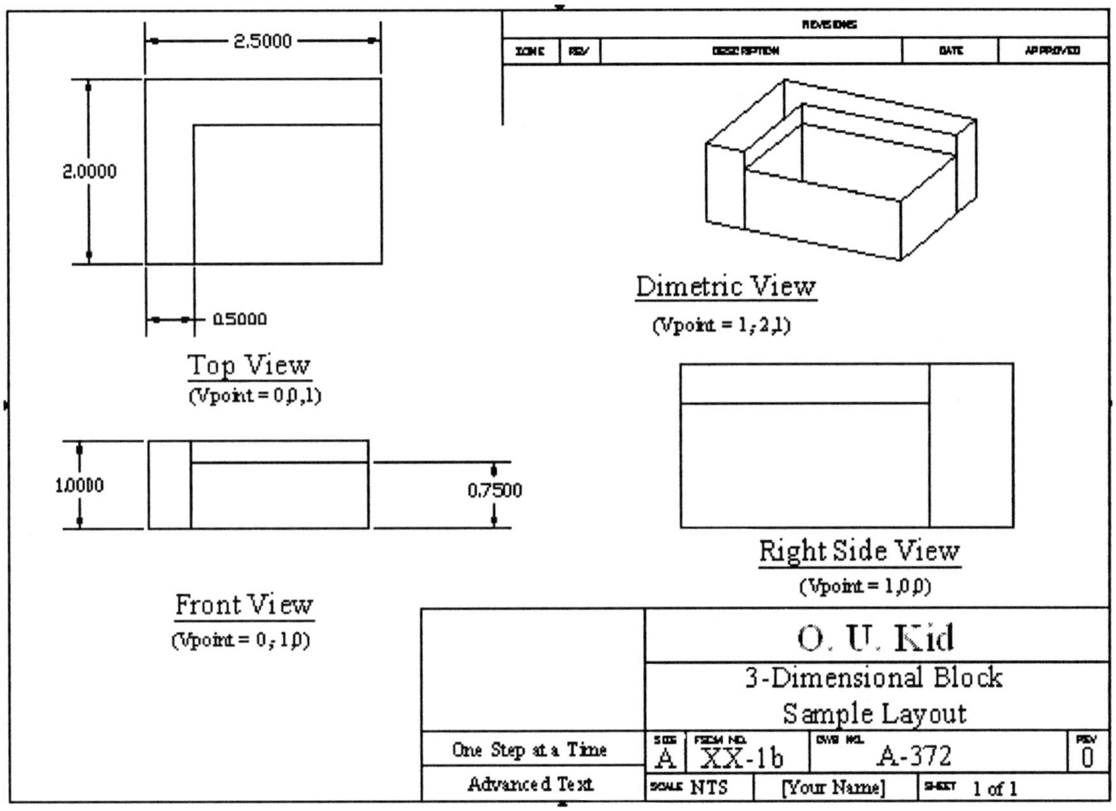

Figure 1.044

11. Create the drawing in Figure 1.045 according to the following parameters.
    11.1. Start the drawing from scratch and use the default settings.
    11.2. Use polylines with 1/16" width for everything except the handle.
    11.3. Use an thickness of 1/16" for all objects.
    11.4. I used a scale of 1:4 in the plan and elevation views but no scale in the isometric.
    11.5. Don't attempt to draw the dimensions yet.
    11.6. Text is in Paper Space and is 3/16" and 1/8".
    11.7. Title block text is ¼", 3/16", and 1/8".
    11.8. Save the drawing as *MyGrill.dwg* in the C:\Steps3D\Lesson01 folder.

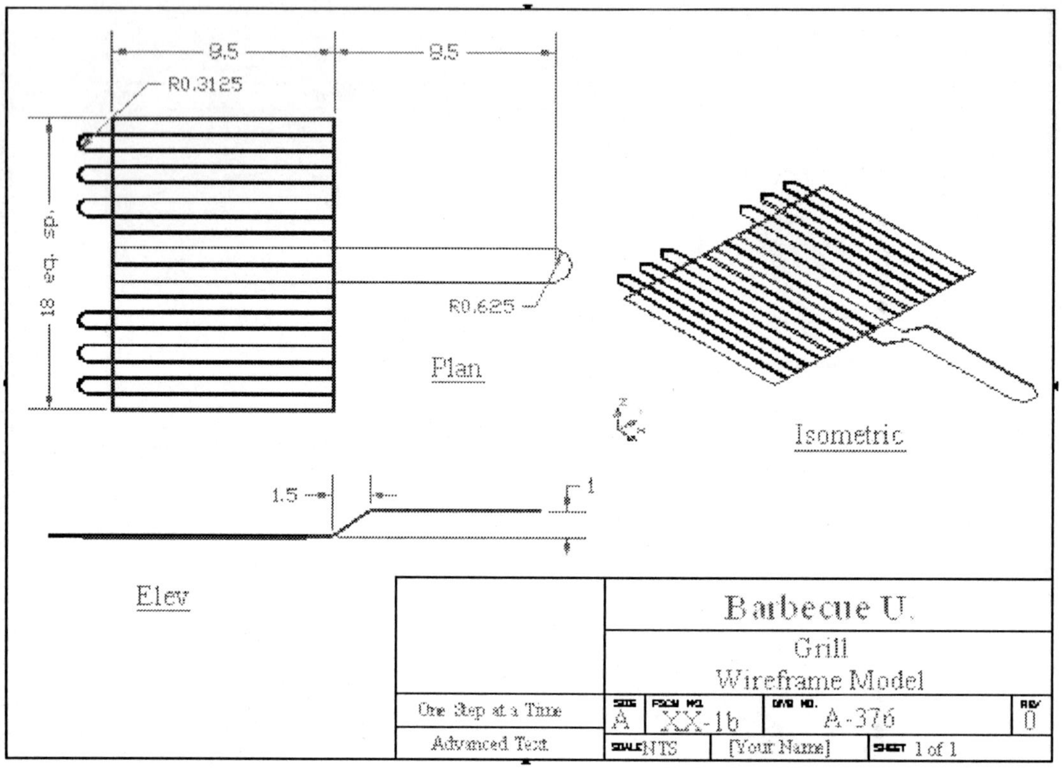

Figure 1.045

12. Create the drawing in Figure 1.046 according to the following parameters:
    12.1. Start the drawing from scratch and use the default settings.
    12.2. I used a scale of 1:2 for all views.
    12.3. Don't attempt to draw the dimensions yet.
    12.4. Text is in Paper Space and is 3/16" and 1/8"
    12.5. Title block text is ¼", 3/16", and 1/8".
    12.6. You'll find it easier to fillet the corners rather than drawing arcs.
    12.7. Save the drawing as *MyBookEnd.dwg* in the C:\Steps3D\Lesson01 folder.

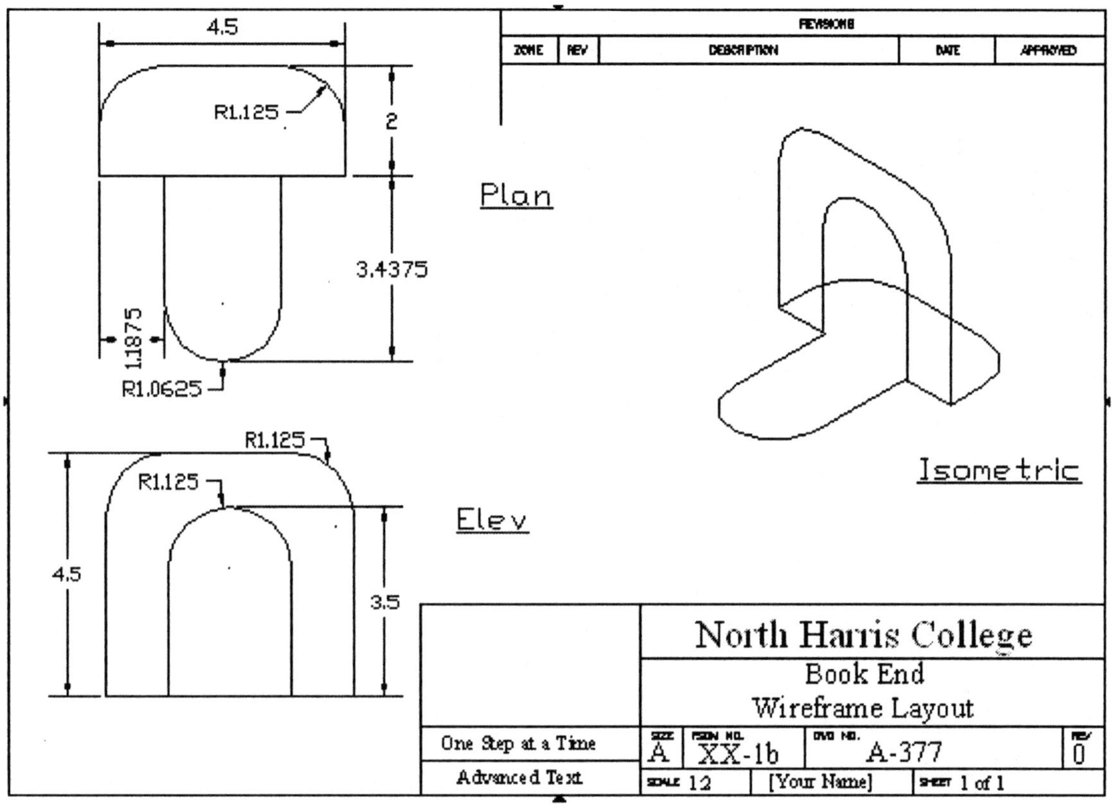

Figure 1.046

13. Create the drawing in Figure 1.047 according to the following parameters:
    13.1. Start the drawing from scratch and use the default settings.
    13.2. Use 0.6mm lineweights for all lines.
    13.3. Don't attempt to draw the dimensions yet.
    13.4. Text is in Paper Space and is 3/16" and 1/8"
    13.5. Title block text is ¼", 3/16", and 1/8".
    13.6. I used ½" diameter circles with ½" thickness for the legs.
    13.7. Save the drawing as *MyMagRack.dwg* in the C:\Steps3D\Lesson01 folder.

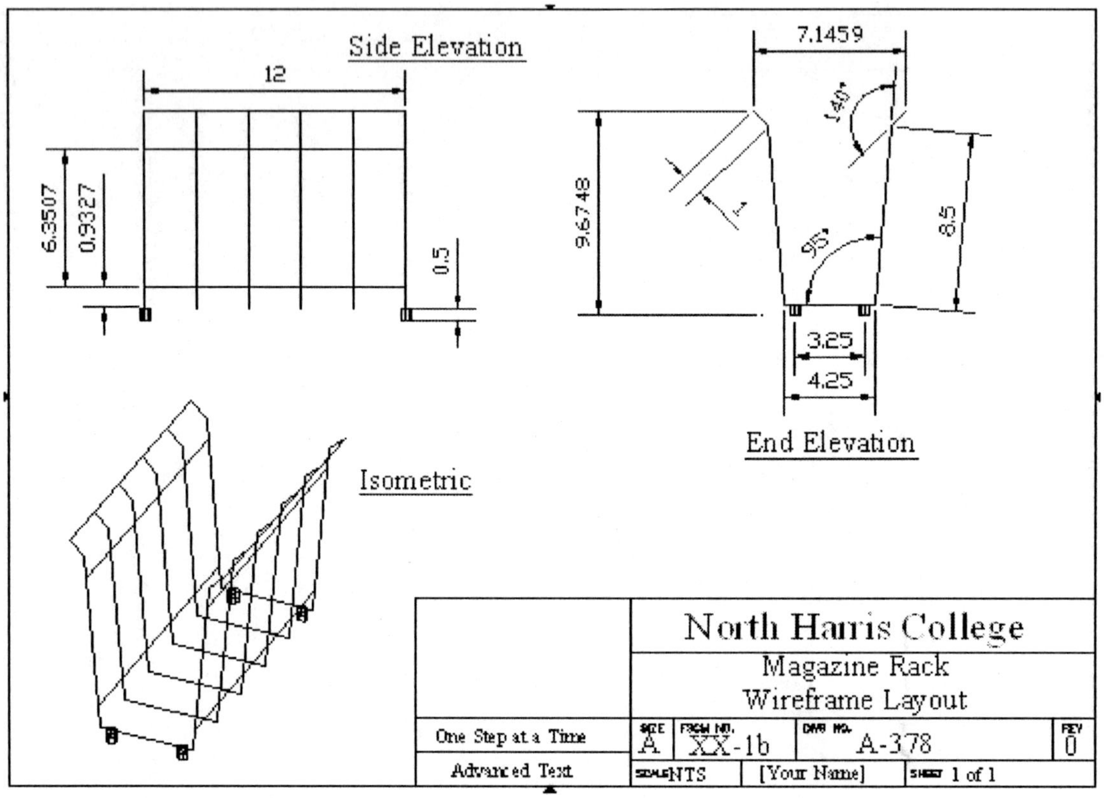

Figure 1.047

14. Create the drawing in Figure 1.048 according to the following parameters:
    14.1. Start the drawing from scratch and use the default settings.
    14.2. Don't attempt to draw the dimensions yet.
    14.3. Text is in Paper Space and is 3/16" and 1/8"
    14.4. Title block text is ¼", 3/16", and 1/8".
    14.5. The viewpoint for each viewport is indicated.
    14.6. Use lines without thickness except when drawing the slot.
    14.7. Save the drawing as *MyBookEnd.dwg* in the C:\Steps3D\Lesson01 folder.

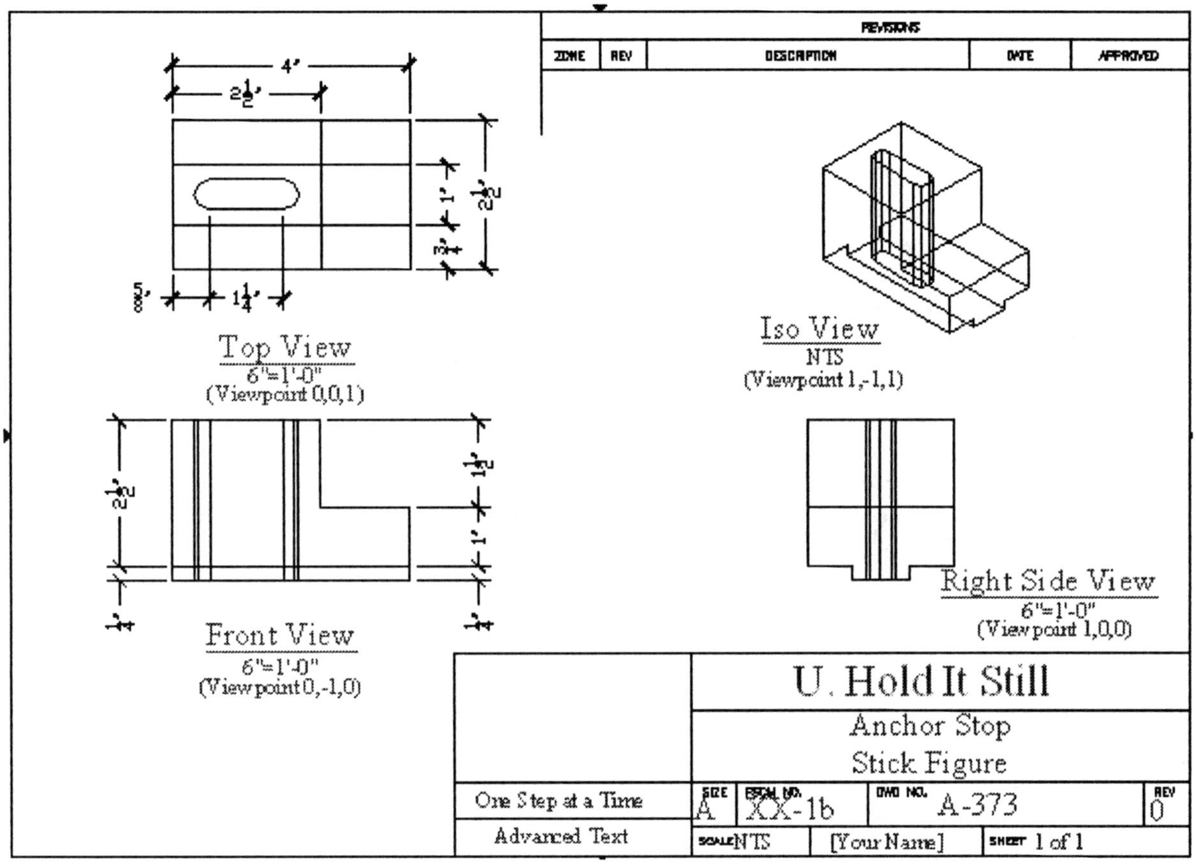

Figure 1.048

15. Here's a challenge. Create the drawing in Figure 1.049 according to the following parameters:
    15.1. Start the drawing using *template #1* found in the C:\Steps3D\Lesson01 folder.
    15.2. Don't attempt to draw the dimensions yet.
    15.3. Text is in Paper Space and is 3/16" and 1/8".
    15.4. Title block text is ¼", 3/16", and 1/8".
    15.5. The crown is made up of solids drawn with thickness.
    15.6. The round pieces are donuts drawn with thickness.
    15.7. The diametric view has been Gouraud shaded.
    15.8. Save the drawing as *MyQueen.dwg* in the C:\Steps3D\Lesson01 folder.

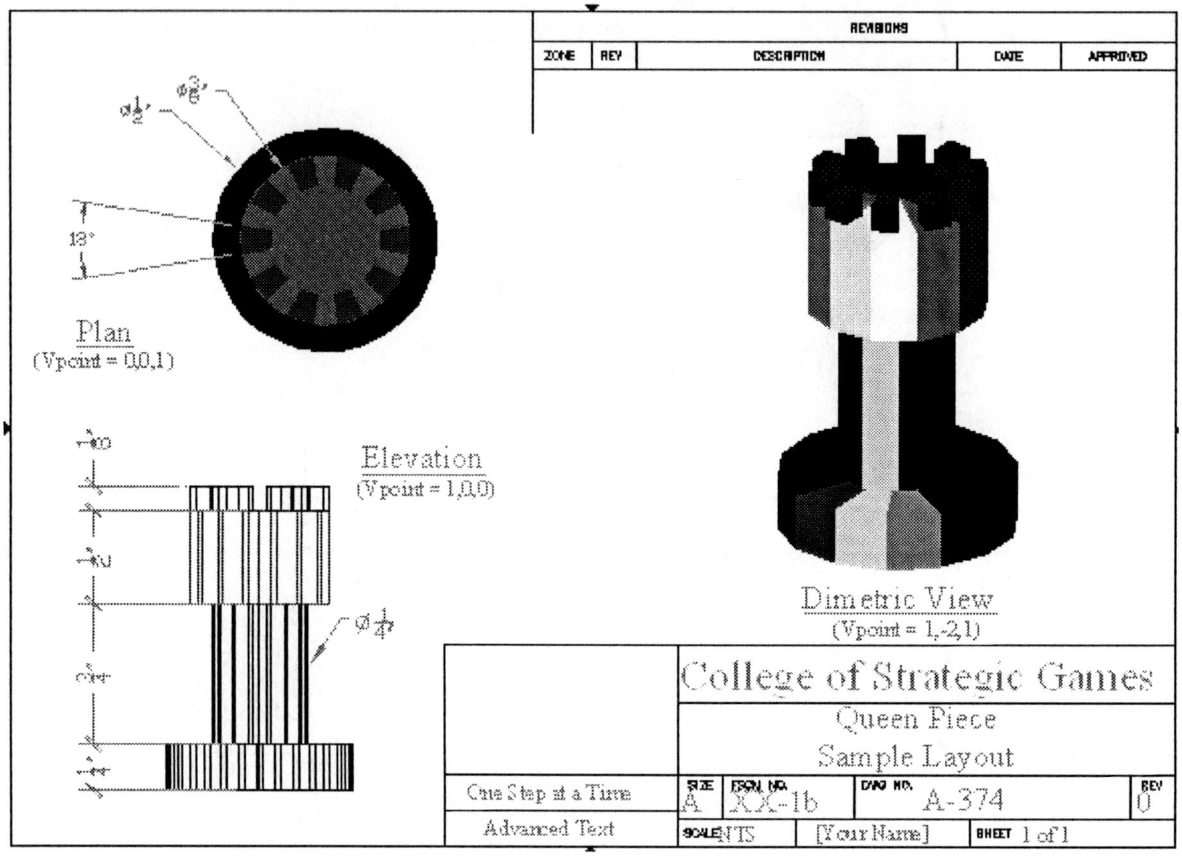

Figure 1.049

16. Here's another challenge. Create the drawing in Figure 1.050 according to the following parameters:
    16.1. Start the drawing using *template #1* found in the C:\Steps3D\Lesson01 folder.
    16.2. Don't attempt to draw the dimensions yet.
    16.3. Text is in Paper Space and is 3/16" and 1/8"
    16.4. Title block text is ¼", 3/16", and 1/8".
    16.5. Polylines have a width of 1/16".
    16.6. Anchors are donuts with an outer diameter of 9/16" and an inner diameter of 3/16".
    16.7. The nut is a six-sided polygon inscribed in a radius of ¼"; its width is 1/16".
    16.8. The isometric view has been Gouraud shaded..
    16.9. Save the drawing as *MyRingStand.dwg* in the C:\Steps3D\Lesson01 folder.

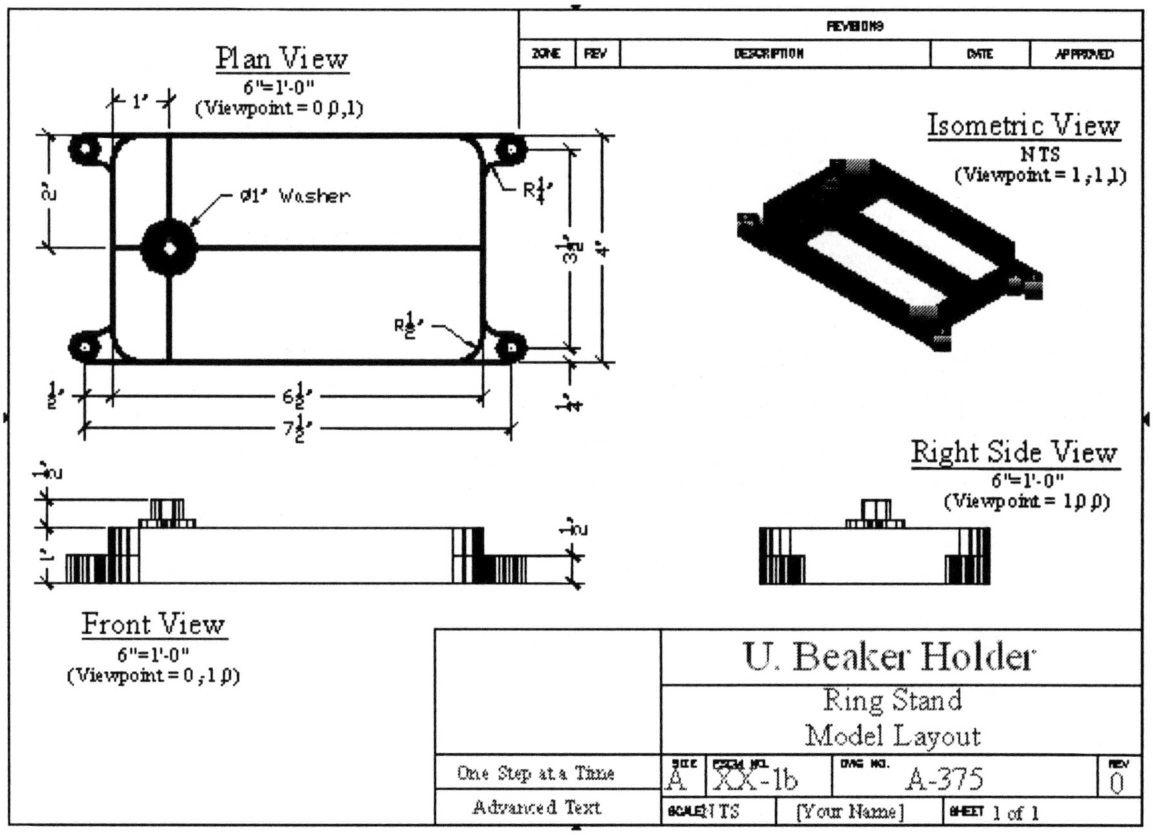

Figure 1.050

1.8	**For Web-Based Review Questions, visit: http://www.uneedcad.com/2006/Files/06R1-3D.pdf**

Lesson 2

Following this lesson, you will:

- ✓ *Understand the differences between the UCS and the WCS*
  - *Know how to use the **UCS** command to create working planes*
  - *Know how to use the UCS Dialog Box (Manager)*
  - *Know how to dimension a three-dimensional drawing*
- ✓ *Be familiar with some advanced viewing techniques*
  - *Be able to use the **3DOrbit** and **3DCOrbit** commands*
  - *Be able to use the **3DPan** and **3DZoom** commands*
  - *Be able to use the **3DDistance** and **3DSwivel** commands*
- ✓ *Be able to use the **3DClip** command*

More of Z Basics

When we first drew our three-dimensional stick figure house, it had no walls and only sticks for a roof. Later we saw that, by using the Thickness and Elevation system variables, we could make our walls solid. But what about the roof? There must be an AutoCAD tool for making it solid as well.

We also discussed the need for point entry precision in Z-space. What about text and dimensions? Must we use coordinates to place them? (Have you tried to dimension a three-dimensional drawing yet?) And what if you want to place text along a slope (like the roof)?

Believe it or not, all of these questions have the same answer! The answer lies in a tool called the User Coordinate System – the UCS.

In Lesson1, you learned how to create simple three-dimensional objects and how to view those objects from different angles. In this lesson, we'll discuss the tools with which you'll work on the different faces of your three-dimensional objects. Then we'll look at some more advanced viewing tools.

Let's start with the UCS.

2.1 WCS vs. UCS

Although you may not be aware of it, you're already familiar with the UCS. You've been using it since you began your study of AutoCAD. However, it's always been aligned with the World Coordinate System – the WCS – so you never noticed it. So what's the difference? You need to understand a little about how AutoCAD works to really understand the UCS.

> To avoid any chance of damage to the WCS, AutoCAD placed it out of your reach.

In the basic text, you learned that all objects in an AutoCAD drawing are defined by information stored in that drawing's database. When you regenerate a drawing, AutoCAD reads this information and restores the drawing accordingly. When you create an attributed block, the attribute information is also stored in the database.

Part of the information stored in a drawing's database is the location and orientation of each object. For consistency (and to avoid a programming nightmare), AutoCAD developed a coordinate system that remains the same throughout the life of the drawing. Thus, the definition of point 0,0,0 will remain the same, and the point will always be in the same place. The X, Y, and Z directions will never change.

This coordinate system is the WCS.

AutoCAD uses the WCS point 0,0,0 much as a mariner uses the North Star. Unchanging in the night sky, Polaris shows the sailor the way home. Unchanging in the WCS, point 0,0,0 shows AutoCAD how to orient any object in the drawing.

The User Coordinate System – or UCS – is what the CAD operator uses to determine up from down and left from right for the immediate task at hand. Until now, the UCS has always been aligned with the WCS – our mariner's ship has always been pointed toward the North Star. Until now, the front of our ship has always been north; the masts have always pointed upward. With a compass, our sailor could find anything on the ship. Until now, we had only two dimensions with which to create a drawing – our sailor had only a single deck on which to work. Until now, the CAD operator had no need to know about the UCS or WCS. Until now…

Sigh. It used to be so easy …

Now our ship has changed directions and acquired new decks. Our sailor with the compass is completely lost. To make it easier for him to understand where things are on the ship, he must change his reference point from the North Star to something on the ship itself. This way, he can always find his way regardless of the direction in which the ship is sailing.

He'll use the mainsail as his reference point (0,0,0). His compass directions will now reference points on the ship – the bow becomes north, the stern becomes south, starboard and port become east and west. The mainsail will always point upward from the main deck (regardless of how violently the sea rocks the ship). So now, using the North Star, our mariner's ship will never be lost, and using his new Mariners Coordinate System, he'll never be lost on the ship.

CAD operators must adjust for Z-space as our mariner adjusted for a change in the ship's direction. Using the 0,0,0 coordinate of the WCS as our North Star, we can always find our way home. And using the User Coordinate System as our sailor uses his Mariner's Coordinate System, we can work on any surface – or working plane – of a three-dimensional model (as our mariner could work on any deck of his ship).

So how do you use the UCS? Where do you begin?

Let's begin by accessing the ***UCS*** command. What do you suppose the command would be?

To keep it simple, AutoCAD calls the command ***UCS***! The command line approach looks like this:

Command: ***ucs***

Current ucs name: *WORLD*

Enter an option [New/Move/orthoGraphic/Prev/Restore/Save/Del/ Apply/?/World] <World>: ***[enter an option]***

Let's look at each option.

> AutoCAD provides several methods of accessing the ***UCS*** command and its options. Easiest, perhaps, is the UCS toolbar. But you can also access the options by cursor or dynamic input menu once you've entered the ***UCS*** command, or by selecting from the Tools pull down menu.

- **Move** allows you to move the UCS without changing its orientation. It prompts

  Specify new origin point or [Zdepth]<0,0,0>:

  Respond by picking a new origin point or by selecting the **Zdepth** option. The **Zdepth** option allows you to move the origin along the Z-axis.
- The **orthoGraphic** option offers a quick way to change the UCS to one of the standard orthographic projections without moving 0,0,0 from the WCS location. It prompts

  Enter an option [Top/Bottom/Front/BAck/Left/Right]<Top>:
- **Prev** restores the previous UCS.
- **Save** allows you to save the current UCS for later retrieval. **Restore** allows you to restore a previously saved UCS. Remember these; they'll save wasted time spent repeating a setup.
- **Del** allows you to delete a stored UCS.
- The **Apply** option allows you to apply the current UCS setting to another viewport(s). This can be a useful timesaver.
- The question mark (**?**) can also be pretty handy. It will provide the names, 0,0,0 locations, and orientations of the X-, Y-, and Z-axes of all the saved UCSs as they relate to the current UCS.
- The **World** option is the most important (so important, in fact, it deserves "Hail Mary" status). It tells AutoCAD to restore the UCS to match the WCS's orientation. In other words, when you're lost, use the World option to reorient yourself.
- The **New** option, of course, allows you to create a new UCS – that is, you can define a new 0,0,0 and new orientations for the X-, Y-, and Z-axes. You can do this in one of several ways depending on the option you select at the prompt:

  Specify origin of new UCS or [ZAxis/3point/OBject/Face/View/X/Y/Z] <0,0,0>:

- Using the **ZAxis** option, you'll define the new UCS by identifying a location for 0,0,0, and then a point on the Z-axis.
- The **3point** option allows you to define the new UCS by identifying a location for 0,0,0, and then points on the X- and Y-axes.
- The **OBject** option allows you to define the new UCS by selecting an existing three-dimensional object. AutoCAD aligns the new UCS with that object.
- The **Face** option allows you to define the new UCS by selecting a face on an existing three-dimensional solid object.
- **View** sets the UCS flat against the screen (the X-axis parallel to the bottom of the screen and the Y axis parallel to the left side of the screen). The 0,0,0 coordinate remains where it is currently located.
- The **X/Y/Z** options allow you to rotate the UCS around the selected axis. They prompt

  Specify rotation angle about X *[or Y or Z]* axis <90>:

Most of this will become clearer with practice. Let's try a get-acquainted exercise.

> Some knowledge of the UCS Icon will help in this exercise. You can access the various options of the ***UCSIcon*** command using right-click cursor menus or the View pull-down menu. Follow this path:
>
> *View – Display – UCS Icon – [option]*

Do This: 2.1.1	**Manipulating the UCS**

I. Open the *ucs practice2.dwg* file in the C:\Steps3D\Lesson02 folder. The drawing looks like Figure 2.001.

II. Follow these steps.

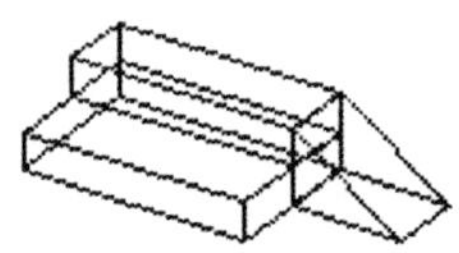

Figure 2.001

TOOLS	COMMAND SEQUENCE	STEPS
No Button Available	**Command:** ***ucsicon*** **Enter an option [ON/OFF/All/ Noorigin/ORigin Properties] <ON>:** ***or***	1. Our first step is one of the most important steps to remember when manipulating the UCS. Set the UCS icon to **ORigin**. This way, you'll always know where 0,0,0 is.
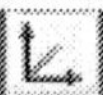 UCS Button	**Command:** ***ucs***	2. Now we'll tell AutoCAD to use the UCS (our mariner must navigate within the ship). Enter the ***UCS*** command. Alternately, you can pick the **UCS** button on the UCS toolbar.

TOOLS	COMMAND SEQUENCE	STEPS
	**Current ucs name: *TOP*** **Enter an option [New/Move/ orthoGraphic/Prev/Restore/Save/ Del/Apply/ ?/World] <World>:** ***n***	3. AutoCAD responds by asking what you would like to do. We'll create a new UCS. Select the **New** option.
	Specify origin of new UCS or [ZAxis/3point/OBject/Face/ View/X/Y/Z] <0,0,0>: ***za***	4. Our first UCS will simply relocate 0,0,0. We'll use the **ZAxis** option.
	Specify new origin point <0,0,0>: **Specify point on positive portion of Z-axis <1.0000,1.0000,1.0000>:**	5. Specify the bottom-left corner of the object and accept the default **Z-axis point**. Notice that the UCS icon moves as shown in Figure 2.002. It's locating 0,0,0 as we told it to do in Step 1.

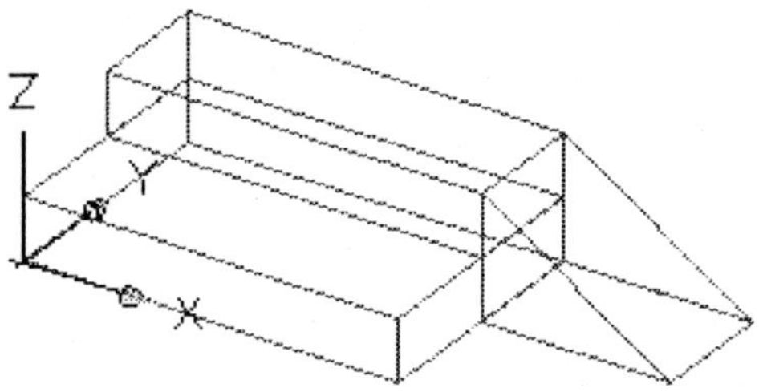

Figure 2.002

TOOLS	COMMAND SEQUENCE	STEPS
	**Command:** ***[enter]*** **Current ucs name: *NO NAME***	6. Now we'll save this UCS for later retrieval. Repeat the ***UCS*** command.
	**Enter an option [New/Move/orthoGraphic/Prev/ Restore/Save/Del/Apply/?/World] <World>:** ***s***	7. Tell AutoCAD you wish to save the UCS …
	Enter name to save current UCS or [?]: ***lower left base***	8. … and call it something appropriate.
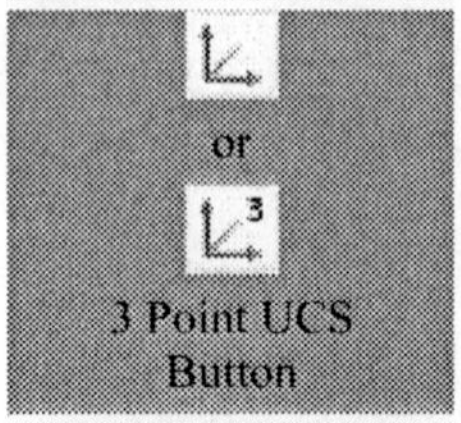	**Command:** ***[enter]*** **Current ucs name: lower left base** **Enter an option [New/Move/ orthoGraphic/Prev/Restore/Save/ Del/Apply/?/World] <World>:** ***n***	9. Now let's create another UCS. Repeat the command and tell AutoCAD to create a new UCS. (Alternately, the **3 Point UCS** button will satisfy Steps 9 and 10.)
	Specify origin of new UCS or [ZAxis/3point/OBject/Face/ View/X/Y/Z] <0,0,0>: ***3***	10. (If you used the 3 Point UCS button in Step 9, skip this step.) Use the **3point** option.

TOOLS	COMMAND SEQUENCE	STEPS
	Specify new origin point <0,0,0>: **Specify point on positive portion of X-axis <1.0000,0.0000,0.0000>:** **Specify point on positive-Y portion of the UCS XY plane <0.0000,1.0000,0.0000>:**	11. Select the points indicated in Figure 2.003.

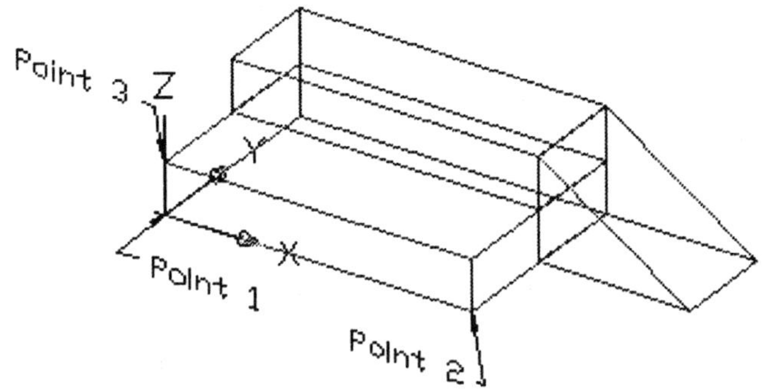

Figure 2.003

TOOLS	COMMAND SEQUENCE	STEPS
		12. The UCS icon now appears like Figure 2.004. Before continuing, stop and ask yourself where the X-, Y-, and Z-axes are located. Use the UCS icon and the Right-Hand Rule to help answer that question.*

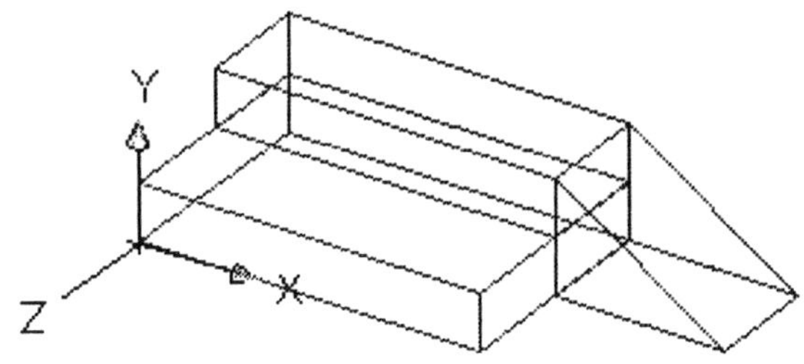

Figure 2.004

TOOLS	COMMAND SEQUENCE	STEPS
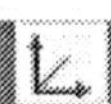	**Command:** ***[enter]***	13. Repeat Steps 6 through 8 to save this UCS as **MyFront**.
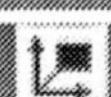 UCS Object Button	**Command:** ***ucs*** **Current ucs name: MyFront** **Enter an option [New/Move/ orthoGraphic/Prev/Restore/Save/ Del/Apply/?/World] <World>:** ***ob***	14. Now let's look at some other ways to set the UCS. You can select the **New** option of the ***UCS*** command for the next several steps, or you can simply pick the button indicated. Let's begin with the **Object** option. Enter the sequence shown or pick the **Object UCS** button on the UCS toolbar.

* The UCS icon indicates the X-, and Y-, and Z-axes.

TOOLS	COMMAND SEQUENCE	STEPS
	Select object to align UCS:	15. Select the "Z" line closest to the lower-left corner of the screen. The UCS icon now looks like Figure 2.005. The 0,0,0 coordinate of the UCS matches the line's end point closest to where you picked. The Y- and Z-axes also line up according to the line's definition.

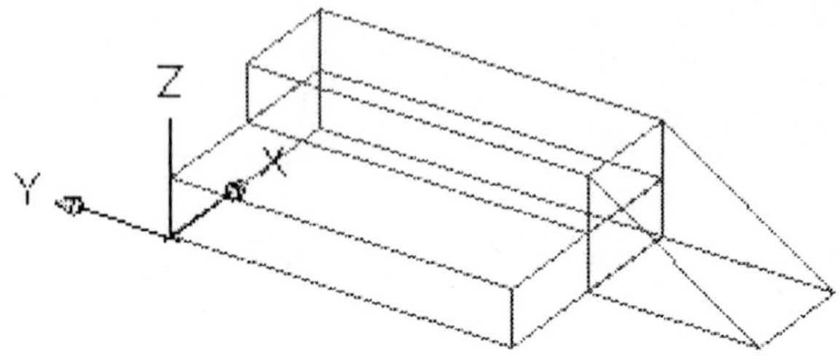

Figure 2.005

TOOLS	COMMAND SEQUENCE	STEPS
 UCS Previous Button	**Command:** ***ucs*** **Current ucs name: *NO NAME*** **Enter an option [New/Move/ orthoGraphic/Prev/Restore/Save/ Del/Apply/?/World] <World>:** ***p***	16. Restore the **Previous** UCS.
 Face UCS Button	**Command:** ***ucs*** **Current ucs name: MyFront** **Enter an option [New/Move /orthoGraphic/Prev/Restore/Save/ Del/Apply/?/World] <World>:** ***fa***	17. Now we'll use the **Face** option. (The wedge piece is a three-dimensional solid – an item we'll discuss in Lesson 7. It was necessary to include it here for demonstration purposes.) Enter the sequence shown or pick the **Face UCS** button on the UCS toolbar.
	Select face of solid object: ***[select the wedge]*** **Enter an option [Next/Xflip/Yflip] <accept>:** ***[enter]***	18. Select the lower-left corner of the front of the wedge. Notice that you can adjust the UCS after selecting the solid. When yours looks like Figure 2.006, accept it.

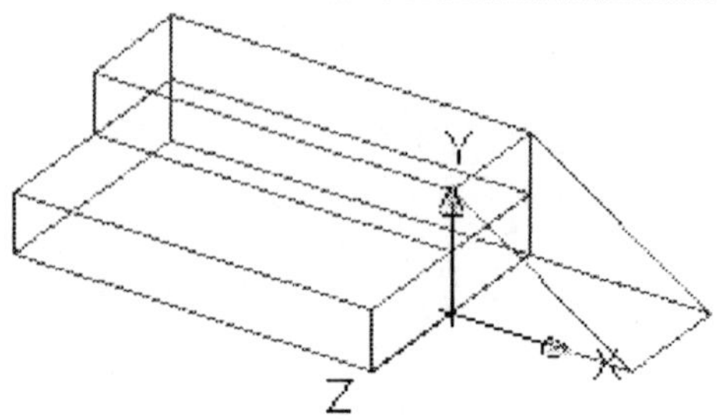

Figure 2.006

TOOLS	COMMAND SEQUENCE	STEPS
 View UCS Button	**Command:** ***ucs*** **Current ucs name: *NO NAME*** **Enter an option [New/Move/ orthoGraphic/Prev/Restore/Save/ Del/Apply/?/World] <World>:** ***v***	19. Now set the UCS according to the current view. Enter the sequence shown or pick the **View UCS** button. Notice the new orientation of the UCS icon. Notice also that, although it changes orientation, it doesn't change position.
	**Command:** ***ucs*** **Current ucs name: *NO NAME*** **Enter an option [New/Move/orthoGraphic/ Prev/Restore/Save/Del/Apply/ ?/World]** **<World>:** ***r*** **Enter name of UCS to restore or [?]:** ***myfront***	20. Restore the *MyFront* UCS. Notice that the UCS icon returns to the location defined by the *MyFront* UCS.
 X Button		21. Let's experiment with the X/Y/Z options. Watch the UCS icon as we rotate the UCS. Select the **X** button.
	Specify rotation angle about X axis <90>: ***[enter]***	22. Accept the **90°** default and watch the UCS icon (Figure 2.007). It can be difficult to follow axial rotations, but that's where the Right-Hand Rule comes in handy. Here's how it works: Point the finger that represents the axis about which you're rotating (in this case, the thumb) directly at your nose. Now rotate your hand 90° (or the desired angle of rotation) counterclockwise. This configuration will match the UCS icon and show you the orientation of your UCS.

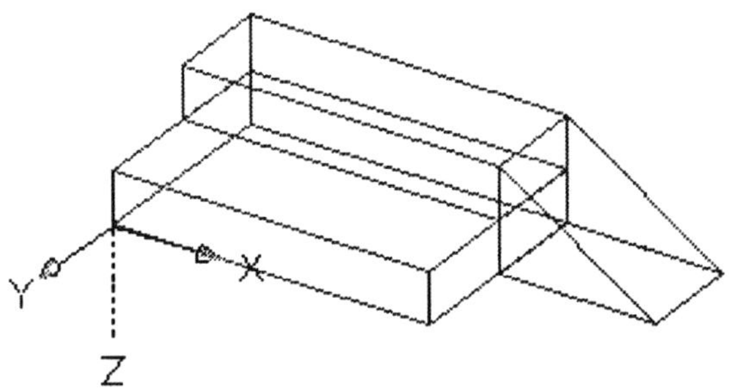

Figure 2.007

Tools	Command Sequence	Steps
		23. Repeat Steps 21 and 22 for the **Y** and **Z** options.
		24. Using any of the tools just discussed, create the new UCS setups identified in Figures 2.008 through 2.013. Save the setups as indicated.

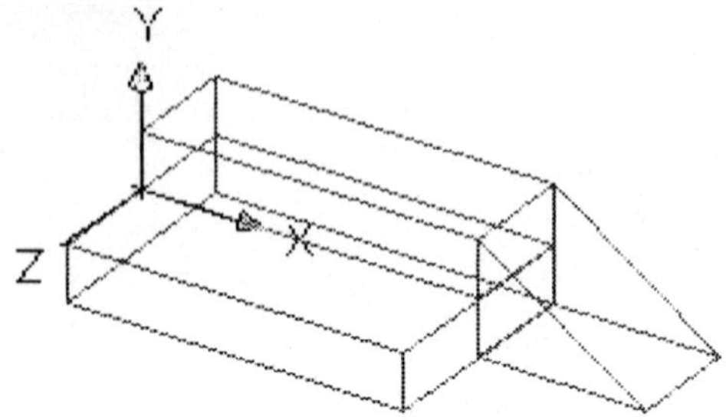

Figure 2.008: *myFront-1*

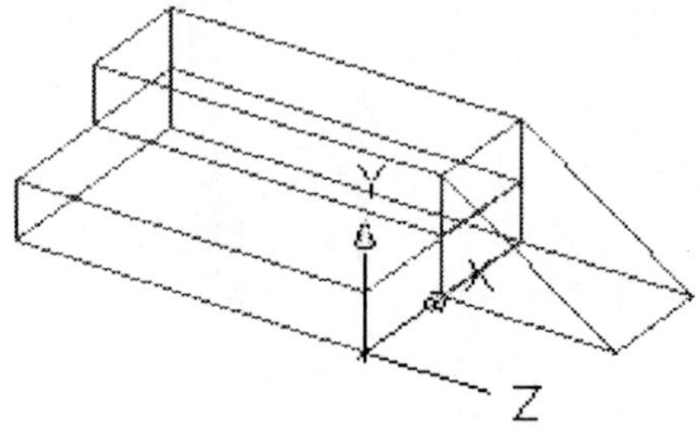

Figure 2.009: *MyRight*

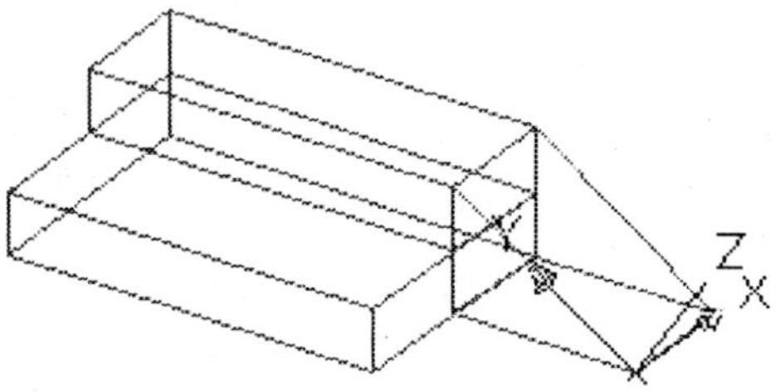

Figure 2.010: *Inclined*

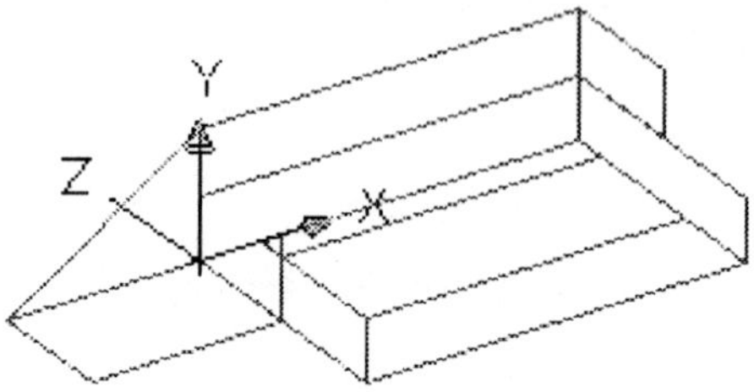

Figure 2.011: *myBack*

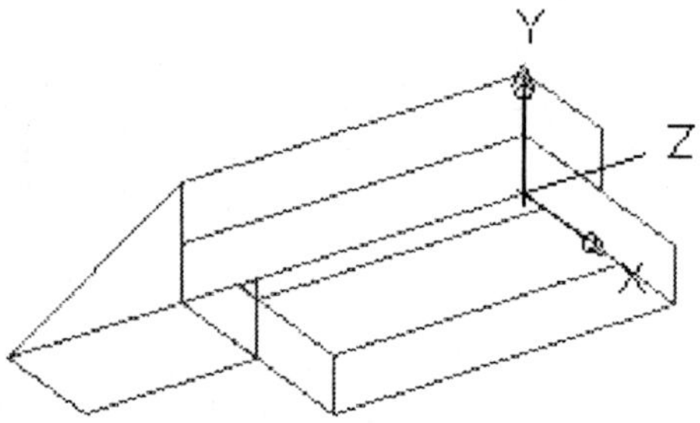

Figure 2.012: *myLeft*

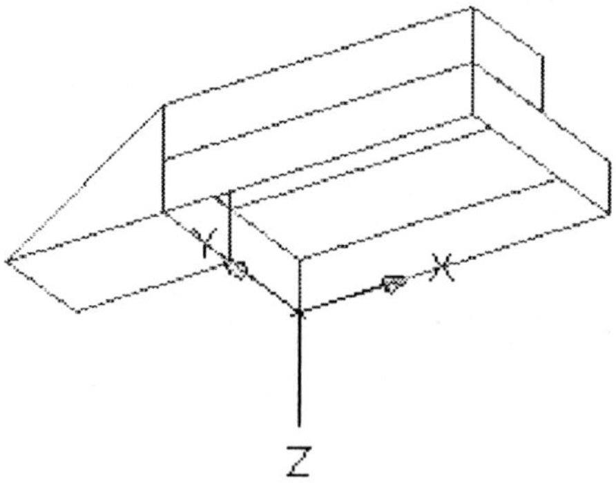

Figure 2.013: *myBottom*

Tools	Command Sequence	Steps
	Command: *qsave*	25. Save the drawing.

You've created several UCS setups using a variety of methods. It's okay to be a bit confused at this point. Let's pause for a moment, catch our breath, and review what we've learned. Here are some important things to remember.

- Despite their similarities, viewpoints, viewports and the UCS are three different things. Remember:

- o Viewpoints are *points* (where you stand) *from which you view* the model.
  - o Viewports are like *port*holes in a ship *through which you view* the model.
  - o The UCS orients you *on* the model itself.

COOL STUFF

- Points used to define viewpoints will always reference the WCS (*not* the UCS). This will make it easier for you to get your bearings.
- The ***View*** command will save viewpoints.
- Each viewport can have a unique viewpoint and/or UCS assigned to it.
- Each – viewpoint, viewport, and UCS – works independently of the other two, but all should be considered as a team to assist you when you work on a three-dimensional model.

Let's take a moment to look at a tool that might make UCS management easier. Then we'll use the working planes to create some lines, text, and dimensions on our stick house.

2.2 The UCS Dialog Box (Manager)

Over the course of creating a drawing – particularly a larger drawing – you may find it necessary to create several UCSs. You have already seen how to save these setups for later retrieval, but where do you keep the list of names you've assigned the UCSs?

AutoCAD makes it simple with the UCS dialog box (aka. UCS Manager – Figure 2.014). This provides a dialog box approach to keeping track of the various setups as well as some additional tools. Access the UCS Manager by entering the ***UCSMan*** command.

Let's take a look.

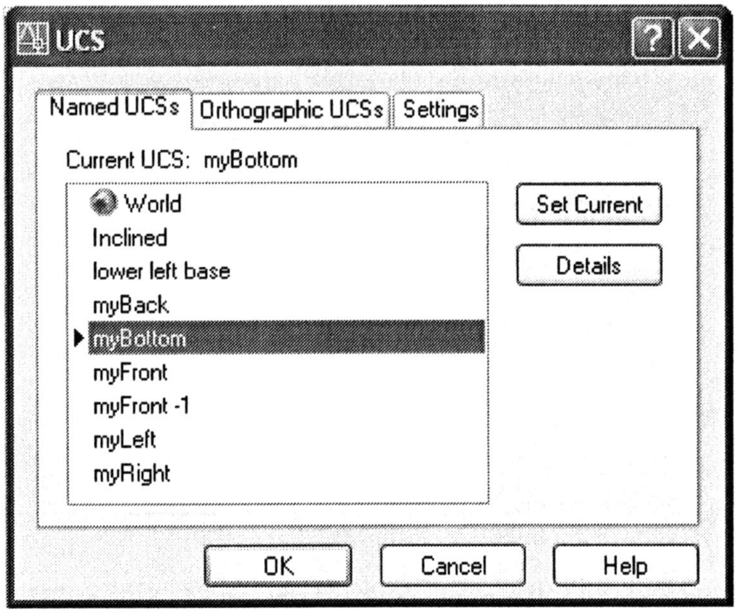

Figure 2.014

- The first tab presents a list of **Named UCSs** as well as **World** and **Previous** options. To set a UCS current, either double-click on its name or select the name and pick the **Set Current** button. Then pick the **OK** button to complete the procedure. It's that simple – no need to remember or to store a list of names! (But it's always a good idea when assigning the names to make them self-explanatory.)

The **Details** button presents a dialog box (Figure 2.015) that indicates the origin location, as well as the orientation of the X-, Y-, and Z-axes for the currently selected UCS. You view the data in relation to the WCS or any other existing UCS by selecting the coordinate system from the **Relative to** control box.

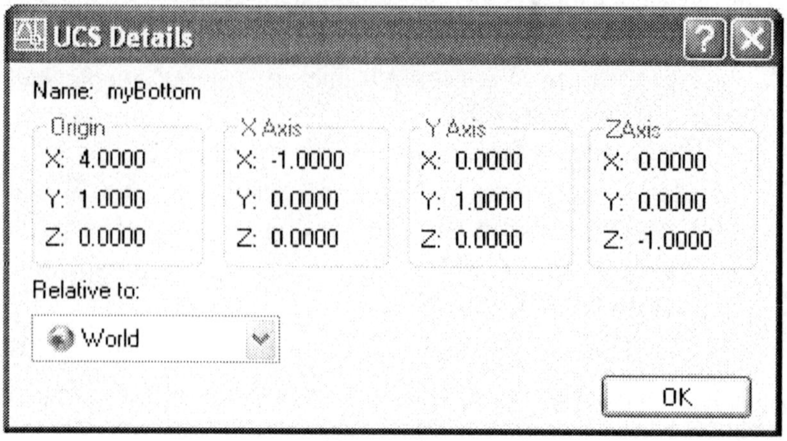

Figure 2.015

- The middle tab (Figure 2.016) – **Orthographic UCSs** – lists several standard UCSs (the same ones available using the **orthoGraphic** option of the ***UCS*** command). Use the same procedure you used on the **Named UCSs** to set one current.

  Notice the **Relative to** control box on this tab. You can set one of the orthographic UCSs relative to the WCS or relative to one of the user-defined UCSs. I suggest leaving this control set relative to the WCS at least for now. Any other setting might make it difficult to orient yourself.

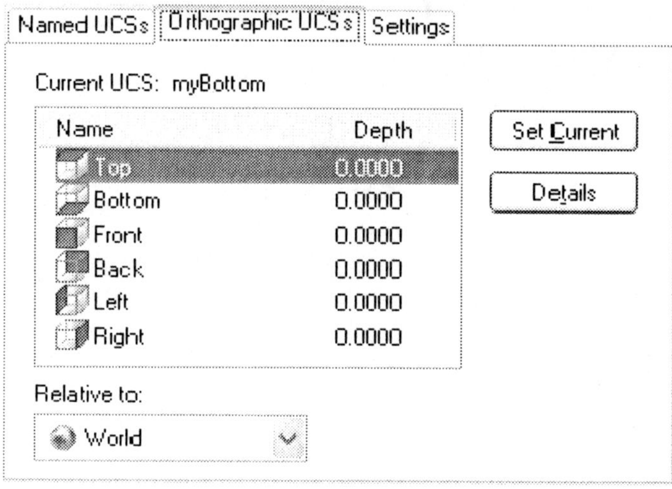

Figure 2.016

- The **UCS icon settings** frame of the **Settings** tab (Figure 2.017) is a dialog box interface for the ***UCSIcon*** command.

  Use the **UCS settings** frame to **Save UCS with viewport** (the default). This means that each viewport can have its own UCS setting. Clear this check and each viewport will reflect the UCS settings of the current viewport.

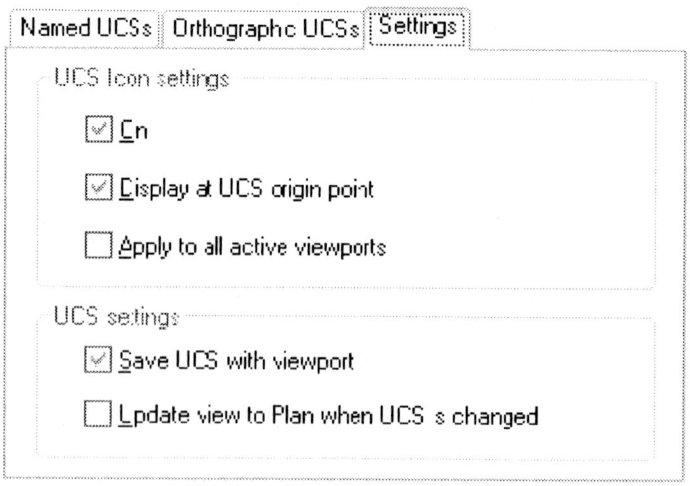

Figure 2.2d

  A check next to **Update view to Plan when UCS is changed** will regenerate the viewport in a plan view of the current UCS whenever the UCS settings are changed. I suggest leaving this box clear since you're more likely to want to keep the view even if you change the UCS.

We'll use the UCS dialog box in our next exercise to help us see how using different UCS settings can benefit us.

2.3 Using Working Planes

We've spent many pages learning to set up User Coordinate Systems. But as yet, we haven't seen how to use the UCS once it's set up. We haven't seen the answers to the questions that began our lesson – How do we make a solid roof? How do we place text and dimensions in a three-dimensional drawing.

Let's do an exercise to put UCSs to practical use.

Do This: 2.3.1 Manipulating the UCS

I. Open the *Stick House 2.dwg* file in the C:\Steps3D\Lesson02 folder. The drawing looks like Figure 2.018.

This drawing has already been set up for you with several UCSs and Paper Space viewports. The house is a stick figure (wireframe) structure. We'll recreate the house using lines with thickness, and we'll create a solid roof. Then we'll add a few dimensions.

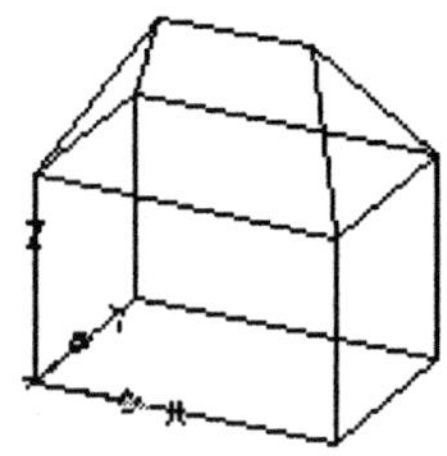
Figure 2.018

II. Be sure **Obj2** is the current layer.

III. Follow these steps.

TOOLS	COMMAND SEQUENCE	STEPS
	Command: *th* **Enter new value for THICKNESS <0.0000>: *8*** **Command: *l***	1. Begin by drawing the walls with 8" thickness. Trace the bottom lines using lines with 8" thickness. You won't need to change the UCS for this step.
	Command: *hide*	2. Remove the hidden lines to help see what you've done. Your drawing looks like Figure 2.019.

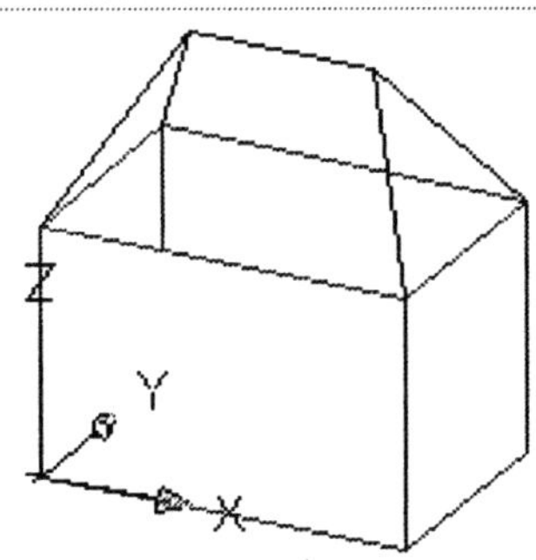
Figure 2.019

TOOLS	COMMAND SEQUENCE	STEPS
	Command: *th* **Enter new value for THICKNESS <8.0000>: *0***	3. Set **Thickness** back to **0**.

Tools	Command Sequence	Steps
	Command: ***so***	4. Try drawing a solid using the points indicated in Figure 2.020.

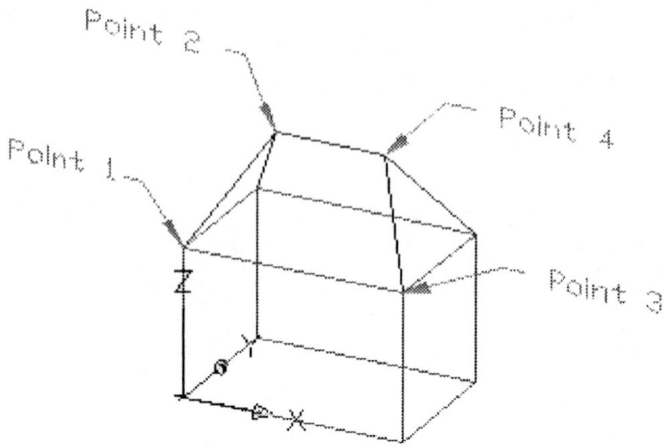

Figure 2.020

Tools	Command Sequence	Steps
	Command: ***e***	5. Notice that the solid was drawn two-dimensionally in the current UCS (Figure 2.021). To use a solid to form the roof, the UCS must be aligned to the side of the roof you wish to draw. Erase the solid.

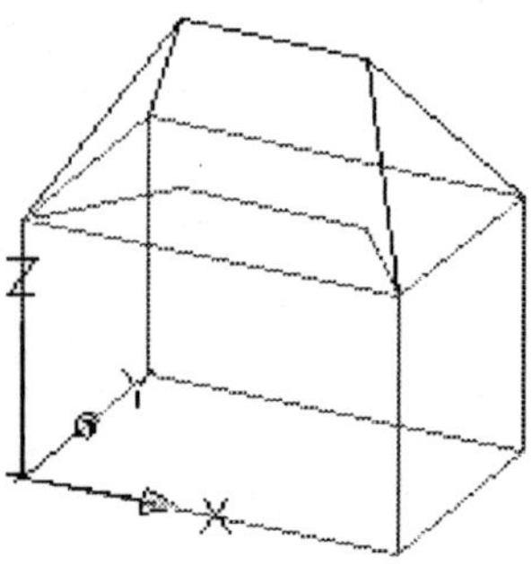

Figure 2.021

Tools	Command Sequence	Steps
Named UCS Button	**Command:** ***ucsman***	6. Call the UCS Manager.
		7. Double-click on the south roof UCS to make it current. Then pick the **OK** button. Notice that the UCS icon moves to align itself with the face of the south roof (see Figure 2.022).

TOOLS	COMMAND SEQUENCE	STEPS

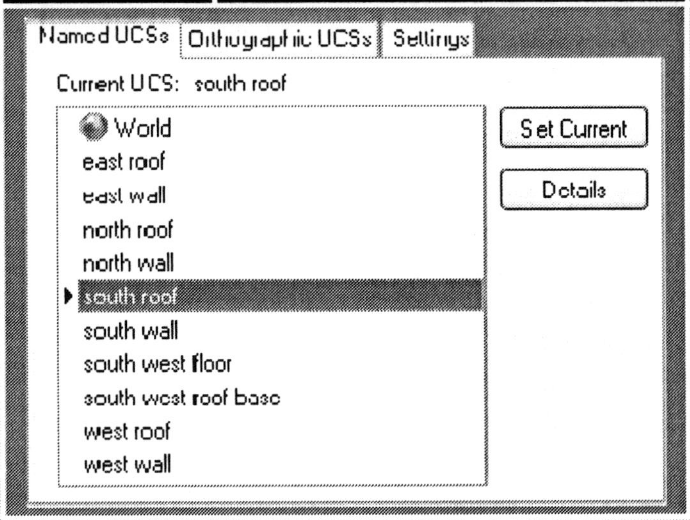

8. Now that the UCS has been properly set, repeat Step 4. Remove the hidden lines (Figure 2.022).

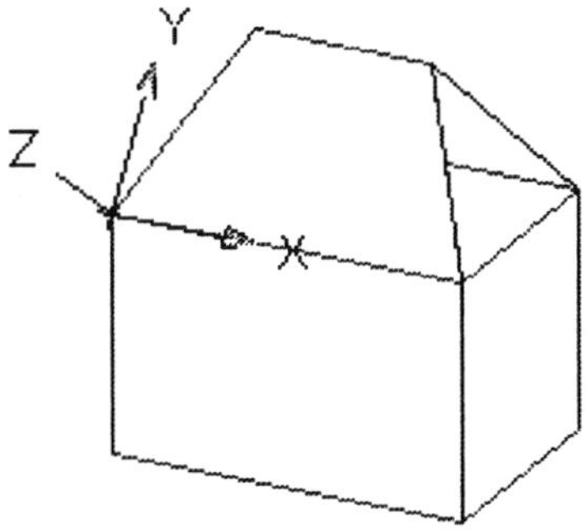

Figure 2.022

9. Using the techniques seen in this exercise, complete the roof using solids (adjust the viewpoint as necessary, then return to the current setting of 1,-2,1).

Your drawing looks like Figure 2.023.

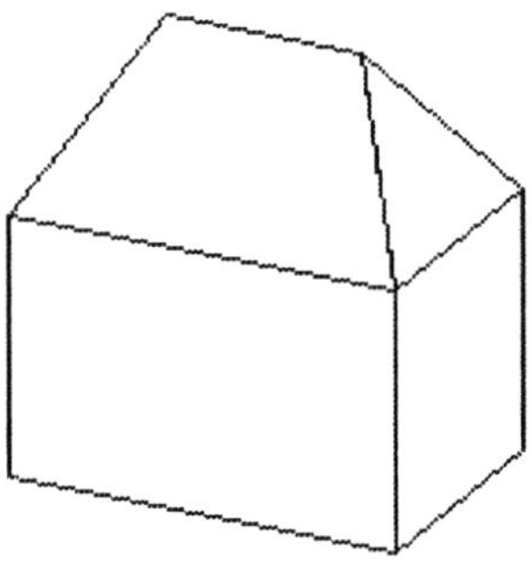

Figure 2.023

Tools	Command Sequence	Steps
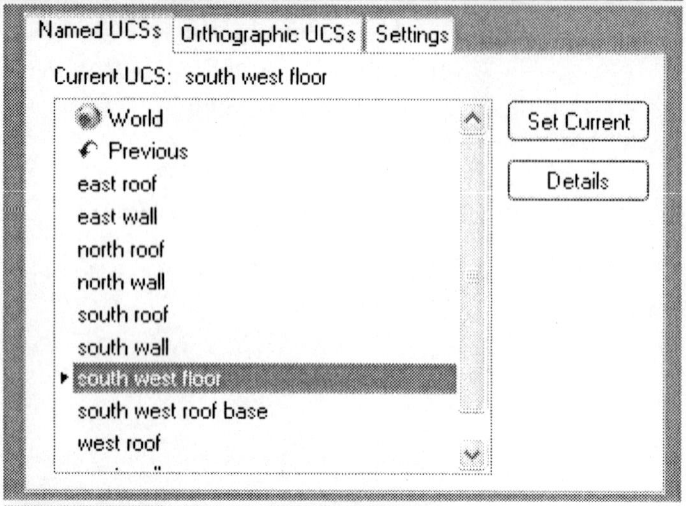		10. Set the current UCS to **south west floor**.
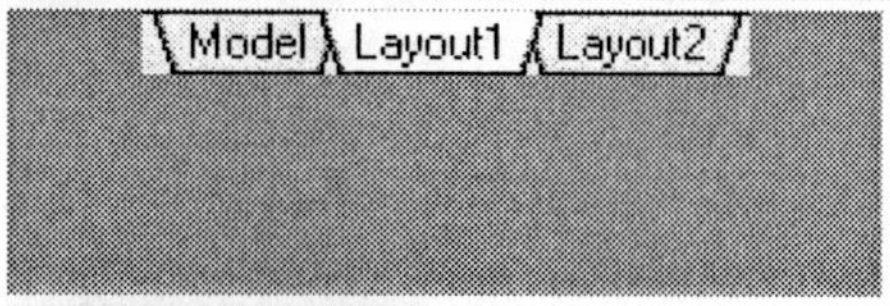		11. Now we'll add some dimensions, but we need to use viewports to this properly. Activate the **Layout1** tab.
	Command: ***ms***	12. Open Model Space. (Refer to Figure 2.024 for Steps 12 – 17.)
		13. Activate the upper-left viewport and do the following: ▪ Create a WCS plan view (set the viewpoint to 0,0,1). ▪ Set the scale for this viewport to 1:8.
		14. Activate the lower-left viewport and do the following: ▪ Create a WCS front view (set the viewpoint to 0,-1,0). ▪ Set the scale for this viewport to 1:8. ▪ Adjust the size of the viewport and the position of the house as necessary to see the entire house.
		15. Activate the lower-right viewport and do the following: ▪ Create a WCS right side view (set the viewpoint to 1,0,0). ▪ Set the scale for this viewport to 1:8. ▪ Adjust the size of the viewport and the position of the house as necessary to see the entire house.

Tools	Command Sequence	Steps
		16. Activate the upper-right viewport and do the following: ▪ Set the scale to 1:8. ▪ Adjust the size of the viewport and the position of the house as necessary to see the entire house.
	Command: ***mvsetup***	17. Using the ***MVSetup*** command to align the plan, front, and side views. Adjust the position of the viewports as necessary for aesthetics. Your drawing looks something like Figure 2.024.

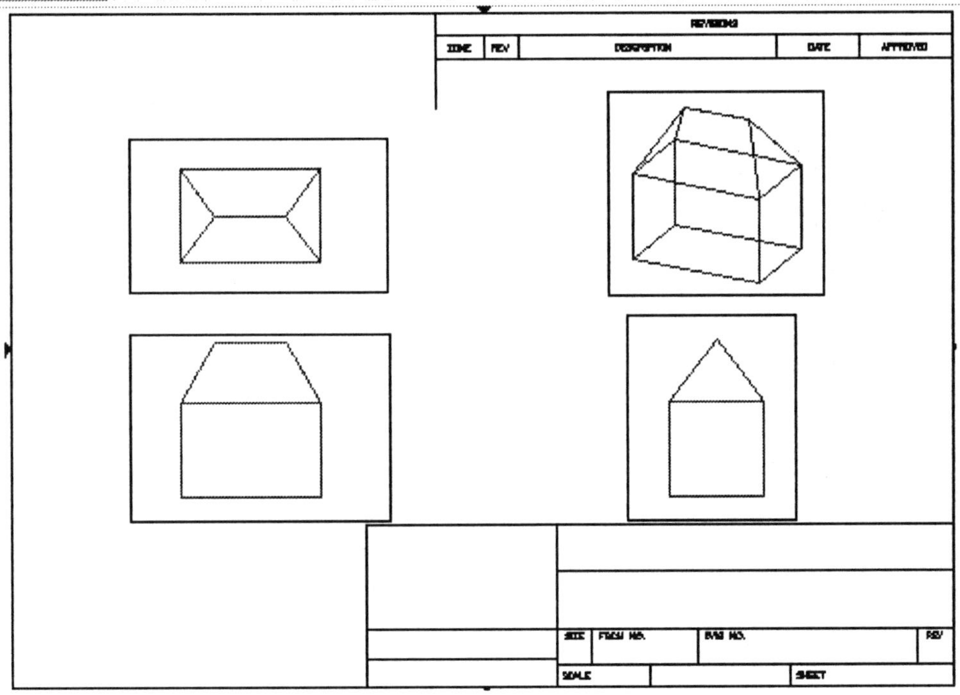

Figure 2.024

Tools	Command Sequence	Steps
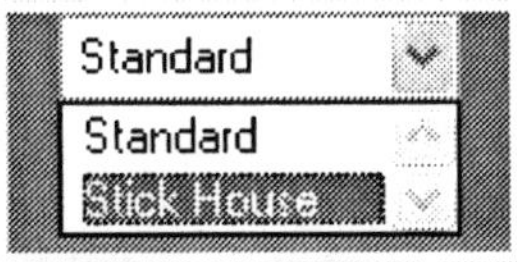		18. Be sure the **Stick House** dimstyle is current. (This style has been set up to work in Paper Space.)
	**Command:** ***dli***	19. On the dims layer, add the dimensions shown in the upper left viewport (Figure 2.025). (Note: *Freeze the* ***obj2*** *layer to make this easier.*)

Tools	Command Sequence	Steps

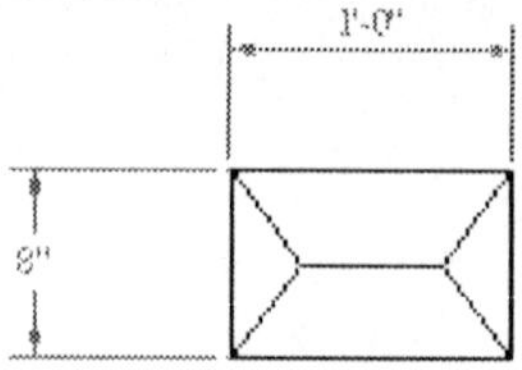

Figure 2.025

Tools	Command Sequence	Steps
		20. Create the dimensions in Figure 2.026.

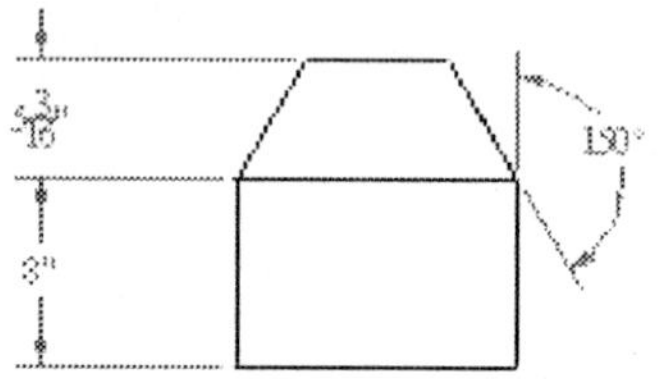

Figure 2.026

Tools	Command Sequence	Steps
		21. Complete the drawing as shown in Figure 2.027. Here are some hints: ▪ Try to hatch a shingle pattern onto the roof – I used the AR_RSHKE pattern at a 1/16" scale. (It helps to freeze the **obj2** layer.) ▪ Like dimensioning, you can only hatch in the current UCS. ▪ Use the **Hidden** option of the ***Shademode*** command in the Dimetric View.

TOOLS	COMMAND SEQUENCE	STEPS

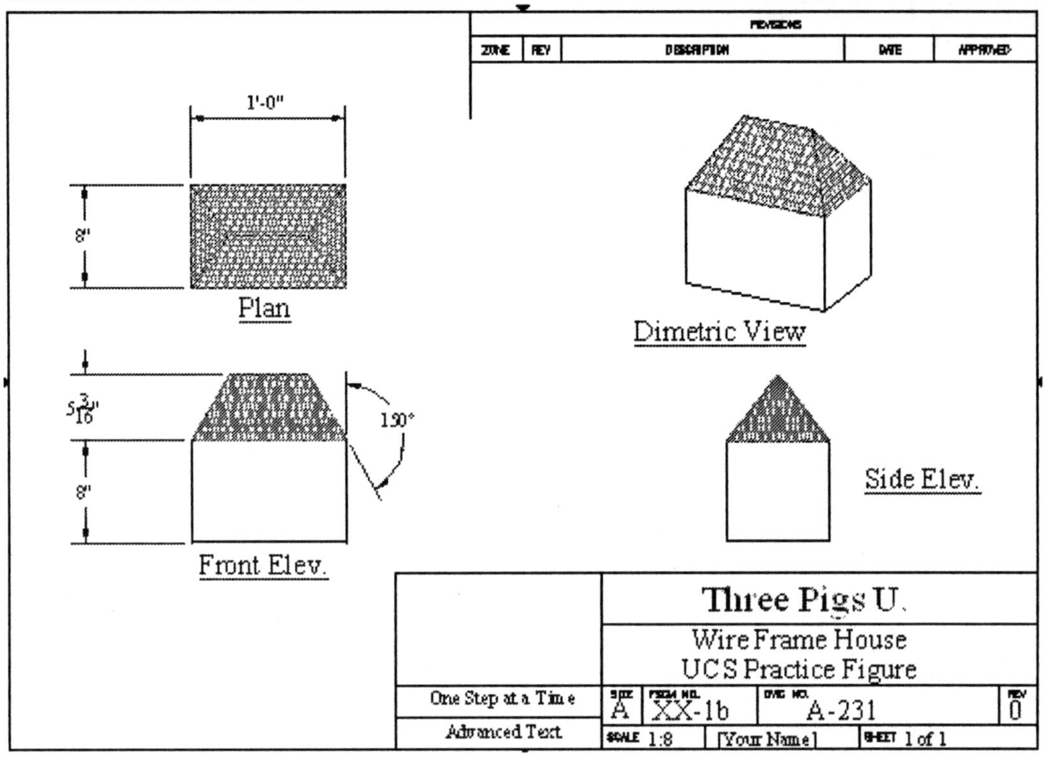

Figure 2.027

2.4 Advanced Viewing Techniques

We discovered some very useful viewing tools and procedures in our last lesson and have used some of them in this lesson. Indeed, we'll continue to use them throughout our text and our computer-drafting career. But consider this scenario.

A builder is creating an object – we'll use the figure we saw in our *UCS Practice 2* drawing (Exercise 2.1.1) as an example. As he adds a piece here or trims a piece there, he holds the object in his hand and rotates it this way and that to get a better understanding … a better feel for its shape. How can you do the same thing with the computer model *before* our builder creates it?

You can use the viewpoint tool we've been using. You can even speed it up slightly by using preset viewpoints on the View toolbar. But let's face it; at best, this tool is too slow and cumbersome for the scenario just discussed. It won't provide the insights our builder will get by rotating the object at different angles.

For this reason, AutoCAD has provided two extraordinary tools – ***3DOrbit*** and ***3DCOrbit*** (3D Continuous Orbit). Let's take a look at each.

2.4.1 3DOrbit

The ***3DOrbit*** command allows you to rotate a model on the computer screen just as the builder did in his hand.

Hands are marvels of engineering. We use them without a second thought – forgetting the time our infant minds struggled to control them.

Orbit!?

3DOrbit is also quite a marvel – and it'll also take some time to learn to control it. But we'll soon come to appreciate it almost as much as a builder appreciates his hands!

Access the 3D Orbit screen with the ***3DOrbit*** command like this:

Command: *3dorbit*
(or *3do* or *orbit*)

Press ESC or ENTER to exit, or right-click to display shortcut-menu.

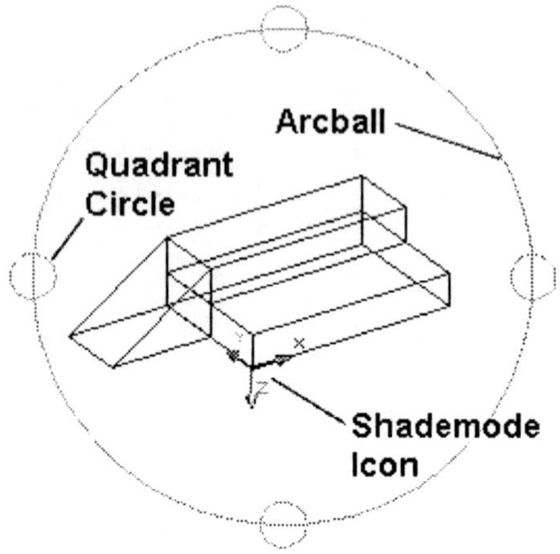

Figure 2.028

AutoCAD displays a sort of three-dimensional compass over the screen (in the current viewport). The compass is called an **Arcball** and will help you control your movements while using the orbiter. Let's examine how it works (refer to Figure 2.028).

- When you enter the orbiter screen, AutoCAD will replace the UCS icon with the Shademode icon. Here you see it located at the last 0,0,0 UCS point identified in Exercise 2.1.1.
- The arcball and smaller quadrant circles allow you to determine how to manipulate the model. It works like this:

When the Cursor Begins:	Dragging the cursor will:	The cursor will be a:
Inside the arcball	Manipulate the model freely about its center point.	
Outside the arcball	Perform a two-dimensional rotation (called a roll) about an axis extending through the center of the arcball (and outward from the screen).	
Inside the east or west quadrant circle	Rotate the model about the north axis of the arcball.	
Inside the north or south quadrant circle	Rotate the model about the east/west axis of the arcball.	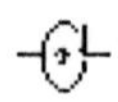

To reset the original view, right-click and select **Reset View** from the cursor menu. (More on the cursor menu after the next exercise.)

Most of this will become clearer with practice. Let's try an exercise.

Do This: 2.4.1.1	Three-Dimensional Orbiting

I. Reopen the *UCS Practice 2.dwg* file in the C:\Steps3D\Lesson02 folder. (If you didn't save your changes in this drawing, open UCS *Practice 2a.dwg* instead.) The drawing looks like Figure 2.029. If it doesn't look like Figure 2.029, restore the *MyBottom* UCS and set the viewpoint to –1,1.5,-1.

II. Follow these steps.

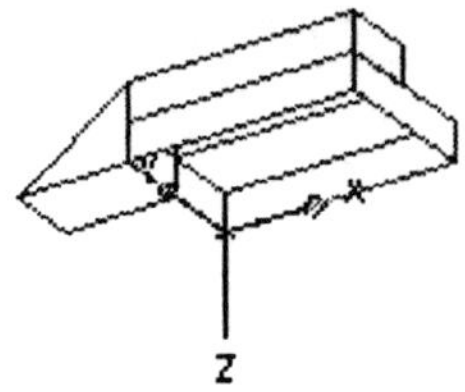

Figure 2.029: Shown with hidden lines removed

TOOLS	COMMAND SEQUENCE	STEPS
 Hidden Button	**Command:** ***shademode***	1. Set the **Shademode** system variable to **Hidden**.
 3DOrbit Button	**Command:** ***3do***	2. Enter the ***3DOrbit*** command. Alternately, you can also select **3D Orbit** from the View pull-down menu or pick the **3D Orbit** button on the 3D Orbit toolbar.
	**Press ESC or ENTER to exit, or right-click to display shortcut-menu.**	3. Place your cursor in the western quadrant circle (refer to Figure 2.028 if you have problems locating it.)
		4. With the left mouse button, click and drag to the opposite quadrant circle. Notice how the model rotates. Release the mouse button. Your drawing now looks like Figure 2.030.

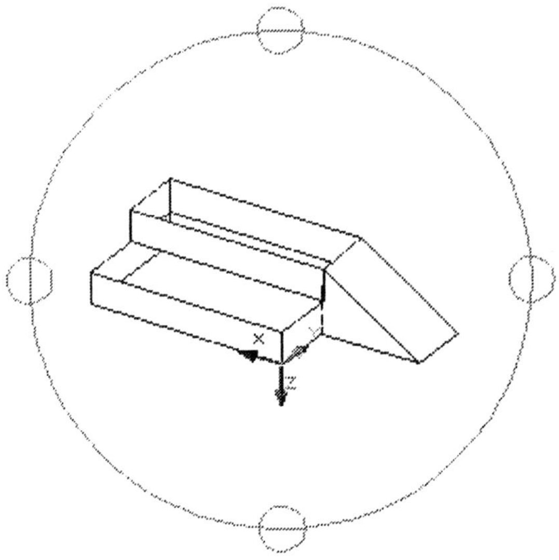

Figure 2.030

Tools	**Command Sequence**	**Steps**
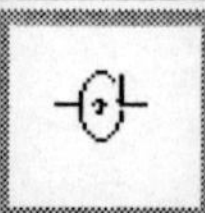		5. Repeat Steps 3 and 4, but this time begin in the upper quadrant circle and end in the lower quadrant circle. Your drawing looks like Figure 2.031. (Notice that the UCS icon continues to orient the model for you.)

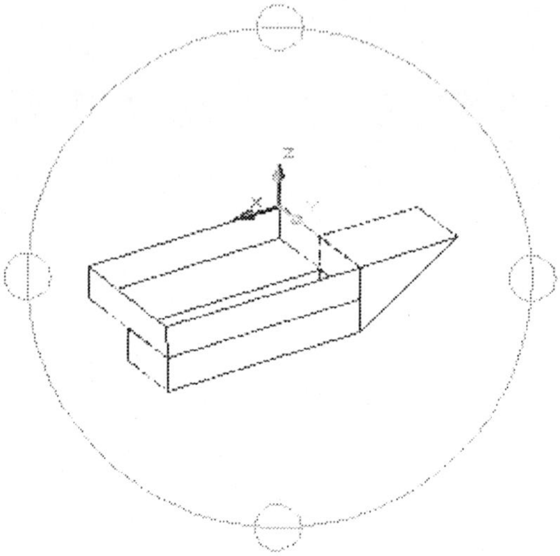

Figure 2.031

Tools	**Command Sequence**	**Steps**
		6. Place your cursor outside the arcball near the western quadrant circle. Pick and drag around the arcball to a similar location outside the eastern quadrant circle. (Note: Don't pass through the arcball, as your cursor and the manipulation procedure will change.) Your drawing looks like Figure 2.032.

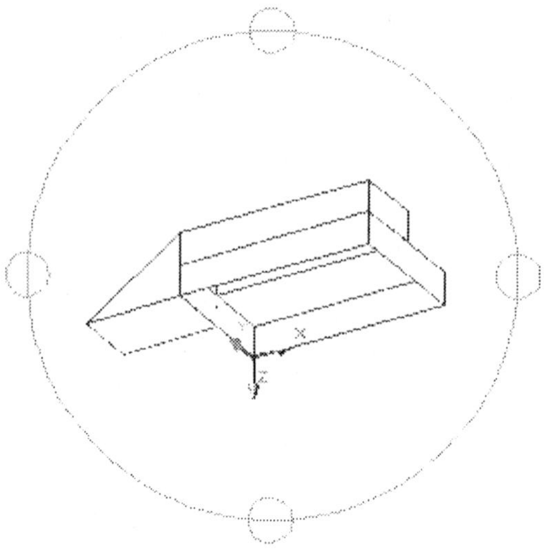

Figure 2.032

TOOLS	COMMAND SEQUENCE	STEPS
		7. Now experiment with the freestyle rotation. Place your cursor inside the arcball and drag it around. Watch and try to control the rotation.
	Press ESC or ENTER to exit, or right-click to display shortcut-menu. ***[enter]***	8. Complete the command.
	**Command:** ***u***	9. Undo your changes, but don't exit the drawing.

> You don't have to drag from quadrant circle to quadrant circle as we did here, but using the quadrant circles makes it a little easier to control what your doing. Feel free, once you've gotten the feel for the ***3DOrbit*** command, to rotate as much or as little as you want.

As if these procedures alone wouldn't make this a priceless tool in a three-dimensional world, notice that the command prompt offers you the opportunity to **right-click to display shortcut-menu.** There's more!

Let's take a look at the cursor menu (shortcut-menu). Refer to Figure 2.033.

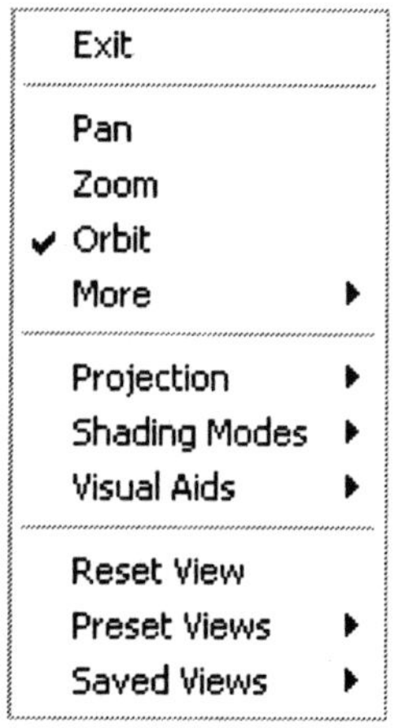

Figure 2.033

- The top frame contains only one option. Select **Exit** to leave the orbiter.
- The first two options in the next frame probably look familiar. But these aren't the ***Pan*** and ***Zoom*** commands you already know and love.
  - Selecting **Pan** is the equivalent of entering the ***3DPan*** command at the command line. This command is very similar to the ***Pan*** command (it even uses the same cursor) except that, with ***3DPan***, you can pan using a single object (or a few objects) as a reference. AutoCAD removes all other objects from the display until you exit the command.

    The trick is to select the objects you want to view *before* entering the ***3DOrbit*** or ***3DPan*** command. AutoCAD then removes all but the selected objects from view during the pan or orbit procedure.
  - Selecting **Zoom** is the equivalent of entering the ***3DZoom*** command at the command line. Like the ***Pan*** and ***3DPan*** commands, ***Zoom Realtime*** and ***3DZoom*** also share the same cursor. Also like ***3DPan***, ***3DZoom*** can view only selected objects if desired (again, if they were selected before entering the command).

    We'll get a chance to use ***3DPan*** and ***3DZoom*** in our next exercise.
  - The **Orbit** option is the one we examined in our last exercise.
  - **More** calls another menu (Figure 2.034).

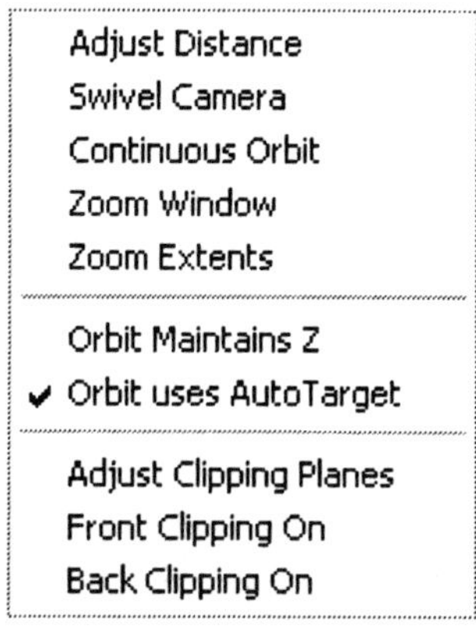

Figure 2.034

    - The **Adjust Distance** option of the **More** cursor menu appears to do the same thing that the ***3DZoom*** command

- did. The difference is that this option calls the ***3DDistance*** command, which actually changes the distance between you and the model. No distortion results from the ***3DDistance*** command as it may from the ***3DZoom*** command. The ***3DDistance*** command even has its own cursor (Figure 2.035).

Figure 2.035

Figure 2.036

- The **Swivel Camera** option calls the ***3DSwivel*** command. This adjusts the view as though a camera, although stationary, is revolving on its tripod. (See Figure 2.036 for the **Swivel Camera** cursor.) This tool comes in handy when viewing an architectural, structural, or piping drawing from inside the model.
- The **Continuous Orbit** option calls the ***3DCOrbit*** command. We'll discuss this command in Section 2.4.2.
- **Zoom Window** and **Zoom Extents** are the standard ***Zoom*** options but allow the command to take place transparently – without leaving the orbiter.
- When checked, **Orbit Maintains Z** forces the 3D orbiter to hold the Z-axis when rotating objects horizontally. You must use the left/right quadrant circles when dragging. This handy setting can save you a lot of grief by preventing the model from accidentally rotating off the screen.
- I suggest keeping the **Orbit Uses Auto Target** option checked (default setting). This ensures that rotations will be about the target point on the objects you're viewing instead of the center of the viewport. Removing this may have some undesirable results.
- **Adjust Clipping Planes** calls the ***3DClip*** command. This command presents a separate window with the model shown at 90° to the current 3D Orbital display (Figure 2.037). The lines through the center of the window are the clipping planes. Pick and drag to adjust their locations.

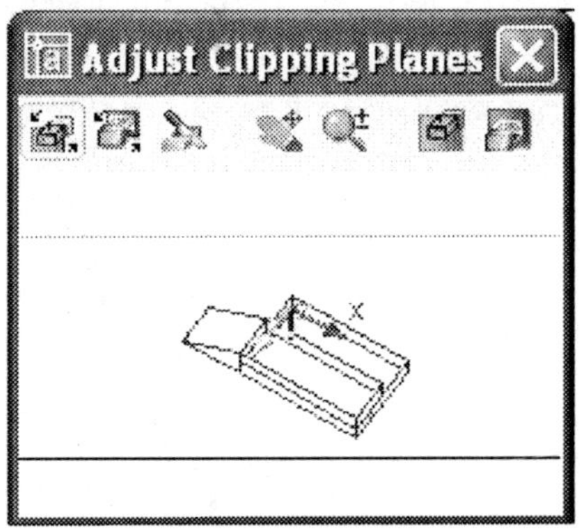

Figure 2.037

  Control what you see with the five buttons along the top of the window. These are (from the left):

  ⇒ **Adjust Front Clipping** allows you to move the front clipping plane up or down.

  ⇒ **Adjust Back Clipping** allows you to move the back clipping plane up or down.

  ⇒ **Create Slice** allows you to move both clipping planes together.

  ⇒ **Pan** calls the ***3DPan*** command within the Adjust Clipping Plane window.

  ⇒ **Zoom** calls the ***3DZoom*** command within the Adjust Clipping Plane window.

  ⇒ **Front Clipping On/Off** toggles front clipping on or off. When **On**, AutoCAD won't display anything in front of (below) the front clipping plane.

  ⇒ **Back Clipping)On/Off** toggles back clipping on or off. When **On**, AutoCAD won't display anything behind (above) the clipping plane.

  ⇒ Toggle both clippings off the view the entire model.

- (Back on the **More** cursor menu – I know this can be difficult to follow, but take your time.) **Front Clipping On** and **Back Clipping On** are toggles that work the same as the buttons in the Adjust Clipping Planes window. (We'll see more on clipping planes in our next exercise.)

- The third frame of the orbiter's cursor menu (remember that one? See Figure 2.033) contains three options – **Projection**, **Shading Modes**, and **Visual Aids**. Each calls a separate menu, but these are much simpler than the **More** menu.
  - The **Projection** option allows you to select either a **Parallel** or a **Perspective** view. In engineering drafting, you'll almost always wish to keep the default **Parallel** view. A **Perspective** is more artistic but also more difficult to use when creating dimensionally accurate drawings.
  - The **Shading Modes** option lists the same options available in the **Shademode** system variable. They're repeated here for convenience.
  - **Visual Aids** include the UCS icon, a two-dimensional grid, or a three-dimensional compass. Although you can use any or all of these tools, I recommend using the UCS icon. The two-dimensional grid is distracting in three-dimensional space and the compass doesn't display well.
- The bottom frame of the Orbiter's cursor menu also presents thee options – **Reset View**, **Preset Views**, or **Saved Views**.
  - **Reset View** behaves like an **Undo** or **Previous** option within the orbiter. It restores the view that was current when the orbiter was displayed. (So does the ***Undo*** command once the ***3DOrbit*** command ends.)
  - **Preset Views** presents a list of AutoCAD's preset viewpoints (again, this one is put here for convenience).
  - **Saved Views** presents a list of user created views.

Whew! A person can drown in that much information all at once! But don't worry; most of it's for reference. There's only one thing we need to look at a bit closer.

Let's look at the orbiter's clipping planes.

Do This: 2.4.1.2	**Using the Orbiter's Clipping Planes**

I. Be sure you're still in the *UCS Practice 2.dwg* file in the C:\Steps3D\Lesson02 folder.

II. Activate the Shademode's **Hidden** option.

III. Follow these steps.

TOOLS	COMMAND SEQUENCE	STEPS
	**Command:** ***3do***	1. Open the orbiter.
		2. Right-click in the graphics area to call the cursor menu.

Tools	**Command Sequence**	**Steps**
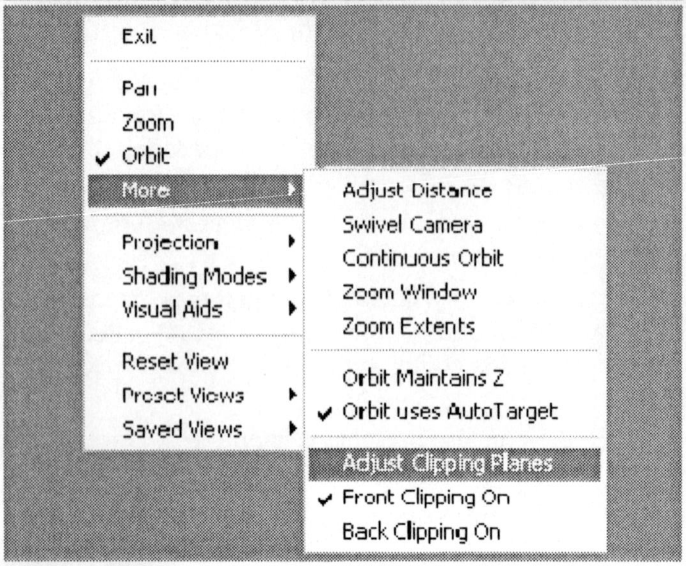		3. Select the **More** option and then **Front Clipping On** on the flyout menu. Then select the **Adjust Clipping Planes** option on the same menu.
 Adjust Front Clipping Button		4. AutoCAD presents the **Adjust Clipping Planes** window with **Adjust Front Clipping** toggled **On** (refer to Figure 2.037).
		5. Move the front clipping plane up and down (pick and drag). Watch the display in the orbiter as you move the plane.
 Adjust Back Clipping Button		6. Repeat Steps 3 through 5 using the **Back Clipping** plane. (Be sure to toggle Back Clipping On.)
 Create Slice Button		7. Move both clipping planes fairly close to each other, then pick the **Create Slice** button.
		8. Repeat Step 5.
		9. Close the Adjust Clipping Planes window.
		10. Rotate the object as you did in the previous exercise and watch the effect clipping has on your view.
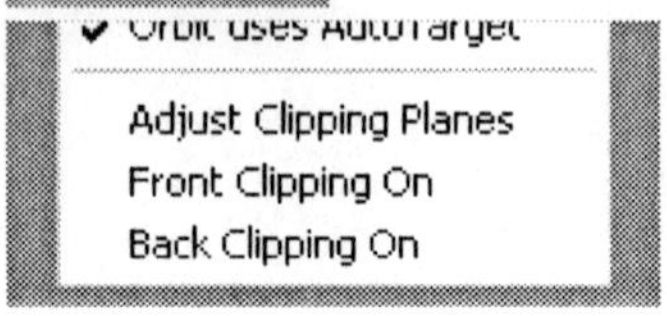		11. To clear the clipping, remove the checks from **Front Clipping On** and **Back Clipping On** on the **More** cursor menu.

Tools	Command Sequence	Steps
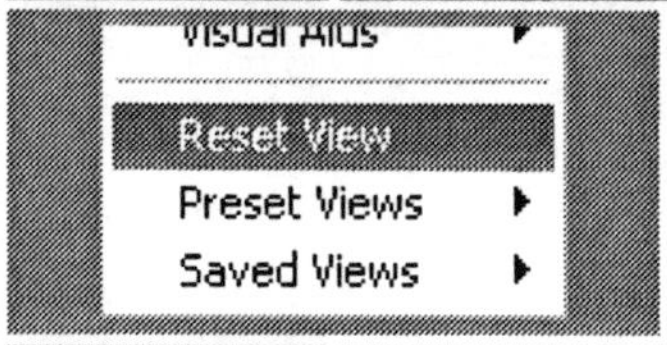		12. **Reset** the view.
	Command: ***vpoint*** ***** Switching to the WCS ***** **Current view direction: VIEWDIR=-11.3354,17.0031,-11.3354** **Specify a view point or [Rotate] <display compass and tripod>:** ***1,-1.5,1***	13. Exit the orbiter and set the view point to 1,-1.5,1.
		14. Without entering a command, select the inclined solid on the east end of the object (Figure 2.038).

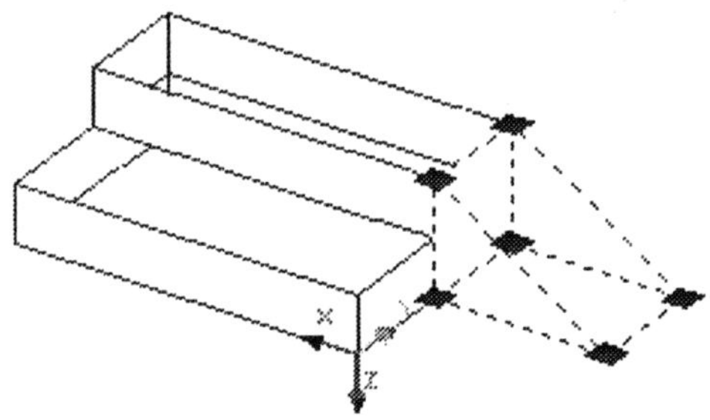

Figure 2.038

Tools	Command Sequence	Steps
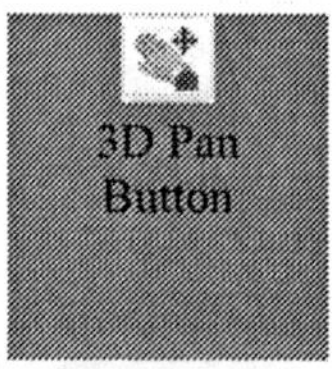	**Command:** ***3dpan***	15. Enter the ***3DPan*** command. Alternately, you can pick the **3D Pan** button on the 3D Orbit toolbar. Notice that only the selected objects appear on your screen.
		16. Hold down the left mouse button and move the object around the screen.
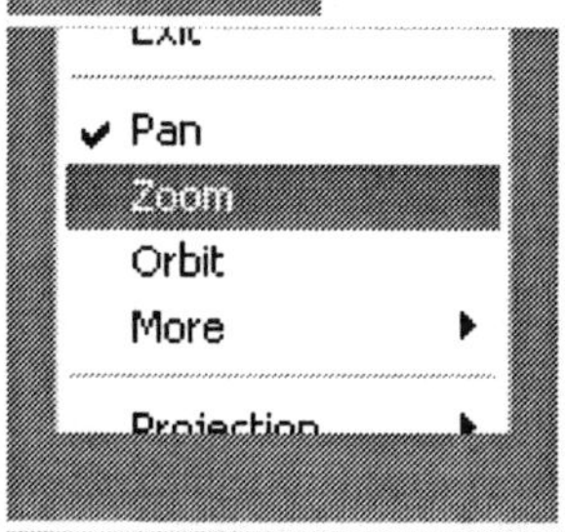		17. Right-click and select **Zoom** from the cursor menu. (If you were at the command prompt, you could pick the **3D Zoom** button on the 3D Orbit toolbar.)
		18. Repeat Step 16, zooming in and out on the object.
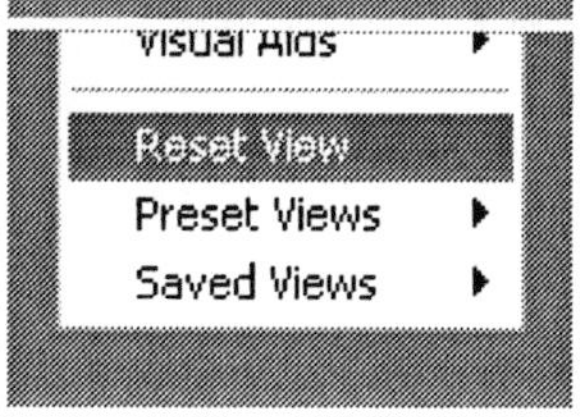		19. Reset the view.
		20. Right-click and select **Exit** from the cursor menu. Notice that all the objects in the drawing return to the display.

2.4.2 A Continuous Three-Dimensional Orbit – *3DCOrbit*

This tool is guaranteed to razzle-dazzle friends and co-workers (and employers) alike! The ***3DCOrbit*** command (3D Continuous Orbit) allows you to begin a rotation – and then remove your hands from the keyboard or mouse and watch as AutoCAD continuously rotates the model on the screen.

To use the ***3DCOrbit*** command, first enter the command or select it from the 3DOrbit toolbar. AutoCAD prompts:

> **Command:** ***3dcorbit***
>
> **Press ESC or ENTER to exit, or right-click to display shortcut-menu.**

Notice that the prompt is the same as the ***3DOrbit*** command prompt. By right-clicking at any time during the 3D Continuous Orbit, you can access the same cursor menu available through any of the ***3DOrbit*** commands. Alternately, you can begin a continual orbit of the model. To do this, pick any point on the screen and drag the cursor in the direction you'd like the model to spin. Notice the cursor changes to one of the 3DOrbit cursors (depending on which direction you drag). Release the mouse button and the model continues to spin on its axis – like a planet. The speed at which you drag the cursor determines the speed of the spin.

To stop the spinning, hit either the ENTER or ESC key on the keyboard or right-click and select **Exit** from the cursor menu. The model stops spinning where it is. Use the ***Undo*** command to return to the orientation it had before the command.

Give it a try.

Do This: 2.4.2.1 Creating a Continuous Orbit

I. Be sure you're still in the *UCS Practice 2.dwg* file in the C:\Steps3D\Lesson02 folder.

II. Follow these steps.

TOOLS	COMMAND SEQUENCE	STEPS
 3D Continuous Orbit Button	**Command:** ***3dcorbit***	1. Enter the ***3DCOrbit*** command at the command line or pick the **3D Continuous Orbit** button on the 3DOrbit toolbar.
	Press ESC or ENTER to exit, or right-click to display shortcut-menu.	2. Pick a point on the right side of the screen and drag to the left. Release the mouse button about halfway across the screen but continue the mouse movement (use some follow-through as though you're hitting a golf ball).
		3. Watch the model spin before your very eyes!
	Press ESC or ENTER to exit, or right-click to display shortcut-menu. ***[enter]***	4. Hit ***enter*** to stop the rotation.
	**Command:** ***u***	5. Undo the rotation.

2.5 Extra Steps

It may take some practice to get the feel of the ***3DCOrbit*** command. Take a few minutes now and repeat the last exercise until you feel comfortable. For a real treat, shade the model (use the Gouraud Shademode) before starting the 3D Continuous Orbiter. Change some of the colors for a flashier shown. Try different speeds of rotation.

Can you see how this tool might be useful?

2.6 What Have We Learned?

Items covered in this lesson include:

- *The differences between the UCS and the WCS*
- *How to use the UCS and different working planes*
- *How to use the UCS Dialog Box (Manager)*
- *How to dimension a three-dimensional drawing*
- *How to use AutoCAD's orbiting tools*
- *Commands:*
    - ***UCS***
    - ***UCSMan***
    - ***3DOrbit***
    - ***3DPan***
    - ***3DZoom***
    - ***3DDistance***
    - ***3DSwivel***
    - ***3DClip***
    - ***3DCOrbit***

My chief Grammar and Usage Editor will say this was another full lesson!

But pat yourself on the back! Having made it to this point is no slight accomplishment. In these two lessons, you've mastered the basics for working in Z-Space. Let's take a minute and think about what you can do now that you couldn't do before beginning this text.

- You can maneuver in a drawing from front to back, side to side, and up and down (Spherical and Cylindrical Coordinate Systems, Point Filters).
- You can see your drawing from any point in the universe (***VPoint*** and ***Plan***).
- You can work on any surface as though it were lying flat on your desk (***UCS***).
- You have a host of new viewing tools to help you see your model from any angle or several angles at one time (***VPoint***, ***VPorts***, 3DOrbit tools, **Shademode**).
- You can create three-dimensional stick figures (wireframes) and even draw with thickness and elevation.

You've come a long way in a short period of time, but there's still far to go. (Oh, the sights still to see …) Most of what you learn about Z-space from here will be tools and techniques to make three-dimensional drawing easier, faster, and prettier!

As always, we should practice what we've learned before continuing our study. Do the exercises and answer the questions. Then proceed to our study of Wireframe and Surface Modeling techniques.

2.7 Exercises

1. to 8. Dimension the drawings you created in Exercises 1 through 8 of Section 1.7 (in Lesson 1). Refer to the drawings in Appendix B as a guide. If these drawings are not available, use the corresponding drawing in the C:\Steps3D\Lesson02 folder.

9. [Refer to Section 1.7 of Lesson 1 – Exercise #9.] Open the *My Twisted Y.dwg* file in the C:\Steps3D\Lesson01 folder. [If the *My Twisted Y.dwg* file isn't available, use the *Twisted Y.dwg* file found in the C:\Steps3D\Lesson02 folder.]

   9.1. Place the dimensions shown in Figure 1.043.

   9.2. Save the drawing to the C:\Steps3D\Lesson02 folder.

10. [Refer to Section 1.7 of Lesson 1 – Exercise #10.] Open the *My Block.dwg* file in the C:\Steps3D\Lesson01 folder. [If the *My Block.dwg* file isn't available, use the *Block.dwg* file found in the C:\Steps3D\Lesson02 folder.]

    10.1. Place the dimensions shown in Figure 1.044.

    10.2. Save the drawing to the C:\Steps3D\Lesson02 folder.

11. [Refer to Section 1.7 of Lesson 1 – Exercise #11.] Open the *MyGrill.dwg* file in the C:\Steps3D\Lesson01 folder. [If the *My Grill.dwg* file isn't available, use the *Grill.dwg* file found in the C:\Steps3D\Lesson02 folder.]

    11.1. Place the dimensions shown in Figure 1.045.

    11.2. Save the drawing to the C:\Steps3D\Lesson02 folder.

12. [Refer to Section 1.7 of Lesson 1 – Exercise #12.] Open the *MyBookEnd.dwg* file in the C:\Steps3D\Lesson01 folder. [If the *MyBookEnd.dwg* file isn't available, use the *BookEnd.dwg* file found in the C:\Steps3D\Lesson02 folder.]

    12.1. Place the dimensions shown in Figure 1.046.

    12.2. Save the drawing to the C:\Steps3D\Lesson02 folder.

13. [Refer to Section 1.7 of Lesson 1 – Exercise #13.] Open the *MyMagRack.dwg* file in the C:\Steps3D\Lesson01 folder. [If the *MyMagRack.dwg* file isn't available, use the *MagRack.dwg* file found in the C:\Steps3D\Lesson02 folder.]

    13.1. Place the dimensions shown in Figure 1.047.

    13.2. Save the drawing to the C:\Steps3D\Lesson02 folder.

14. [Refer to Section 1.7 of Lesson 1 – Exercise #14.] Open the *My Anchor Stop.dwg* file in the C:\Steps3D\Lesson01 folder. [If the *My Anchor Stop.dwg* file isn't available, use the *Anchor Stop.dwg* file found in the C:\Steps3D\Lesson02 folder.]

    14.1. Place the dimensions shown in Figure 1.048.

    14.2. Save the drawing to the C:\Steps3D\Lesson02 folder.

15. [Refer to Section 1.7 of Lesson 1 – Exercise #15.] Open the *My Queen.dwg* file in the C:\Steps3D\Lesson01 folder. [If the *My Queen.dwg* file isn't available, use the *Queen.dwg* file found in the C:\Steps3D\Lesson02 folder.]

    15.1. Place the dimensions shown in Figure 1.049.

    15.2. Save the drawing to the C:\Steps3D\Lesson02 folder.

16. [Refer to Section 1.7 of Lesson 1 – Exercise #16.] Open the *My Ring Stand.dwg* file in the C:\Steps3D\Lesson01 folder. [If the *My Ring Stand.dwg* file isn't available, use the *Ring Stand.dwg* file found in the C:\Steps3D\Lesson02 folder.]

    16.1. Place the dimensions shown in Figure 1.050.

16.2. Save the drawing to the C:\Steps3D\Lesson02 folder.

17. Create the drawing in Figure 2.039 according to the following parameters:
    17.1. Start the drawing using *template #2* found in the C:\Steps3D\Lesson02 folder.
    17.2. The Paper Space text is 3/16" and 1/8".
    17.3. Title block text is ¼", 3/16", and 1/8".
    17.4. Use the **Hidden** option of the ***Shademode*** command where necessary to achieve the views shown.
    17.5. Text on the blocks uses either AutoCAD's standard text style or a type using the Time New Roman font. Text heights are ¼".
    17.6. Save the drawing as *My Angled Blocks.dwg* in the C:\Steps3D\Lesson02 folder.

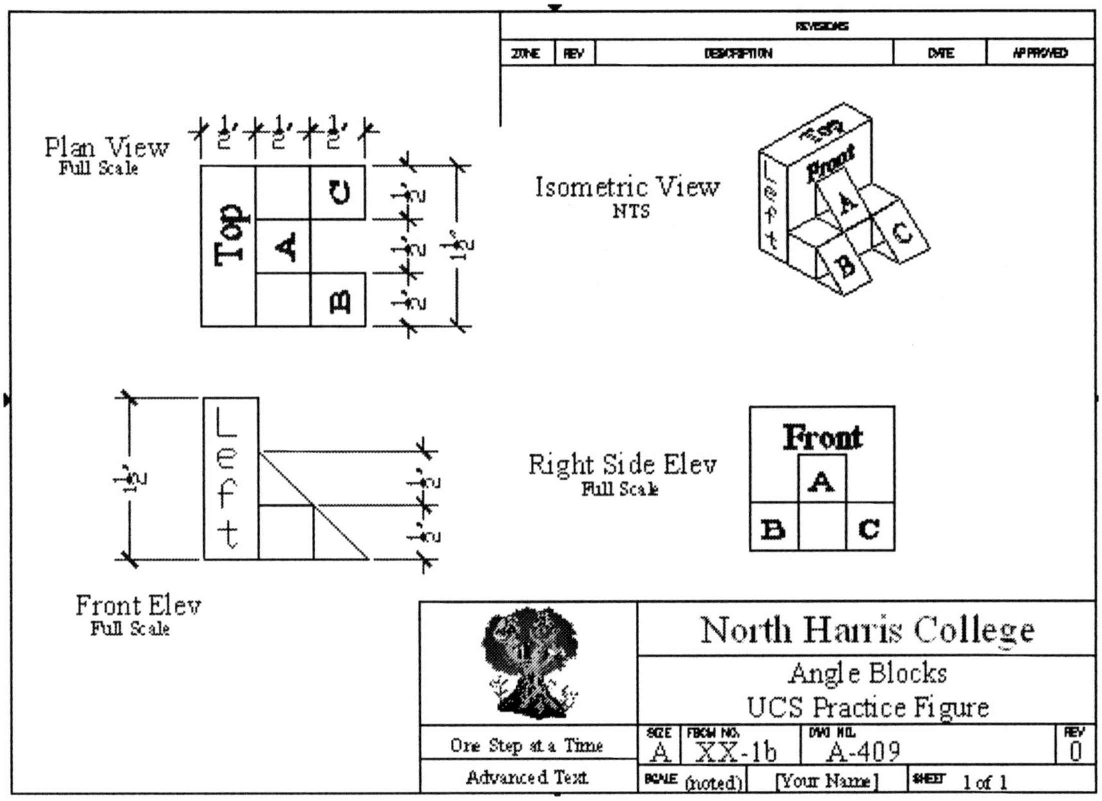

Figure 2.039

18. Create the drawing in Figure 2.040 according to the following parameters:
    18.1. Start the drawing using *template #1* found in the C:\Steps3D\Lesson01 folder.
    18.2. The Paper Space text is 3/16" and 1/8".
    18.3. Title block text is ¼", 3/16", and 1/8". The font is Times New Roman.
    18.4. Use the **Hidden Shademode** system variable where necessary to achieve the views shown.
    18.5. Text on the blocks uses AutoCAD's standard text style. Text height is 3/16".
    18.6. Save the drawing as *My Corner Steps.dwg* in the C:\Steps3D\Lesson02 folder.

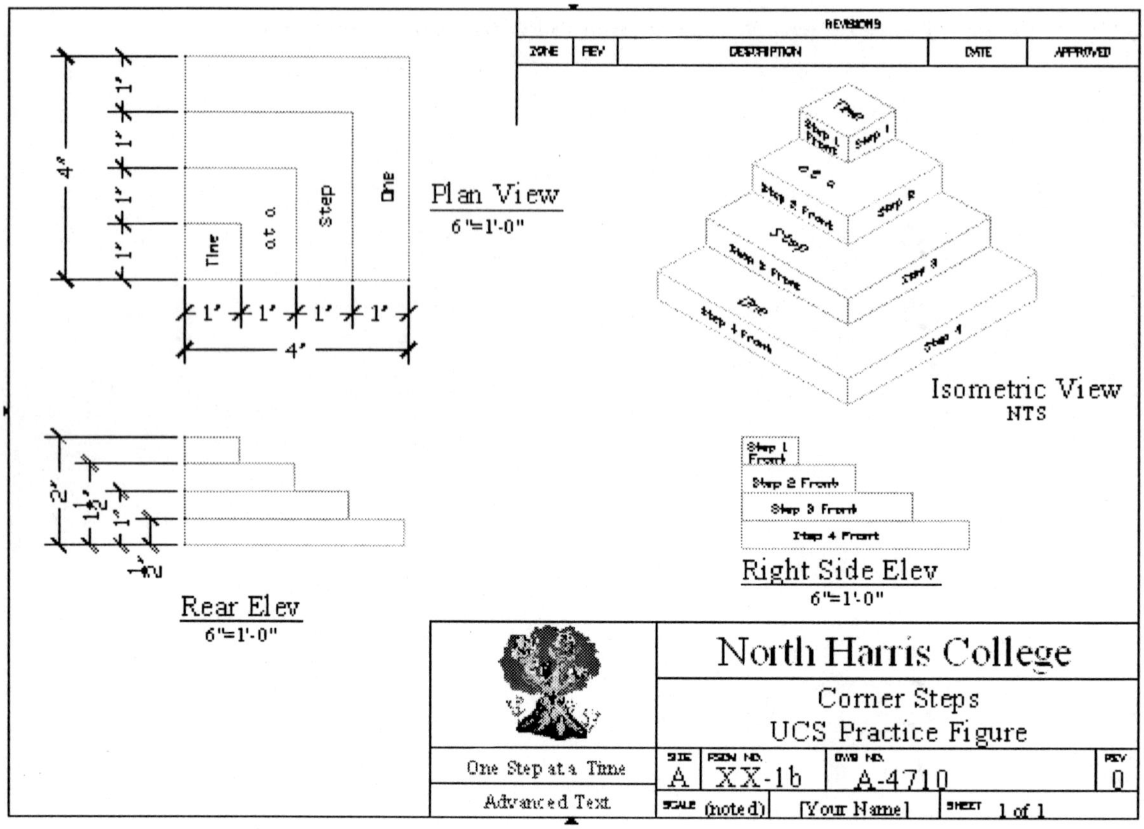

Figure 2.040

19. The drawing in Figure 2.041 is a game of UCS manipulation and the ***Array*** and ***Mirror*** commands. Create it according to the following parameters:
    19.1. The tabletop is a 5" width x 5" long polyline drawn with a ½" thickness.
    19.2. Each of the eight legs is made up of eight lines. Each line is 5" long. The original was drawn at 30° in the XY-lane and 60° from the XY-plane. The lines were then arrayed in a 1/16" circle.
    19.3. The eight "feet" are ¼" radius circles. Again, they were arrayed (eight circles in each array).
    19.4. The center ball (at the intersection of the legs) is also an arrayed circle (eight in all) – this one with a ½" radius.
    19.5. The upper legs are mirrored from the bottom.
    19.6. Have fun!
    19.7. Save the drawing as *My Table.dwg* in the C:\Steps3D\Lesson02 folder.

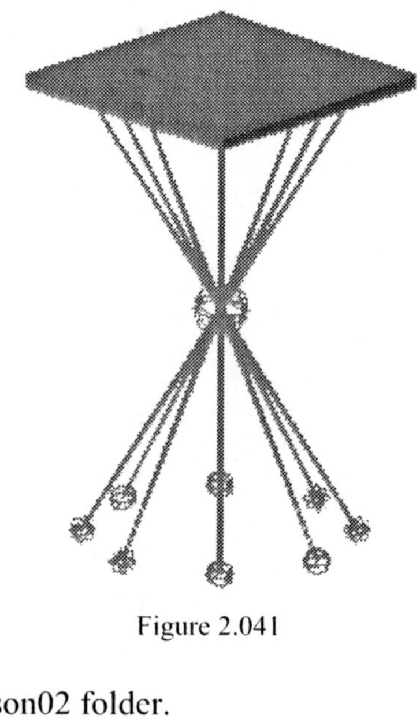

Figure 2.041

20. Create the drawing in Figure 2.042 according to the following parameters:
    20.1. This is a B-size (11" x 17") layout. Use the appropriate AutoCAD title block/border.
    20.2. The Paper Space text is 3/16" and 1/8".
    20.3. Title block text is ¼", 3/16", and 1/8". The font is Times New Roman.
    20.4. The top is a single polyline drawn with ½ thickness.
    20.5. Save the drawing as *My Other Table.dwg* in the C:\Steps3D\Lesson02 folder.

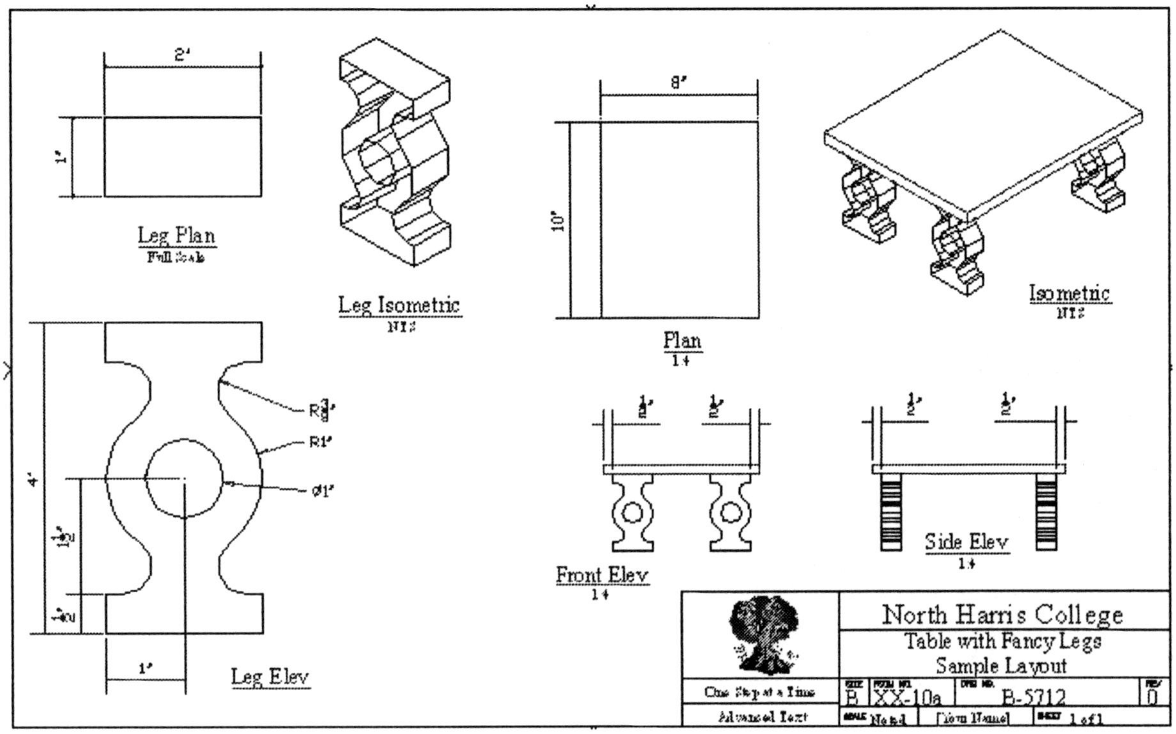

Figure 2.042

21. Create the drawing in Figure 2.043 according to the following parameters:
    21.1. This is a B-size (11" x 17") layout. Use the appropriate AutoCAD title block/border.
    21.2. The Paper Space text is 3/16" and 1/8".
    21.3. Title block text is ¼", 3/16", and 1/8". The font is Times New Roman.
    21.4. This is a wireframe drawing – use thickness only on the arcs/circles.
    21.5. Save the drawing as *My Corner Bracket.dwg* in the C:\Steps3D\Lesson02 folder.

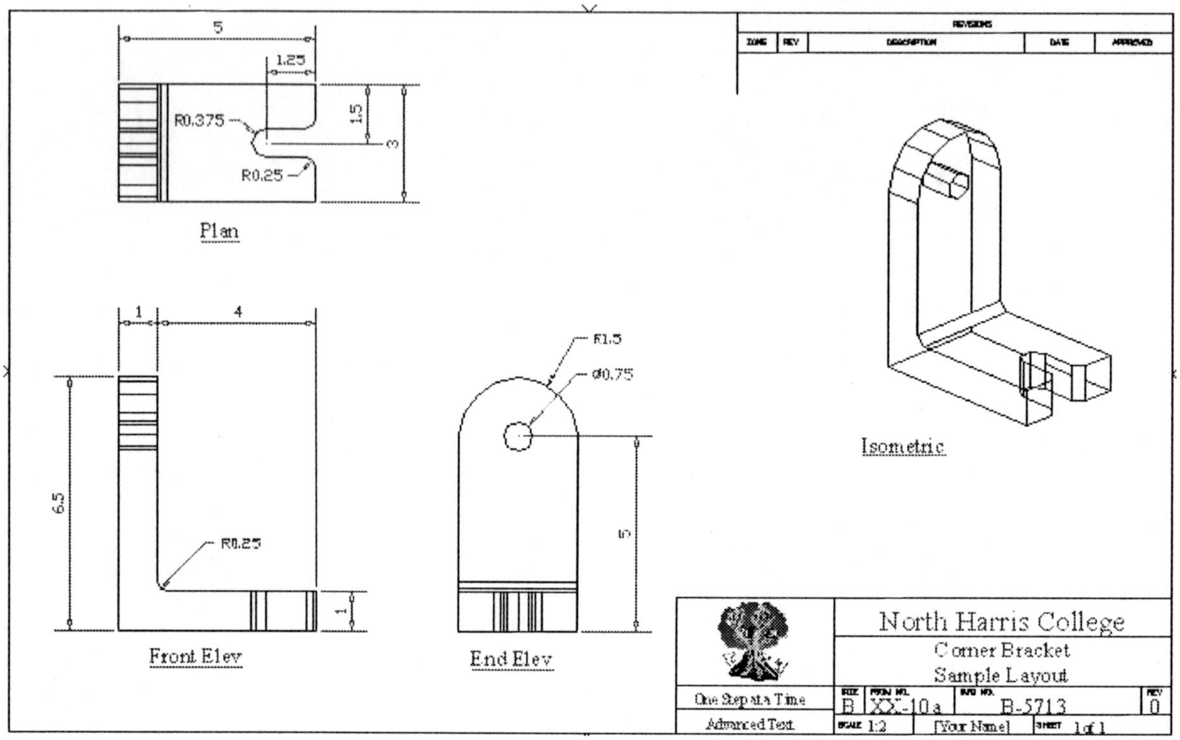

Figure 2.043

22. Create the drawing in Figure 2.044 according to the following parameters:
    22.1. This is an A-size (8½" x 11") layout. Use the appropriate AutoCAD title block/border.
    22.2. The Paper Space text is 3/16" and 1/8".
    22.3. Title block text is ¼", 3/16", and 1/8". The font is Times New Roman.
    22.4. This is a wireframe drawing – don't use thickness.
    22.5. You'll need to use the ***3DClip*** procedures to clean up the views.
    22.6. Save the drawing as *My Phone Plug.dwg* in the C:\Steps3D\Lesson02 folder.

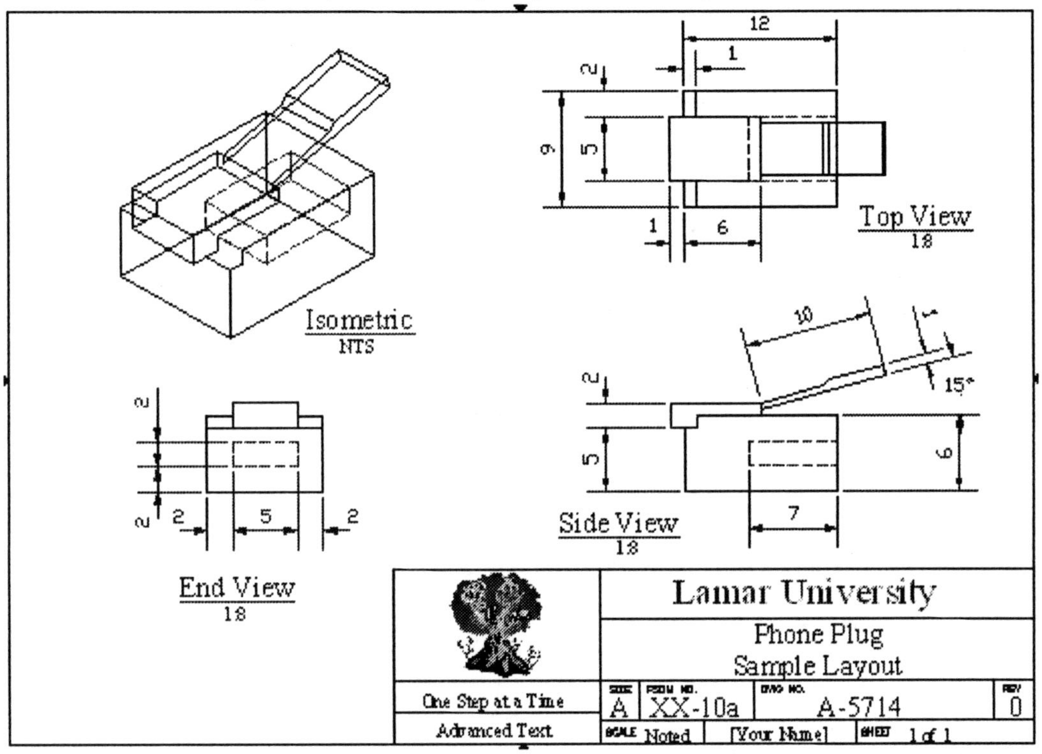

Figure 2.044

23. Create the drawing in Figure 2.045 according to the following parameters:
    23.1. This is a B-size (11" x 17") layout. Use the appropriate AutoCAD title block/border.
    23.2. The Paper Space text is 3/16" and 1/8".
    23.3. Title block text is ¼", 3/16", and 1/8". The font is Times New Roman.
    23.4. This is a wireframe drawing – use thickness only on the arcs/circles.
    23.5. You'll need to use the ***3DClip*** procedures to clean up the views.
    23.6. Save the drawing as *My Other Corner Bracket.dwg* in the C:\Steps3D\Lesson02 folder.

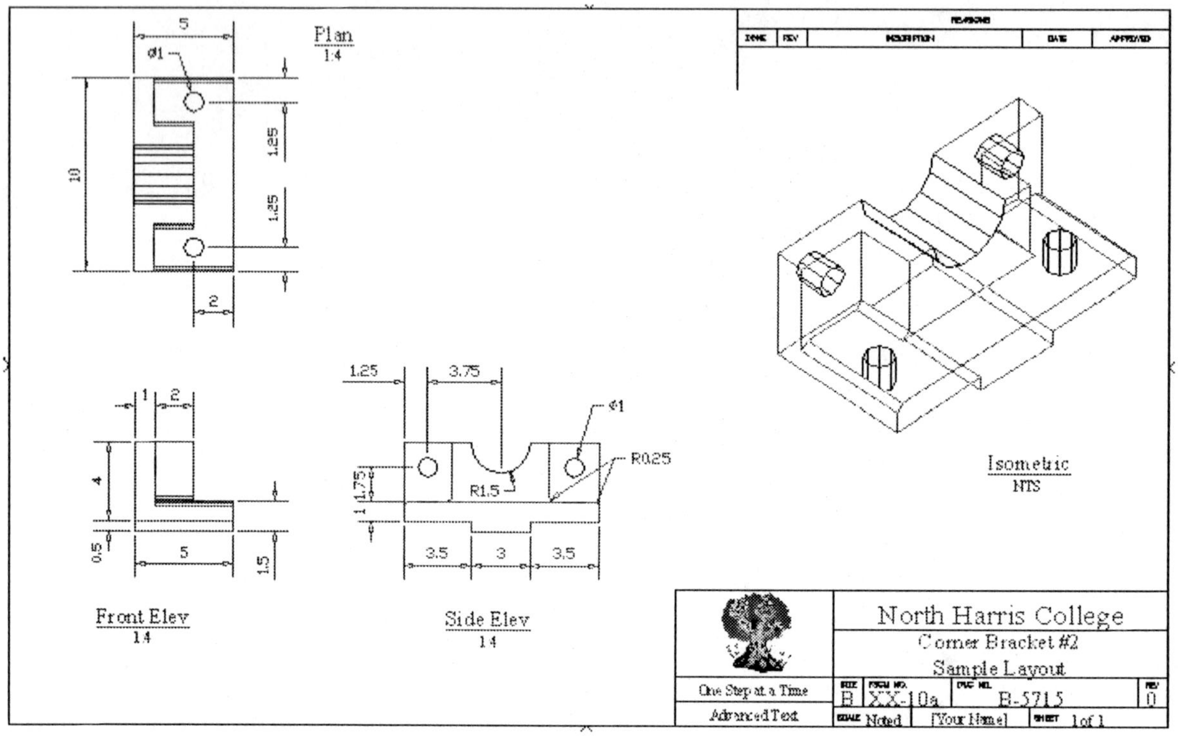

Figure 2.045

2.8	**For Web-Based Review Questions, visit: http://www.uneedcad.com/2006/Files/06R2-3D.pdf**

Lesson 3

Following this lesson, you will:

- ✓ *Know the differences between a polyline and a three-dimensional polyline*
- ✓ *Know how to project a curved surface in three dimensions*
- ✓ *Know how to create a three-dimensional face (**3DFace**)*
  - o *Know how to make the edges of a face visible or invisible*
- ✓ *Know the differences between solids and regions*
  - o *Know how to create a region with the **Region** command*
  - o *Know how to create a region with the **Boundary** command*
- ✓ *Know how to use the **Subtract** command to remove one region from another*

Wireframes and Surface Modeling

Most textbooks separate Wireframe Modeling and Surface Modeling into two distinct chapters. But the inevitable result is confusion. The two are so closely related that distinguishing between them often causes more bewilderment than just teaching them as they are – two sides of the same coin.

Let me make the distinction as simple as possible.

A wireframe model (what we've called a stick figure up until now) is a skeleton drawing. It has all the necessary parts – but no flesh. Drawing a wireframe model is relatively fast (compared to a surface model), but it provides little more than an outline of the model. When used, wireframes generally lead the three-dimensional design process (in the layout stage). Fleshing out the wireframe – turning it into a surface model – comes when the layout is accepted, and you want to turn the skeleton into a production or display drawing.

A surface model essentially stretches some skin over the skeleton. We've seen that Shademode and Hide have no effect on wireframe models. Fleshing out the skeleton makes it possible to see surfaces (hence, the name).

In this lesson, we'll learn how to create more complex wireframe models. Then we'll look at some ways to create surfaces.

3.1 *3DPoly* vs. *PLine*

Now that you're working with Z-coordinates, you may have noticed a certain limitation in polylines – polylines are two-dimensional creatures. True, you can give a polyline thickness and elevation, but you can't draw a polyline using different points on the Z-axis. That is, when prompted to **Specify next point**, your selection will use the same point on the Z-axis as the first point you identified regardless of any three-dimensional coordinate you give it!

But that doesn't mean that you have to sacrifice the benefits of a polyline … well, not entirely anyway. AutoCAD provides a three-dimensional version of the polyline call the *3DPoly*. When you need a multi-segmented polyline drawn in Z-space, simply use the ***3DPoly*** command instead of the ***PLine*** command.

There are, however, some restrictions to the 3DPoly. Chief among these is that a 3D polyline can't contain width. AutoCAD hasn't added this useful property yet. Additionally, you can't draw a 3D polyline using arcs or linetypes other than continuous.

> Bear in mind that, while use of the polyline is restricted in Z-space, use of the spline isn't. You can use the ***Spline*** command when you want to draw curved lines in three dimensions. However, most surfaces created for a surface model will have flat edges and won't be able to lie flat against a spline.

The benefits of the 3DPoly include the ability to draw it in three dimensions, to edit it with the ***PEdit*** command, and to spline it.

We'll use the ***3DPoly*** command in our first exercise.

3.2 Drawing in Three Directions at Once – Point Projection

You may think that wireframe modeling is a fairly easy thing to do. After all, a wireframe model is just stick figures, right?

Of course, you're absolutely right. Stick figures are quite simple to draw – as long as the model you want to draw uses nice straight sticks. But consider the curved panel roof in Figure 3.001. Using the UCS procedures you learned in Lesson 2, you can easily draw the arcs and rooflines. But how would you draw the joint between the roofs? (Uh, oh! Here we go with the hard questions again.)

To draw in three directions at once as this joint requires – (front to back, side to side, and up and down – means that you must identify a series of points where the two roofs intersect. To do this, you must project points – that is, you must identify points by intersection of lines in Z-space.

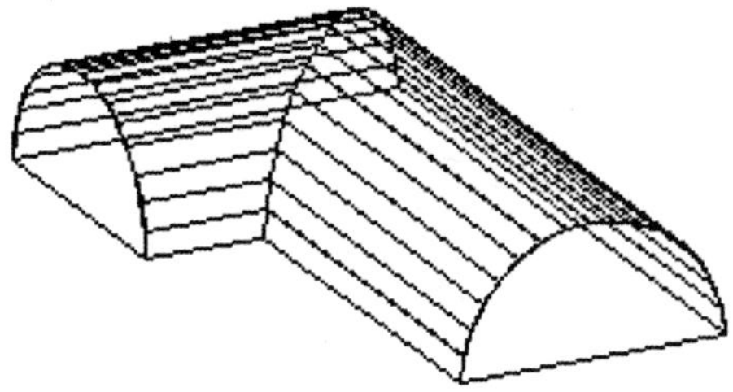

Figure 3.001

This isn't as difficult as it sounds. In fact, it's a lot like duck hunting (or skeet shooting for those with weaker stomachs).

To hit a moving target, you must lead it a bit so that your bullet and the bird (or clay pigeon) arrive at the same place at the same time. What you do when you project points is simply identify where the bird and the bullet will meet. You do this by projecting one line along the bird's flight path and another along the barrel of your rifle. The intersection of the two lines is your actual target – or in drafting terms, your projection point.

Let's see how this works. We'll use our projection technique to identify the intersecting arc of the two roofs and then draw the arc using the 3D polyline.

> Remember that all three-dimensional drafting requires precise point identification using one of the methods we've already discussed. You can't pick an "about here" point and get away with it as you might have in two-dimensional drafting.

Do This: 3.2.1	Projections and 3D polylines

I. Open the *Cabin.dwg* file in the C:\Steps3D\Lesson03 folder. The drawing looks like Figure 3.002.

II. Follow these steps.

Figure 3.002

TOOLS	COMMAND SEQUENCE	STEPS
	Command: ***div*** **Select object to divide:** **Enter the number of segments or [Block]:** ***16***	1. Use the ***Divide*** command to divide each of the arcs into 16 segments. (If the divisions aren't clearly marked, set the **PDMode** system variable to **3** and regenerate the drawing.) Notice that the nodes appear in the UCS that was current when the arcs were drawn.
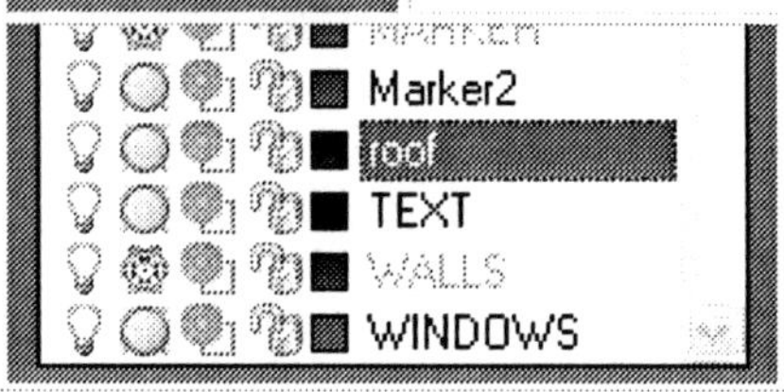		2. Set the **roof** layer current.

TOOLS	COMMAND SEQUENCE	STEPS
	**Command:** ***l***	3. Draw lines between the corresponding nodes and endpoints as shown in Figure 3.003. (The lines represent the duck's flight path and the barrel of our rifle.) (Hint: You may find it easier to draw one line in each direction and then copy it to each of the nodes/endpoints.)

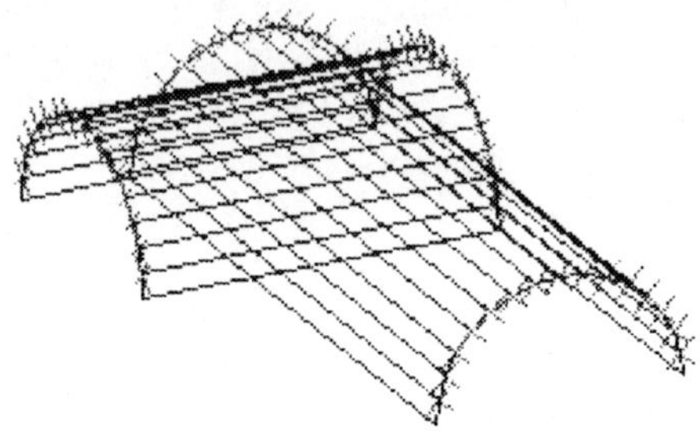

Figure 3.003

TOOLS	COMMAND SEQUENCE	STEPS
		4. Freeze the **Marker2** layer.
	**Command:** ***tr***	5. Carefully trim the extra portions of the lines. Start with the nearest intersection – lowest north-south line with lowest east-west line – and work back. (Hint: You may fine it easier to do this from the plan view. But watch the first and last endpoints.) Return to this view (VPoint 2,-1,1) when you've finished. Your drawing looks like Figure 3.004.

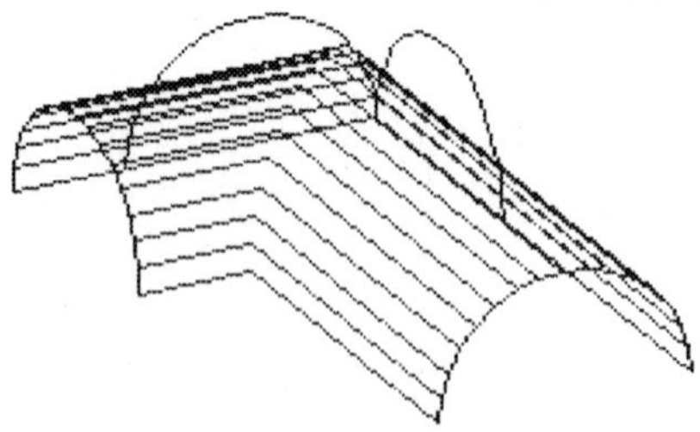

Figure 3.004

TOOLS	COMMAND SEQUENCE	STEPS
	**Command:** ***e***	6. Erase the arcs in the back.

TOOLS	COMMAND SEQUENCE	STEPS
No Button Available	**Command:** ***3dpoly*** **Specify start point of polyline:** **Specify endpoint of line or [Undo]:** **Specify endpoint of line or [Undo]:** **Specify endpoint of line or [Close/Undo]:**	7. Draw a 3D polyline connecting the intersections of the extension lines as shown in Figure 3.005. Use OSNAPs! (It might be easier to do this in plan view.)
	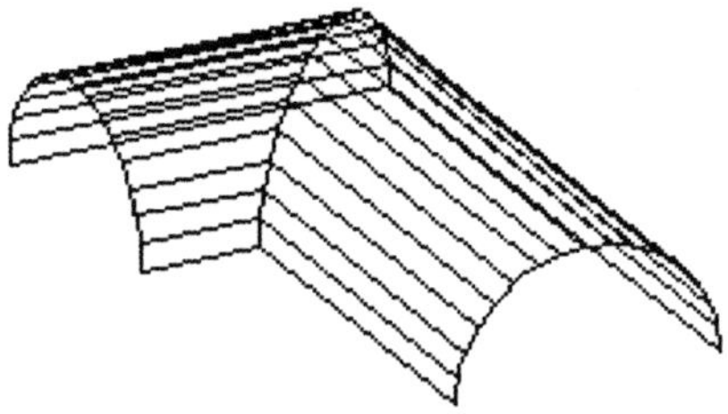 Figure 3.005	
	Command: ***qsave***	8. Save the drawing but don't exit.

Try the ***Hide*** command on the drawing. (Go ahead; experiment!) If you do, you'll notice that it has no effect. Remember that we've created a *wireframe model* – a skeleton of the roof we want.

Look closely at the 2D polyline you created. Notice that it's a series of straight lines, not curved like the roof. We might have used a spline instead of the 3D polyline and achieved a nice soft curve, but the tool we'll use to "stretch the skin around our skeleton" doesn't allow for curves. So we're better off using the straighter 3D polyline (as we'll soon see).

3.3 Adding Surfaces – Regions, Solids, and 3D Faces

AutoCAD provides three methods for creating surface models – the ***Region***, ***Solid***, and ***3DFace*** commands. The three are so closely related that it's often difficult to tell the difference:

- You'll draw each as a two-dimensional object.
- Each becomes a 3D solid when extruded (more on the ***Extrude*** command in Lesson 7).
- Each creates an opaque (or solid) surface.

But despite their similarities, each has its place.

- The ***Region*** command converts a closed object (polygon, circle, two-dimensional spline, etc.) into a surface. A region can't have thickness.
- The ***Solid*** command fills an area only in the current UCS, but it can have thickness.
- The ***3DFace*** command draws a true three-dimensional surface (in all of the X-, Y-, and Z-planes). A 3D face can't have thickness.

When would you use one instead of the other two? Let's take a look at each and see.

3.3.1 Creating 3D Faces

Of the three surfacing methods, the ***3DFace*** command is the most versatile. However, it's also more difficult to use when cutouts are involved. The ***3DFace*** command makes no allowance for removal of part of the surface (as the ***Trim*** command allows you to remove part of a line or circle). You can, however, draw a 3D face without concern for the current UCS (provided coordinate entry is precise).

The command sequence looks like this:

> **Command:** ***3DFace***
> **Specify first point or [Invisible]:** ***[select the first corner point]***
> **Specify second point or [Invisible]:** ***[select the second corner point]***
> **Specify third point or [Invisible] <exit>:** ***[select the third corner point]***
> **Specify fourth point or [Invisible] <create three-sided face>:** ***select the fourth point or hit enter to create a 3D face from the three points already selected]***
> **Specify third point or [Invisible] <exit>:** ***[you can continue selecting points or hit enter to exit the command]***

The only option available – **Invisible** – isn't one you really want to use. When creating 3D faces, it'll occasionally be necessary to hide one of the edges (or make it invisible). (This will become apparent in the next exercise.) The command line procedure for doing this involves typing an ***I*** before the first point selection that defines the edge to be hidden. The edge drawn between the two points that follow the ***I*** will be invisible. This becomes a real chore when two or more edges must be hidden. We'll look at an easier approach to hiding the edges of 3D faces following the next exercise.

First, let's use the ***3DFace*** command to place a surface – with a window in it – on one of the walls of our cabin.

> You can also access the ***3DFace*** command by using the **3D Face** button on the Surfaces toolbar or by selecting it from the Draw pull-down menu. Follow this path:
> *Draw – Surfaces – 3D Face*

Do This: 3.3.1.1	Surfaces with 3D Faces

I. Be sure you're still in the *Cabin.dwg* file in the C:\Steps3D\Lesson03 folder. If not, please open it now.

II. Set UCS = WCS.

III. Thaw the **WALLS** and **MARKER** layers; freeze the **roof** layer. The drawing looks like Figure 3.006.

IV. Follow these steps.

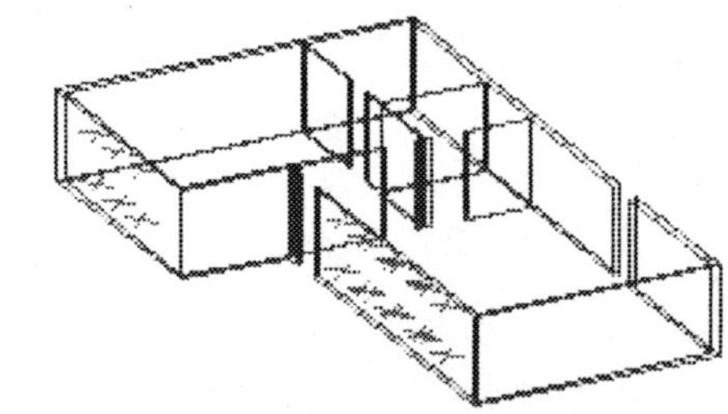

Figure 3.006

TOOLS	COMMAND SEQUENCE	STEPS
	**Command:** *z*	1. Zoom in around the protruding part of the cabin – the wall with six nodes (refer to Figure 3.007).
	**Command:** *props*	2. The front wall consists of two lines drawn with thickness. We can't put a window in these objects, so we'll remove the thickness and replace the lines with 3D faces. Begin by changing the thickness of both lines (inner and outer walls) to **0**.

TOOLS	COMMAND SEQUENCE	STEPS
		(Use the Properties Palette.)
3D Face Button (Surfaces Toolbar)	**Command:** ***3dface*** **Specify first point or [Invisible]:** ***[select Point 1]***	3. (Refer to Figure 3.007.) Begin the ***3DFace*** command. Pick Point 1 as your first point.

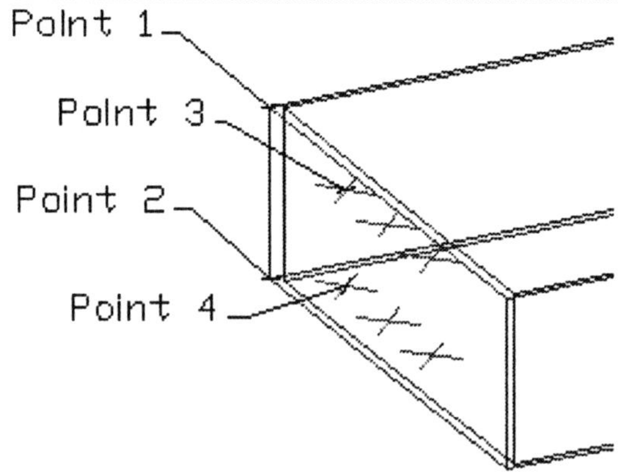

Figure 3.007

	Specify second point or [Invisible]: ***[select Point 2]***	4. Pick Point 2 as the second point.
	Specify third point or [Invisible] <exit>: ***.xy***	5. We'll use point filters to locate the third point. Tell AutoCAD to use the X and Y values of Point 3 …
	of (need Z): ***[select Point 2]***	6. … and the Z value of Point 2.
	Specify fourth point or [Invisible] <create three-sided face>: ***.xy***	7. Again, we'll use point filters to locate the fourth point. Tell AutoCAD to use the X and Y values of Point 4 …
	of (need Z): ***[select Point 1]***	8. … and the Z value of Point 1.
	Specify third point or [Invisible] <exit>: ***[enter]***	9. Complete the command. Your drawing looks like Figure 3.008.

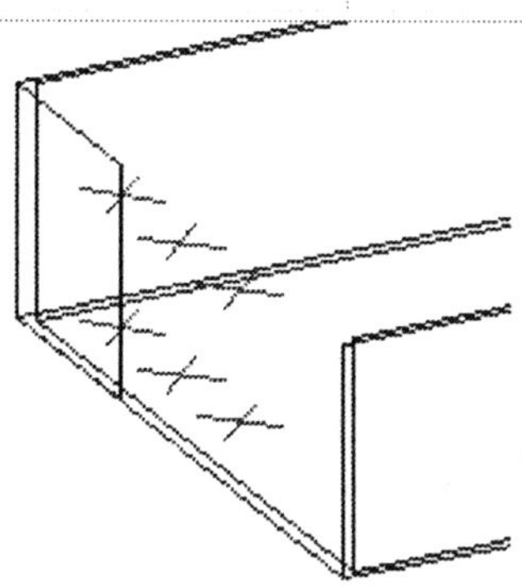

Figure 3.008

Tools	Command Sequence	Steps
	Command: ***3dface***	10. Repeat Steps 4 through 10 to draw the other end of the wall and the walls above and below the window. Your drawing will look like Figure 3.009.
	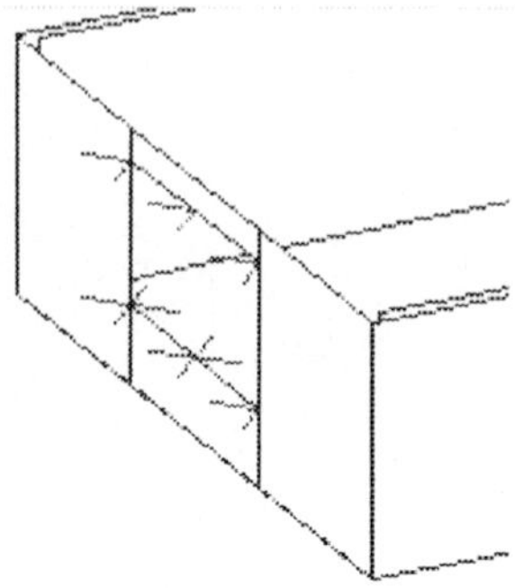 Figure 3.009 (Hidden lines removed for clarity)	
	Command: ***qsave***	11. Save the drawing but don't exit.

You should've noticed three things about the last exercise.

- Although quite useful, the skeleton (wireframe model) isn't required when drawing a surface model.
- The ***3DFace*** command has left lines (edges) above and below the window that don't normally appear on a model (or a wall).
- The current UCS didn't affect placement of the 3D faces.

Let's take a look at those edges.

3.3.2 Invisible Edges in 3D Faces – SPLFrame and the *Edge* Command

Although you can make 3D face edges invisible as you draw them, it's a tedious, time-consuming, and error-prone task. You'll find it much easier to draw the 3D face and then use the ***Edge*** command to hide the edges that you don't want to see.

The command sequence looks like this:

> **Command:** ***edge***
>
> **Specify edge of 3dface to toggle visibility or [Display]:**

Simply select the edge you want to make invisible. How much easier can they make it?

The only option (**Display**) prompts like this:

> **Enter selection method for display of hidden edges [Select/All] <All>:**

Respond by selecting a 3D face whose invisible edges you'd like to see or by hitting ***enter*** to accept the **All** option. The **All** option means that AutoCAD will show all the invisible edges in the display.

Use of the **Display** option allows you to *temporarily* display all invisible edges. Once the command ends, invisible edges are once again invisible.

To display all of the 3D face edges in a drawing, whether visible or not, set the **SPLFrame** system variable to **1**.

Let's make the necessary edges invisible on our new surfaces.

You can also access the ***Edge*** command by using the **Edge** button on the Surfaces toolbar or by selecting it from the Draw pull-down menu. Follow this path:

Draw – Surfaces – Edge

Do This: 3.3.2.1	**Invisible Edges**

I. Be sure you're still in the *Cabin.dwg* file in the C:\Steps3D\Lesson03 folder. If not, please open it now.

II. Follow these steps.

TOOLS	COMMAND SEQUENCE	STEPS
 Edge Button	**Command:** ***edge***	1. Enter the ***Edge*** command.
	Specify edge of 3dface to toggle visibility or [Display]:	2. Select the edges below the window and the edges above the window. Also select the upper and lower window edges.
	Specify edge of 3dface to toggle visibility or [Display]: ***[enter]***	3. Complete the command.
	**Command:** ***hide***	4. Freeze the **Marker** layer and remove the hidden lines. Your drawing looks like Figure 3.010.

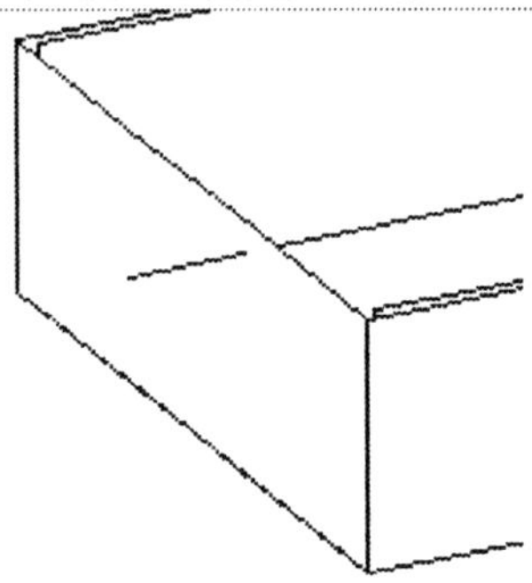

Figure 3.010

TOOLS	COMMAND SEQUENCE	STEPS
No Button Available	**Command:** ***splframe*** **Enter new value for SPLFRAME <0>:** ***1***	5. Now we'll draw the actual window. First we need to find the edges of our 3D face. Do this by setting the **SPLFrame** system variable to **1**. (Regen the drawing to see the results.)
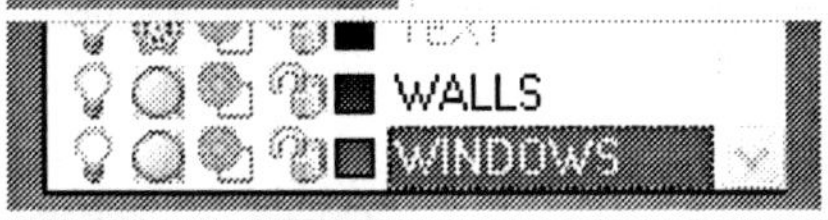		6. Set **Windows** as the current layer.
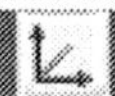	**Command:** ***ucs***	7. Change the UCS to match the front of the wall (Figure 3.011).

TOOLS	COMMAND SEQUENCE	STEPS

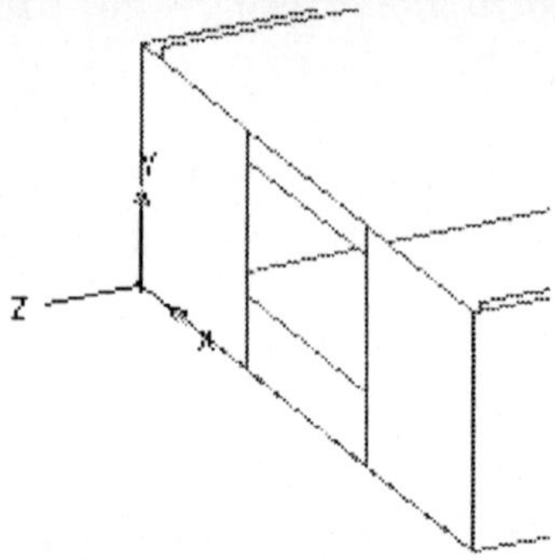

Figure 3.011

	Command: ***th*** **Enter new value for THICKNESS <0'-0">:** ***-5.5***	8. Set **Thickness** to ***-5.5***.
	**Command:** ***l***	9. Draw the window frame (Trace the opening.)
	Command: ***splframe*** **Enter new value for SPLFRAME <1>:** ***0***	10. Return the **SPLFrame** system variable to zero and remove the hidden lines. Your drawing looks like Figure 3.012.

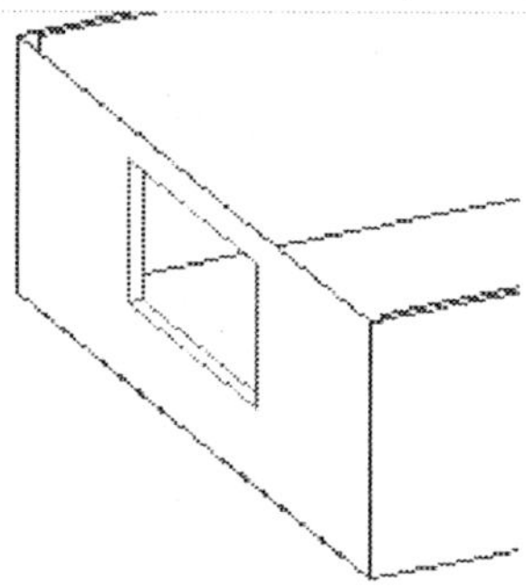

Figure 3.012

	**Command:** ***qsave***	11. Save the drawing but don't exit.

Of course, you've only drawn the outer surface of the wall. If you'd like, you can repeat the exercise to draw the inner surface as well. Then we can look at another method of Surface Modeling – Regions.

3.3.3 Solids and Regions

Some confusion inevitably arises between the terms *solid* and *3D solid*. Let me clarify the distinction.

A *solid* is a two-dimensional, filled polygon. Use solids when you need to highlight a particular object, building, or area on a drawing. Create a solid using the ***Solid*** command.

AutoCAD has no ***3DSolid*** command. The term 3D solid refers to a family of objects (including spheres, cones, boxes, and more). Although similar objects can be created using surface modeling techniques, those techniques won't create *solid* objects. We'll discuss techniques to create 3D solid objects in Lessons 7 and 8.

There are two real differences between solids and regions. The first is that a solid is a filled two-dimensional polygon (although it can have thickness) while a region is an actual surface. The second is that you can create a solid from scratch while creating a region requires an existing object.

A general rule of thumb to follow when considering solids and regions is this: use a region for a surface and a solid to fill in a two-dimensional object.

Since we looked at solids in the basic text and will spend several lessons on Solid Modeling, we'll concentrate here on regions.

What is a region? Without getting too technical, a region is a two-dimensional surface. It looks very much like a 3D face, but you don't have to hide the edges. You'd use a region anywhere an arc, circle, or hole is required. (Imagine trying to show a round hole with the ***3DFace*** command. Remember that you're restricted to straight edges!)

Let's compare 3D faces with regions.

3D FACE	REGION
Can be drawn in three dimensions.	Selected objects must be coplanar (share the same UCS). The ***Boundary*** command will only create regions in the current UCS.
Can't use to show curves or arcs.	Can use to show curves and arcs.
May need to make some edges invisible.	No need to worry about edges.
Can be extruded into a 3D solid.	Can be extruded into a 3D solid.
Create using the ***3DFace*** command.	Create using either the ***Region*** command or the ***Boundary*** command.

As you can see, there are two ways to create a region. Let's consider both.

- The ***Region*** command converts an existing closed object into a region. The objects you can convert include: closed lines, polylines, arcs, circles, splines, and ellipses. Once converted, the original object(s) exists as a region – it's no longer a line, polyline, etc. AutoCAD will create a region from a selected object regardless of the object's relation to the current UCS.

  The ***Region*** command looks like this:

  Command: ***region*** **(or** ***reg*****)**

  Select objects: ***[select the closed object you want to convert]***

  Select objects: ***[hit enter to complete the selection]***

  1 Region created. ***[AutoCAD tells you how many regions it has created]***

- The ***Boundary*** command uses boundaries to create a region. (We discussed boundaries as part of hatching in the basic text.) No objects are lost or converted with this approach and AutoCAD uses a dialog box to assist you. For the ***Boundary*** command to work properly, the objects forming the boundary must be on the zero coordinate of the Z-axis in the current UCS.

We'll use both of these methods to add some more windows in our cabin.

You can also access the ***Region*** command by using the **Region** button on the Draw toolbar. Alternately, you can access both the ***Region*** and ***Boundary*** commands by selecting them from the Draw pull-down menu.

Do This: 3.3.3.1	Creating Regions

I. Be sure you're still in the *Cabin.dwg* file in the C:\Steps3D\Lesson03 folder. If not, please open it now.

II. Set the viewpoint to 1,-1,1 (the SE Isometric view).

III. Follow these steps.

TOOLS	COMMAND SEQUENCE	STEPS
	Command: *z*	1. Zoom in around the other front wall as shown in Figure 3.013.

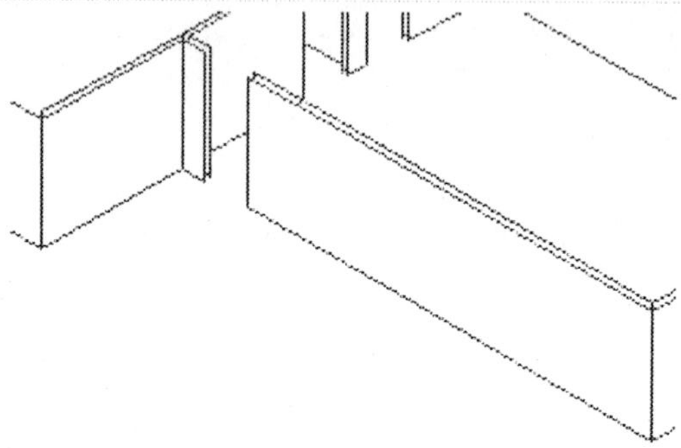

Figure 3.013

TOOLS	COMMAND SEQUENCE	STEPS
	Command: *props*	2. Change the **Thickness** to zero for the lines forming the inner and outer walls (on both sides of the door opening).
		3. Thaw the **MARKER** layer. A group of nodes appears on your screen.
	Command: *pl*	4. Use the nodes as a guide to draw four closed-polyline windows as shown in Figure 3.014. (The thickness should already be set to –5.5 and the UCS should already be the SE Isometric. If not, please make these corrections before doing this step.)

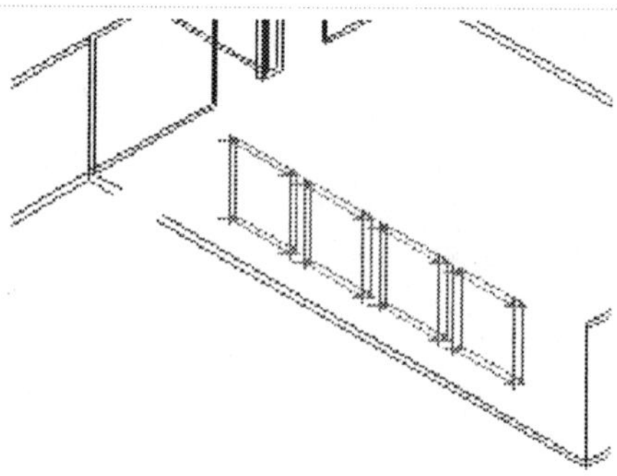

Figure 3.014

TOOLS	COMMAND SEQUENCE	STEPS
	Command: ***th*** **Enter new value for THICKNESS <-0'-5 1/2">:** ***0***	5. Reset **Thickness** to zero.
		6. Set the **Walls** layer current.
	Command: ***pl***	7. Use a polyline to draw the outline of the outer walls as shown in Figure 3.015 (the height of the door opening is 6'-8"). Be sure to close the polyline.

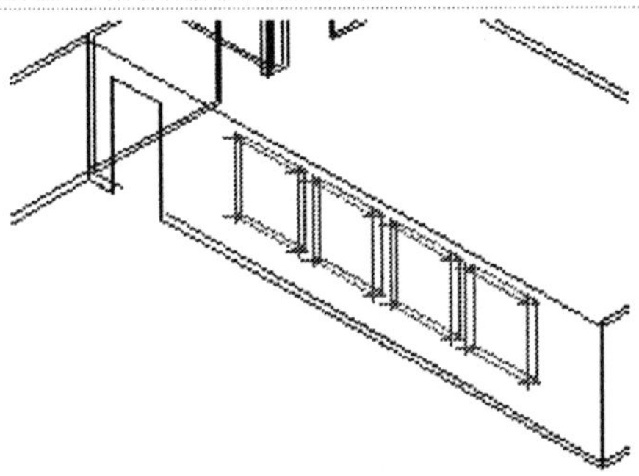

Figure 3.015

TOOLS	COMMAND SEQUENCE	STEPS
Region Button (Draw Toolbar)	**Command:** ***reg***	8. Now we'll create our first region. Enter the ***Region*** command. (If you haven't entered the command previously, it may take a moment to load.)
	Select objects: **Select objects:** ***[enter]*** **1 loop extracted.** **1 Region created.**	9. Select the polyline that defines the wall. AutoCAD tells you that it has created a region.
	Command: ***hide***	10. Freeze the **Marker** layer and then remove hidden lines. Your drawing looks like Figure 3.016. Notice that you can't see through the windows. This is because the wall is a region and there are, as yet, no openings for the windows. We'll deals with that now, first by creating window regions and then by removing the window regions from the wall region.

Tools	**Command Sequence**	**Steps**

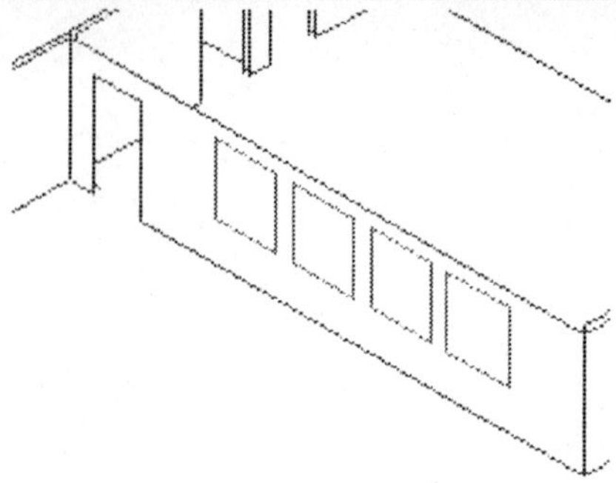

Figure 3.016

 UCS Origin Button	**Command:** ***ucs***	11. Move the current UCS to the lower left corner of the wall as shown in Figure 3.017.

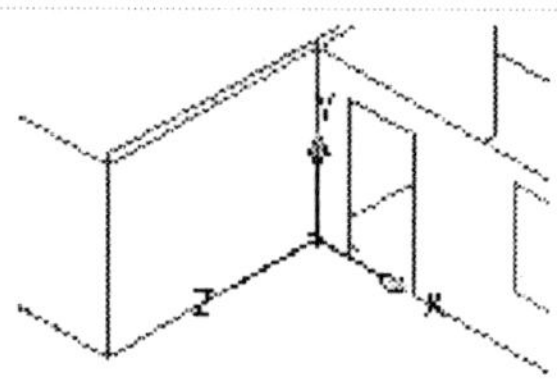

Figure 3.017

No Button Available	**Command:** ***bo***	12. Enter the ***Boundary*** command. AutoCAD presents the Boundary Creation dialog box. (You're familiar with the options from your study of hatching in the basic text.)

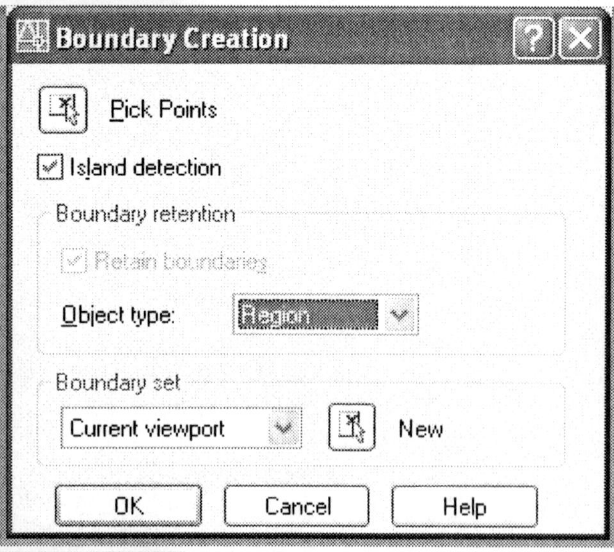

13. Set the control box in the **Object type** frame to **Region** as indicated. Then pick the **Pick Points** button. AutoCAD returns to the graphics screen.

	Select internal point:	14. Select points inside each of the four windows as shown in Figure 3.018.

Tools	Command Sequence	Steps
	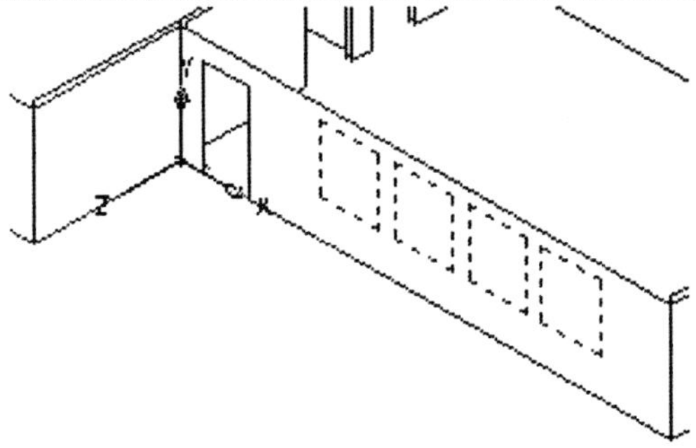 Figure 3.018	
	Select internal point: ***[enter]*** **4 loops extracted.** **4 Regions created.** **BOUNDARY created 4 regions**	15. Complete the command. You can see that the new regions were created without converting the polylines. (Regions are on the **Walls** layer while the polylines are still on the **Windows** layer).
	Command: ***qsave***	16. Save the drawing but don't exit.

You've used two methods to create your regions. The first – the ***Region*** command – converted the polyline outlining the wall to a region. That polyline doesn't exist anymore. The second – the ***Boundary*** command – created regions within the defined boundaries without changing the boundaries themselves (the polylines defining the windows).

But what you haven't done is to use the regions created with your window boundaries to cut holes in the wall for the windows. Right now, you simply have four window regions sitting on top of a wall region. Let's take a look at how we can cut those holes.

We'll use a tool with which we'll become considerably more familiar when we study Solid Modeling. In fact, the tool is one of several modifying tools shared by regions and solid models. The tool we'll use here is the ***Subtract*** command. It looks like this:

> **Command:** ***subtract*** **(or** ***su*****)**
>
> **Select solids and regions to subtract from ..**
>
> **Select objects:** ***[select the region or solid from which you'll subtract – in our exercise, this would be the wall]***
>
> **Select objects:** ***[hit enter to complete the selection]***
>
> **Select solids and regions to subtract ..**
>
> **Select objects:** ***[select the regions you wish to remove]***
>
> **Select objects:** ***[enter to complete the command]***

> You can also begin the ***Subtract*** command by using the **Subtract** button on the Solids Editing toolbar. Alternately, you can access the ***Subtract*** command (and explore other tools shared by regions and solid models) by selecting it from the Modify pull-down menu. Follow this path:
>
> *Modify – Solids Editing – Subtract*

Let's finish our wall.

Do This: 3.3.3.2	Using Regions to Create Holes

I. Be sure you're still in the *Cabin.dwg* file in the C:\Steps3D\Lesson03 folder. If not, please open it now.

II. Follow these steps.

TOOLS	COMMAND SEQUENCE	STEPS
 Subtract Button	**Command:** ***su***	1. Enter the ***Subtract*** command.
	SUBTRACT Select solids and regions to subtract from .. **Select objects:**	2. AutoCAD needs to know from which surface you'll subtract. Select the wall.
	Select solids and regions to subtract .. **Select objects:**	3. Now AutoCAD needs to know what to subtract. Select the windows. (Be sure to select the regions and not the polylines.)
	Command: ***hide***	4. Remove the hidden lines. Your drawing looks like Figure 3.019. (Starting to look pretty good, don't you think?)

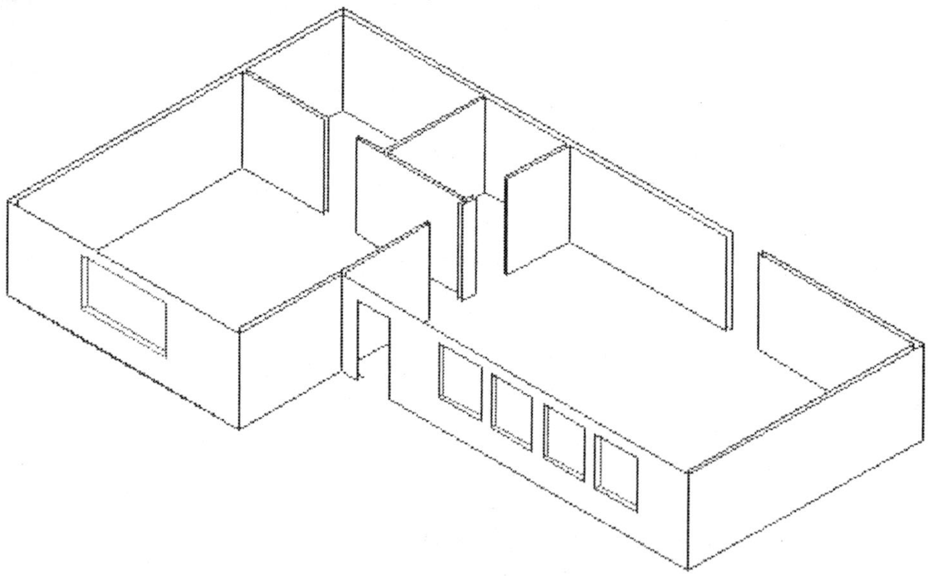

Figure 3.019

	**Command:** ***qsave***	5. Save the drawing.

3.3.4	Which Method Should You Use?

Which method of Surface Modeling do you prefer – 3D Face or Regions? Believe it or not, you'll need both.

Consider the model in Figure 3.020. Take a moment to consider each surface. Ask yourself which method of surface modeling you would use to create it. Then (more importantly) ask yourself why you'd use that method.

Once you've examined each surface, continue to the following explanations.

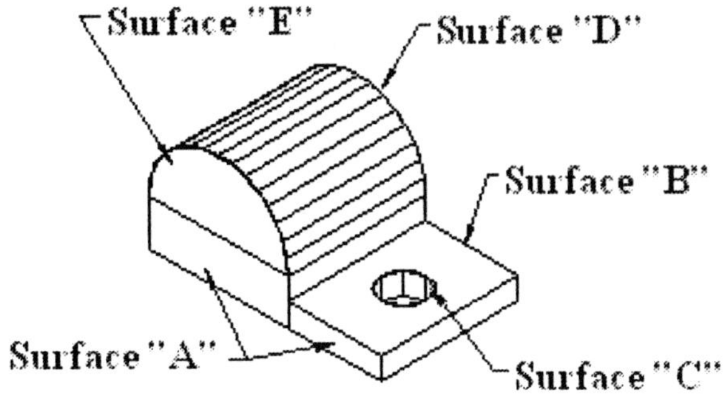

Figure 3.020

- You can easily draw **Surface "A"** using lines with thickness. This brings us to the first rule of three-dimensional work: *Never draw a surface when a line will do.* Surfaces are complex objects and take up more drawing memory than simple lines.
- **Surface "B"** has a round hole in it. 3D faces create flat edges so they won't work here. A region will serve best, but which method should you use? Using the Boundary method to create a region, you can pick a point inside the rectangular area (but outside the circle) and let AutoCAD do the rest. This is the best approach.
- **Surface "C"** is the inside surface of the hole. The only method we've seen to create this surface is **Thickness**. But you can't use a circle to create a hole (refer to Figure 1.034 in Lesson 1). That will result in a closed cylinder (a drum). Use two arcs with thickness to create the hole.
- **Surface "D"** looks very much like the roof of our cabin. Indeed, it's the same type of construction. You can use a region or solid to create the surfaces over the wireframe, but that would involve setting the UCS flat against the surface to be created for each roof section. There are 16 sections, so you'd have to set the UCS and then draw the surface 16 times. Alternately, you can draw 16 3D faces.

  But look at the surface again. There's an easier way that will provide a more rounded surface. How about drawing an arc with thickness in the proper UCS? This will work on this surface, as there are no intersections with which you have to contend. (Rule #2 of three-dimensional work then might be: *Think about it twice; draw it once.*)
- **Surface "E"** has an odd shape to it. Like Surface "B", the odd shape gives away the answer. Another rule of three-dimensional surface work is: *When faced with an unusual shape or holes in a surface, use the boundary approach to create a region.*

Use Figure 3.020 as a guide in your first steps toward creating three-dimensional surface models. (We'll draw Figure 3.020 in our exercises. Then you can plot it and hang it on your monitor!) Memorize the explanation of each surface and consider each point when determining how to draw a surface on your model.

3.4 Extra Steps

Did you use the **Subtract** button on the Solids Editing toolbar in Exercise 3.3.3.2? If you did, you might have noticed that it was grouped with two other buttons – **Union** and **Intersect**. These three buttons work on regions as well as 3D solids. Can you tell from their symbols what they'll do?

We'll discuss them in more detail in Lesson 8, but that doesn't mean that you can't experiment with them now.

Open the *solids & regions.dwg* file in the C:\Steps3D\Lesson03 folder. (It looks like Figure 3.021). Experiment with each of these commands on the objects shown. When you've finished, see Figure 3.022 for the results.

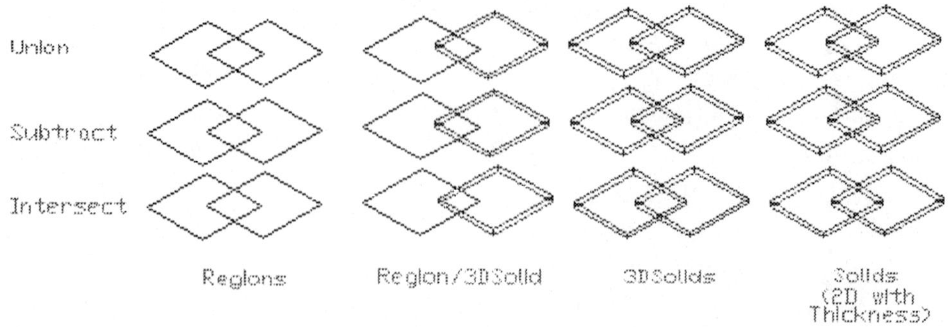

Figure 3.021

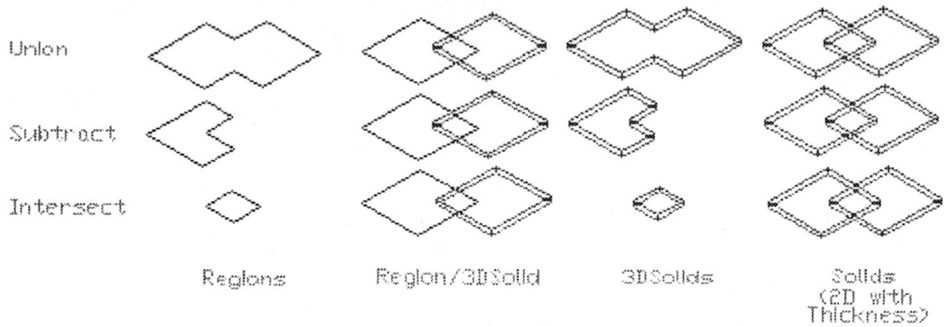

Figure 3.022

You'll notice that these commands work on regions or 3D solids. You can't, however, use them to subtract or join (and so forth) different types of objects with each other. Additionally, you'll notice that they don't work at all on simple solids. This is why you should generally use regions and 3D faces when creating surface models.

Let's do one more thing while we're still in the *solids & regions.dwg* file. Use the Gouraud Shademode to shade the drawing. Now use the ***3DOrbit*** command to flip the view (from bottom to top). Notice anything (Figure 3.023)?

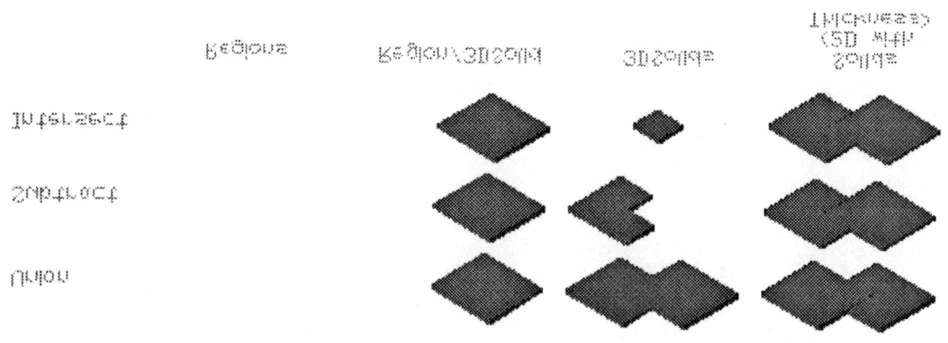

Figure 3.023

You should notice that all of the objects shaded nicely on top and bottom – *except* the regions. This brings up an interesting quirk about regions – you can see the shading on a region only from the top! Does that mean that you can't use a region for a bottom surface?

No! It means that you must draw the bottom region of an object *upside down!* How will you do that? Well, you won't actually draw upside down (face it, there are some things that just weren't meant to be done upside down). You'll use the ***UCS*** command to tell AutoCAD that down is up (tell AutoCAD to work upside down!). In other words, use the ***UCS*** command to flip the positive and negative directions of the Z-axis. (Remember this when drawing the cinder block in the Exercises.)

3.5 What Have We Learned?

Items covered in this lesson include:

- *The differences between 3D faces, solids, 3D solids, and regions*
- *The differences between polylines and 3D polylines*
- *Projecting points in three dimensions*
- *The two approaches to creating regions*
    - ***Region***
    - ***Boundary***
- *Commands*
    - ***3DPoly***
    - ***3DFace***
    - ***Region***
    - ***Edge***
    - ***SPLFrame***
    - ***Boundary***
    - ***Subtract***

It's taken a few lessons to get comfortable with the basics of Z-space wireframe and surface modeling. At this point, you're either anxious to continue or feeling somewhat overwhelmed by it all (probably a little of both).

I can't overemphasize the importance of practice – if you're uncomfortable with the material thus far, go back and do it again. You shouldn't feel that you're the only person who ever found Z-space difficult to master. But that's the benefit of computer labs and a good textbook! (If you've already changed the files that came from the web, just reload them to start over!)

Are you ready for an easy lesson? Our next chapter – "Predefined Surface Models" – will show you how to create some more complex objects easily. So do the problems that follow … review as necessary to get comfortable … then forward – ever forward!

3.6 Exercises

1. through 8. Add surfaces to the drawings you created in Exercises 1 through 8 of Section 1.7 (in Lesson 1). Refer to the drawings in Appendix B as a guide. If these drawing aren't available use the corresponding drawing in the C:\Steps3D\Lesson\Steps\Lesson03 folder.
9. Open the *My Twisted Y.dwg* file you created in Section 2.7 – Exercise 9 (C:\Steps3D\Lesson02 folder). (If that file isn't available, use the *Twisted Y-3.dwg* file in the C:\Steps3D\Lesson03 folder.) Convert the drawing into a surface model by placing thickness, 3D faces, or regions on the necessary surfaces. See Figure 3.024 for the completed drawing.
10. Open the *My Block.dwg* file you created in Section 2.7 – Exercise 10 (C:\Steps3D\Lesson02 folder). (If that file isn't available, use the *Block-3.dwg* file in the C:\Steps3D\Lesson03 folder.) Convert the drawing into a surface model by placing thickness, 3D faces, or regions on the necessary surfaces. See Figure 3.025 for the completed drawing.

Figure 3.024

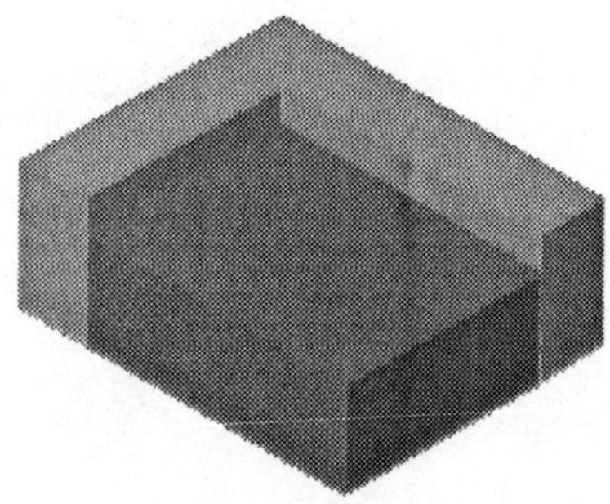

Figure 3025

11. Open the *MyBookEnd.dwg* file you created in Section 2.7 – Exercise 12 (C:\Steps3D\Lesson02 folder). (If that file isn't available, use the *Book End-3.dwg* file in the C:\Steps3D\Lesson03 folder.) Convert the drawing into a surface model by placing thickness, 3D faces, or regions on the necessary surfaces. See Figure 3.026 for the completed drawing.
12. Open the *My Anchor Stop.dwg* file you created in Section 2.7 – Exercise 14 (C:\Steps3D\Lesson02 folder). (If that file isn't available, use the *Anchor Stop-3.dwg* file in the C:\Steps3D\Lesson03 folder.) Convert the drawing into a surface model by placing thickness, 3D faces, or regions on the necessary surfaces. See Figure 3.027 for the completed drawing.

Figure 3.026

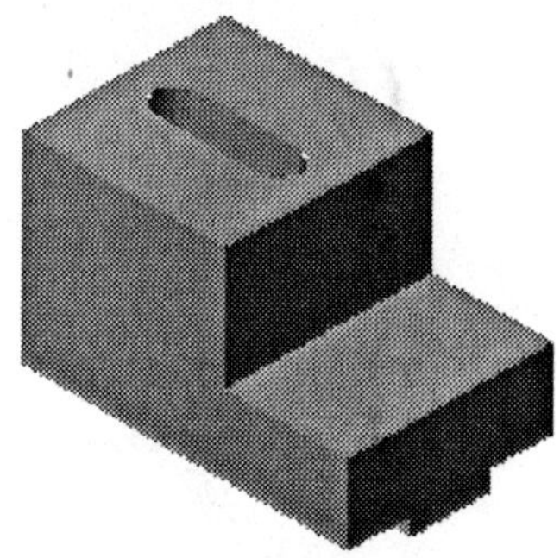

Figure 3.027

13. Open the *MyOtherTable.dwg* file you created in Section 2.7 – Exercise 20 (C:\Steps3D\Lesson02 folder). (If that file isn't available, use the *Other Table-3.dwg* file in the C:\Steps3D\Lesson03 folder.) Convert the drawing into a surface model by placing thickness, 3D faces, or regions on the necessary surfaces. See Figure 3.028 for the completed drawing.
14. Open the *My Corner Bracket.dwg* file you created in Section 2.7 – Exercise 21 (C:\Steps3D\Lesson02 folder). (If that file isn't available, use the *corner bracket-3.dwg* file in the C:\Steps3D\Lesson03 folder.) Convert the drawing into a surface model by placing thickness, 3D faces, or regions on the necessary surfaces. See Figure 3.029 for the completed drawing.

Figure 3.028

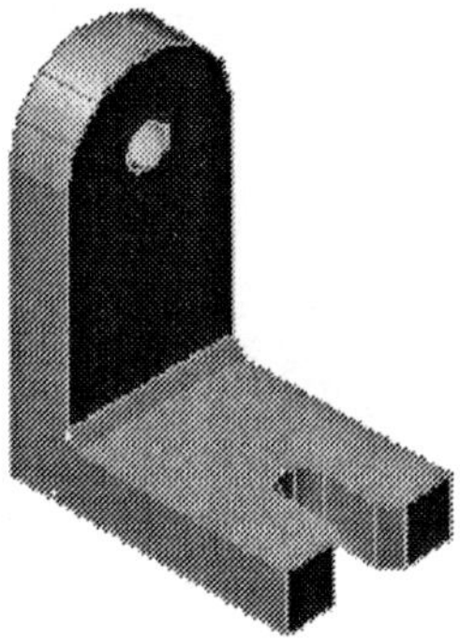

Figure 3.029

15. Open the *My Phone Plug.dwg* file you created in Section 2.7 – Exercise 22 (C:\Steps3D\Lesson02 folder). (If that file isn't available, use the *phone plug-3.dwg* file in the C:\Steps3D\Lesson03 folder.) Convert the drawing into a surface model by placing thickness, 3D faces, or regions on the necessary surfaces. See Figure 3.030 for the completed drawing.
16. Open the *My Other Corner Bracket.dwg* file you created in Section 2.7 – Exercise 23 (C:\Steps3D\Lesson02 folder). (If that file isn't available, use the *other corner bracket-3.dwg* file in the C:\Steps3D\Lesson03 folder.) Convert the drawing into a surface model by placing thickness, 3D faces, or regions on the necessary surfaces. See Figure 3.031 for the completed drawing.

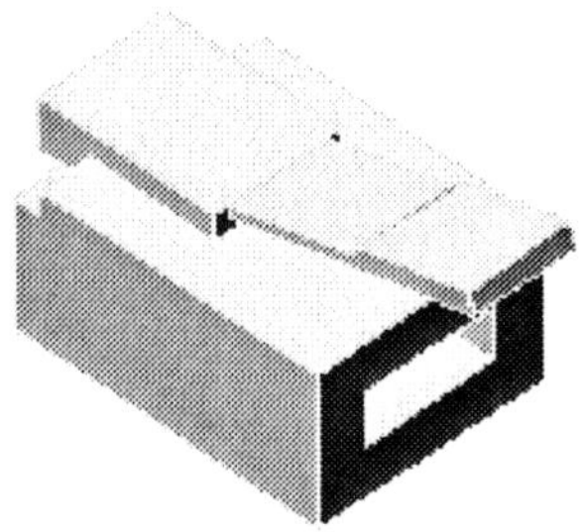

Figure 3030

Figure 3.031

17. Open the *projection.dwg* file in the C:\Steps3D\Lesson03 folder. Create the saddle tee shown in Figure 3.032. Save the drawing as *My Saddle Tee.dwg* in the C:\Steps3D\Lesson03 folder.

Figure 3.032

18. Create the cinder block shown in Figure 3.033. The following details will help:
    18.1. The overall dimensions of a cinder block are 8" x 8" x 16".
    18.2. The holes are 7" x 6½" and are evenly spaced.
    18.3. The fillets in the holes have a 1" radius.
    18.4. Color 164 looks like a cinder block.
    18.5. Remember that a region can only be seen from what was the positive Z-axis direction at the time of its creation.

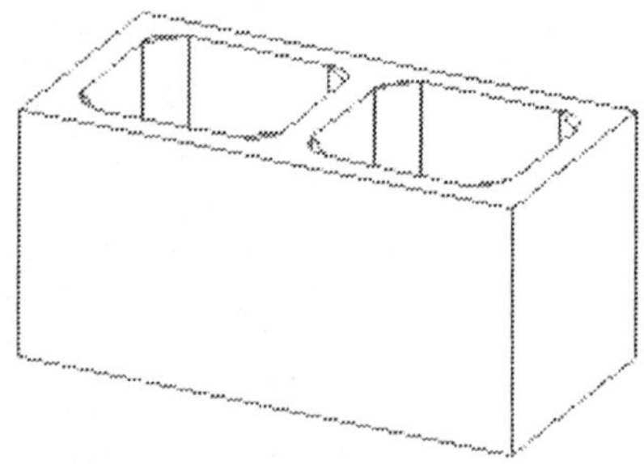

Figure 3.033

    18.6. Save the drawing as *My Cinder Block.dwg* in the C:\Steps3D\Lesson03 folder.

19. Draw the demo model in Figure 3.034 (refer to Figure 3.020). Use the following details as a guide.
    19.1. Label the types of surfaces as explained in Section 3.3.4.
    19.2. Save the drawing as *Surfaces Model.dwg* in the C:\Steps3D\Lesson03 folder.
    19.3. Plot the drawing to fit on about one quarter of a standard 8½" x 11" sheet of paper. Cut it out and tape it to the side of your monitor as a reference.
    19.4. You don't have to dimension the drawing.

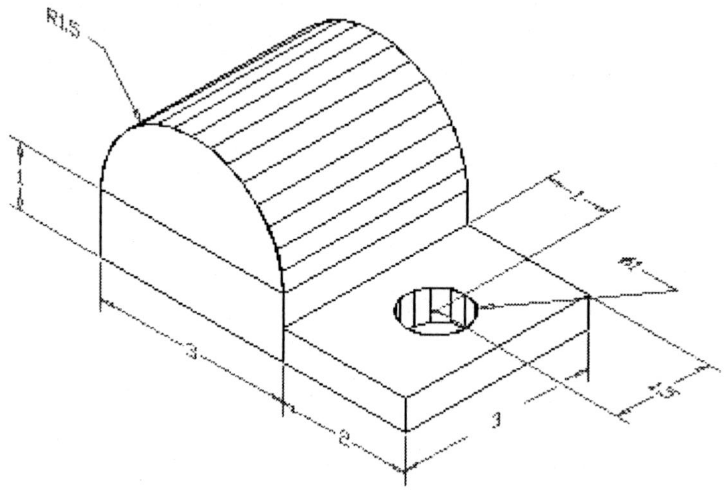

Figure 3.034

20. Finish the Cabin drawing we started in this lesson. The final drawing is shown in Figure 3.035. Use these details as a guide.
    20.1. All the windows are the same size.
    20.2. All the windows are 12" from the top of the walls.
    20.3. Spaced windows are 12" apart.
    20.4. Change the layer for all inner walls to a new **Inner Walls** layer. Then freeze the layer for clarity.
    20.5. Use the ***3DClip*** command for clarity on the elevations.
    20.6. The scale in each viewport is 1/16"=1'-0" (this layout is set up for an 8½" x 11" sheet).
    20.7. Use the title block of your choice, and plot the drawing.
    20.8. Save the drawing in the C:\Steps3D\Lesson03 folder.

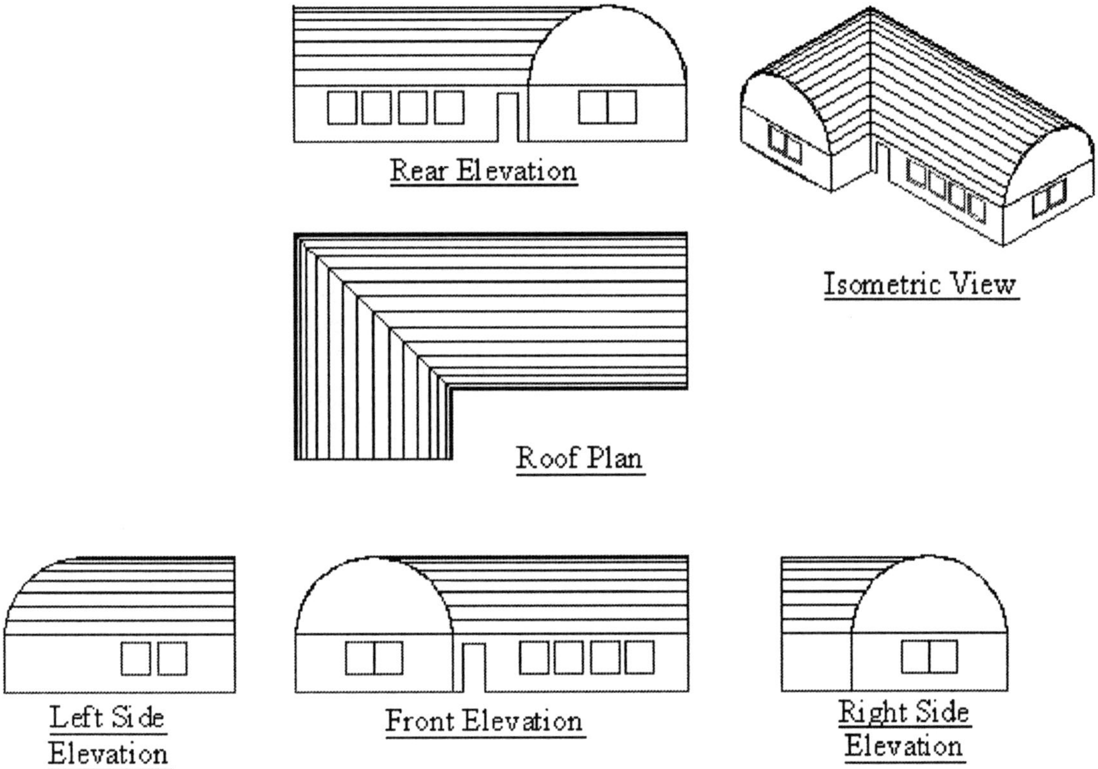

Figure 3.035

21. Reopen the *Cabin.dwg* file you completed in Exercise 20. We'll add a different roof and a chimney. Follow these guidelines. (The final drawing is shown in Figure 3.036).

21.1. Freeze the **Roof** layer.

21.2. Add two new layers – **Roof2** and **Chimney**.

21.3. On the **Roof2** layer, add hexagons at the end walls of the roofs (in the same location where the arcs were found in Exercise 3.2.1).

21.4. Add the 3D polyline and 3D faces to complete the roof.

21.5. Place a 4' x 2' rectangle, with a thickness of 24', at coordinates 33',22',0.

21.6. Use construction lines and UCS manipulation to locate where the chimney penetrates the roof. Draw a polyline around the penetration.

21.7. Finish drawing the chimney using wireframe techniques. Erase the rectangle.

21.8. Add 3D faces to the chimney.

21.9. Hatch everything as shown.

21.10. Save and plot the drawing.

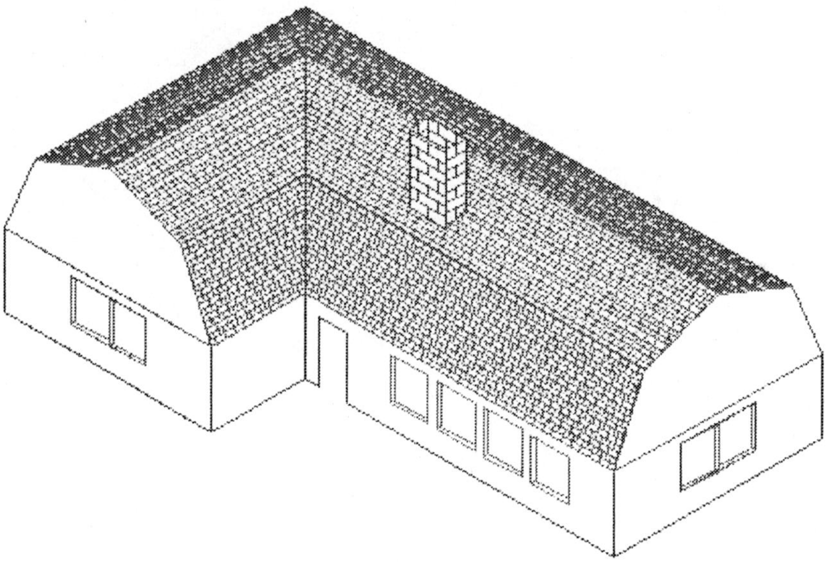

Figure 3.036

3.7	**For Web-Based Review Questions, visit: http://www.uneedcad.com/2006/Files/06R3-3D.pdf**

Lesson 4

Following this lesson, you will:

- ✓ *Know how to build and use AutoCAD's predefined surface models*
  - ***Box***
  - ***Wedge***
  - ***Pyramid***
  - ***Cone***
  - ***Sphere***
  - ***Dome***
  - ***Dish***
  - ***Torus***
- ✓ *Know how to use the **3D** command*

Predefined Surfaces

In our next two lessons, we'll look at some tools that should greatly simplify your work with surface models.

First, in this lesson, we'll learn to use AutoCAD's predefined surface models. Draftsmen at all levels of development can quickly and easily learn to use these remarkable timesaving devices. In fact, I'd be surprised if you hadn't discovered them in your three-dimensional explorations already.

Then in Lesson 5, you'll discover some surprisingly simple tools that you can use to create elaborate, non-uniform surface models.

Let's begin with an overview of the predefined surface modeling tools AutoCAD has provided.

4.1 What Are Predefined Surface Models?

Simply put, predefined surface models are standard geometric shapes that AutoCAD creates for you with a minimal amount of user input. The shapes include (Figure 4.001) a box, a wedge, four types of pyramid, two types of cone, a dome, a dish, a sphere, and a torus. User input for each commonly includes length, width, height, radius, and rotation angle definitions.

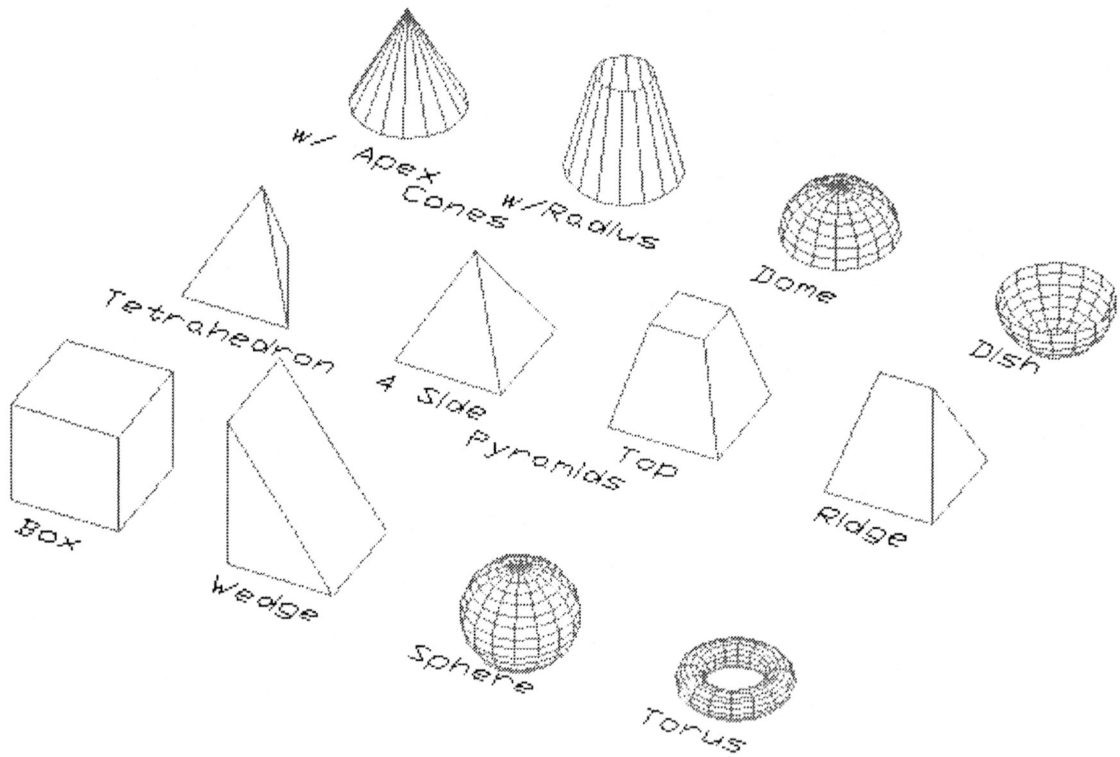

Figure 4.001

You can create each of these models with little effort – simply follow AutoCAD's prompts.

AutoCAD creates each surface model as a 3D mesh (more on 3D meshes in Lesson 5). You can edit a 3D mesh with the ***PEdit*** command (as we'll see in Lesson 6). Additionally, you can explode the model into a series of 3D faces. This enables you to remove part of the model (with the ***Erase*** command) or hide some of the edges (with the ***Edge*** command) without affecting the rest of the faces.

4.2 Drawing Predefined Surface Models

AutoCAD has included a command that enables you to access predefined surface models without difficulty. The command is, simply enough, ***3D***. It looks like this:

Command: ***3d***

Enter an option [Box/Cone/DIsh/DOme/Mesh/Pyramid/ Sphere/Torus/ Wedge]:

Respond with the type of object you wish to draw. But if the keyboard is too much trouble, AutoCAD provides a toolbar – Surfaces – to make accessing its predefined surface modeling tools even easier.

Let's examine the procedures for drawing each of the predefined surface models.

> You can also select a predefined surface model from a dialog box using the **3D Surfaces ...** selection on the Draw pull-down menu. Follow this path:
>
> *Draw – Surfaces – 3D Surfaces ...*
>
> You'll notice an additional tool both on the slide options and in the command – **Mesh**. We'll discuss this tool in Lesson 5.

4.2.1 Box

Use the **Box** option of the ***3D*** command to draw any six-sided box whose sides, top, and bottom are parallel or perpendicular to the current UCS. The command sequence looks like this:

Command: ***3d***

Enter an option [Box/Cone/DIsh/DOme/Mesh/Pyramid/Sphere/Torus/Wedge]: ***b***

Specify corner point of box: ***[identify the first corner of the box]***

Specify length of box: ***[tell AutoCAD how long to make the box]***

Specify width of box or [Cube]: ***[tell AutoCAD how wide to make the box or type C for a cube (all sides equal to the value entered for length)]***

Specify height of box: ***[tell AutoCAD how tall to make the box – AutoCAD skips this prompt if you opt for Cube in the last step]***

Specify rotation angle of box about the Z axis or [Reference]: ***[tell AutoCAD how to orient the box using either mouse pick or reference angle]***

Let's try one.

> You'll notice that, when using the buttons on the Surfaces toolbar, AutoCAD doesn't go through the ***3D*** command. Instead, it uses an older approach – **AI_[command]** (as in ***AI_Box***). You can use this approach at the command prompt as well – if you don't mind the typing! Both approaches produce the same result.

Do This: 4.2.1.1	Creating a 3D Surfaced Box

I. Start a new drawing using the *lesson 04 template* file located in the C:\Steps3D\Lesson04 folder.

II. Follow these steps.

TOOLS	COMMAND SEQUENCE	STEPS
Box Button	**Command:** ***3d*** **Enter an option [Box/Cone/DIsh/DOme/Mesh/ Pyramid/Sphere/Torus/Wedge]:** ***b***	1. Enter the ***3D*** command and select the **Box** option. Alternately, you can pick the **Box** button on the Surfaces toolbar.
	Specify corner point of box: ***1,1***	2. Specify the start point of the box as indicated.
	Specify length of box: ***4***	3. Specify the **length** of the box.
	Specify width of box or [Cube]: ***2*** **Specify height of box:** ***1***	4. We won't draw a cube this time, so enter the width and height as indicated.
	Specify rotation angle of box about the Z axis or [Reference]: ***0***	5. And give the box a zero rotation angle. Your box looks like Figure 4.002.

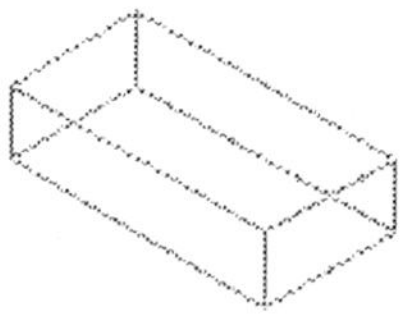

Figure 4.002

	Command: *[enter]* **Enter an option [Box/Cone/DIsh/DOme/Mesh/ Pyramid/Sphere/Torus/Wedge]:** ***b***	6. Let's place a cube atop the box. Repeat the command.
	Specify corner point of box: ***1,1,1***	7. Place the first corner atop the first corner of the box.
	Specify length of box: ***2***	8. Give our cube a **length** of 2 …
Cube	**Specify width of box or [Cube]:** ***C***	9. … and tell AutoCAD to use the length dimension for the width and height as well by selecting the **Cube** option.
	Specify rotation angle of box about the Z axis or [Reference]: ***0***	10. Give the cube the same rotation angle as our first box. Your drawing looks like Figure 4.003.

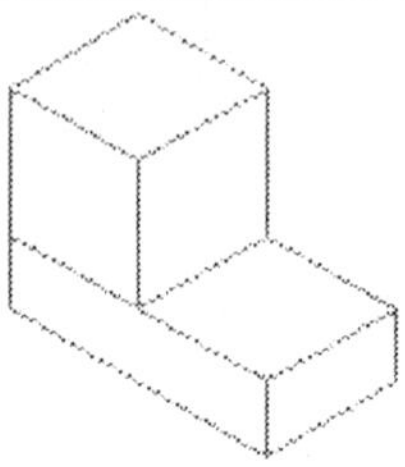

Figure 4.003 (Hidden lines removed for clarity)

TOOLS	COMMAND SEQUENCE	STEPS
	Command: ***save***	11. Save the drawing as *MyBoxes* in the C:\Steps3D\Lesson04 folder.
	Command: ***li***	12. Perform the ***List*** command on the upper box. Notice that each vertex is identified as a polyline mesh (Figure 4.004).

```
                 POLYLINE   Layer: "obj1"
                            Space: Model space
                  Handle = b6
Polyface mesh

                 VERTEX     Layer: "obj1"
                            Space: Model space
                  Handle = b7
Polyface vertex
               at point, X=    1.0000  Y=    1.0000  Z=    0.0000

                 VERTEX     Layer: "obj1"
                            Space: Model space
                  Handle = b8
```

Figure 4.004

TOOLS	COMMAND SEQUENCE	STEPS
	Command: ***x***	13. Explode the upper box.
	Command: ***li***	14. Repeat Step 12. Notice the difference (Figure 4.005). Now you can use the ***Edge*** command to hide edges.

```
              3D FACE   Layer: "obj1"
                        Space: Model space
               Handle = df
          first point, X=    1.0000  Y=    1.0000  Z=    1.0000
         second point, X=    3.0000  Y=    1.0000  Z=    1.0000
          third point, X=    3.0000  Y=    1.0000  Z=    3.0000
         fourth point, X=    1.0000  Y=    1.0000  Z=    3.0000
```

Figure 4.005

TOOLS	COMMAND SEQUENCE	STEPS
	Command: ***qsave***	15. Save and close the drawing.

Wasn't that easier than drawing a stick figure and stretching skin around it? (I love easy!)

Let's try another predefined surface model.

4.2.2 Wedge

You won't find many differences between the ***3D*** command's **Wedge** and **Box** options. In fact, without the **Cube** option, the **Wedge** option is actually easier! It looks like this:

> **Command:** ***3d***
>
> **Enter an option [Box/Cone/DIsh/DOme/Mesh/Pyramid/Sphere/Torus/Wedge]:** ***w***
>
> **Specify corner point of wedge:** ***[identify the first corner of the wedge]***
>
> **Specify length of wedge:** ***[tell AutoCAD how long to make the wedge]***
>
> **Specify width of wedge:** ***[tell AutoCAD how wide to make the wedge]***

Specify height of wedge: ***[tell AutoCAD how tall to make the wedge]***

Specify rotation angle of wedge about the Z axis: ***[tell AutoCAD how to orient the wedge]***

Does it look familiar? Let's draw one for practice.

Do This: 4.2.2.1	Creating a 3D Surfaced Wedge

I. Start a new drawing using the *lesson 04 template* file located in the C:\Steps3D\Lesson04 folder.

II. Follow these steps.

TOOLS	COMMAND SEQUENCE	STEPS
 Wedge Button	**Command:** ***3d*** **Enter an option [Box/Cone/DIsh/DOme/Mesh/ Pyramid/Sphere/Torus/Wedge]:** ***w***	1. Enter the ***3D*** command and select the **Wedge** option. Alternately, you can pick the **Wedge** button on the Surfaces toolbar.
	Specify corner point of wedge: ***1,1***	2. Identify the starting point of the wedge as indicated.
	Specify length of wedge: ***4*** **Specify width of wedge:** ***2*** **Specify height of wedge:** ***1*** **Specify rotation angle of wedge about the Z axis:** ***0***	3. Specify the **length**, **width**, **height**, and **rotation angle** of the wedge as indicated. Your drawing looks like Figure 4.006.

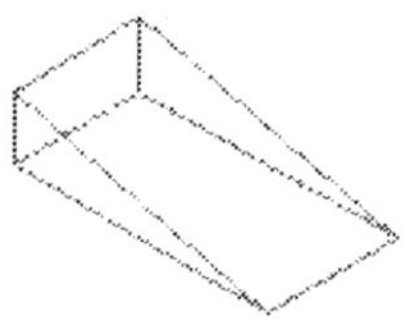

Figure 4.006

TOOLS	COMMAND SEQUENCE	STEPS
	**Command:** ***save***	4. Save the drawing as *MyWedge* in the C:\Steps3D\Lesson folder.

Remember that you can change the UCS prior to creating the wedge to help control the direction of the slope. You can't, however, use negative numbers.

Let's look at something more complex than simple boxes and wedges. Let's look at the **Pyramid** option.

4.2.3 Pyramid

Did you know that there's more than one kind of pyramid?

Technically speaking, a pyramid is a polyhedron – a multi-triangular structure (and you thought pyramid was hard to spell!). The common idea of a pyramid comes from the Egyptian model with a rectangular base (or four triangular sides on a rectangular base). Most of the Egyptian pyramids come to a point at the top.

But there are other pyramids. In fact, the largest pyramid in the world isn't Egyptian at all! Look for it just outside Mexico City. And it has a flat top!

Some pyramids even have triangular bases – or three triangular sides on a triangular base. These are called tetrahedrons. (Sounds like something that ate the bad guys in *Jurassic Park XXIV*.)

You probably didn't realize how complicated the world of pyramids was! But not to worry – AutoCAD provides for drawing each within one simple 3D command option – **Pyramid**. Its command sequence looks like this:

> **Command:** ***3d***
>
> **Enter an option [Box/Cone/DIsh/DOme/Mesh/Pyramid/Sphere/Torus/Wedge]:** ***p***
>
> **Specify first corner point for base of pyramid:**
>
> **Specify second corner point for base of pyramid:**
>
> **Specify third corner point for base of pyramid:**
>
> **Specify fourth corner point for base of pyramid or [Tetrahedron]:** ***[if the base has four corners, enter the fourth point; otherwise, enter T for a three-sided base (tetrahedron)]***
>
> **Specify apex point of pyramid or [Ridge/Top]:** ***[enter a point in Z-space to identify the point at the top of the pyramid, or tell AutoCAD you wish to create a ridge (2-point) or top (3 or 4 points, depending on the base)]***

It really isn't as complicated as it looks. Let's draw some pyramids and see.

Do This: 4.2.3.1	Creating 3D Surfaced Pyramids

I. Start a new drawing using the *lesson 04 template* file located in the C:\Steps3D\Lesson04 folder.

II. Follow these steps.

TOOLS	COMMAND SEQUENCE	STEPS
 Pyramid Button	**Command:** ***3d*** **Enter an option [Box/Cone/DIsh/DOme/Mesh/Pyramid/Sphere/Torus/Wedge]:** ***p***	1. Enter the ***3D*** command and select the **Pyramid** option. Alternately, you can pick the **Pyramid** button on the Surfaces toolbar.
	Specify first corner point for base of pyramid: ***1,1*** **Specify second corner point for base of pyramid:** ***5,1*** **Specify third corner point for base of pyramid:** ***5,5*** **Specify fourth corner point for base of pyramid or [Tetra]:** ***1,5***	2. We'll start with a simple, four-sided pyramid that comes to a point at the top. Enter the base coordinates indicated.
	Specify apex point of pyramid or [Ridge/Top]: ***3,3,4***	3. An apex point (sharp point at the top) is the default, so enter the three-dimensional coordinate indicated. Your first pyramid looks like Figure 4.007 (hidden lines removed for clarity).

TOOLS	**COMMAND SEQUENCE**	**STEPS**

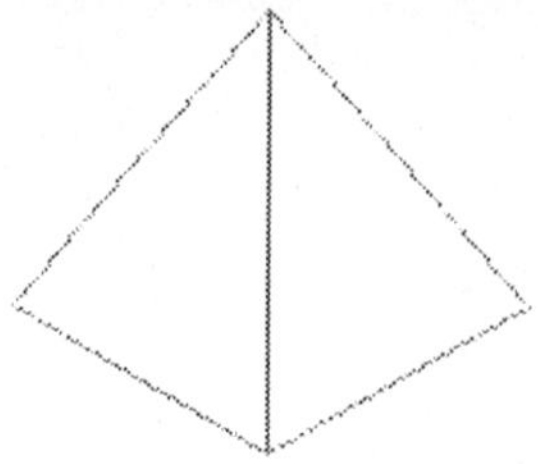

Figure 4.007

TOOLS	**COMMAND SEQUENCE**	**STEPS**
	Command: ***[enter]*** **Enter an option [Box/Cone/DIsh/DOme/Mesh/ Pyramid/Sphere/Torus/Wedge]:** ***p***	4. Let's draw a three-sided pyramid – a tetrahedron. Repeat the command.
	Specify first corner point for base of pyramid: ***11,1*** **Specify second corner point for base of pyramid:** ***11,5*** **Specify third corner point for base of pyramid:** ***7,3***	5. We'll draw this one next to the first. Enter the coordinates indicated for the three corners of the base.
Tetrahedron	**Specify fourth corner point for base of pyramid or [Tetrahedron]:** ***T***	6. When AutoCAD asks for a fourth corner, select the **Tetrahedron** option.
	Specify apex point of tetrahedron or [Top]: ***9,3,4***	7. And place the apex point as indicated. Your drawing looks like Figure 4.008.

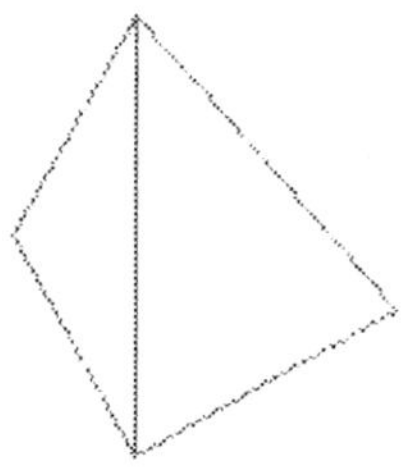

Figure 4.008

TOOLS	**COMMAND SEQUENCE**	**STEPS**
	Command: ***[enter]*** **Enter an option [Box/Cone/DIsh/DOme/Mesh/ Pyramid/Sphere/Torus/Wedge]:** ***p***	8. Now let's draw a four-sided pyramid with a flat top. (Erase the previous pyramids or freeze their layer and set a different one current.) Repeat the command.
	Specify first corner point for base of pyramid: ***1,1*** **Specify second corner point for base of pyramid:** ***5,1*** **Specify third corner point for base**	9. First, draw the base just as you did for the first pyramid.

TOOLS	COMMAND SEQUENCE	STEPS
	of pyramid: ***5,5*** **Specify fourth corner point for base of pyramid or [Tetrahedron]:** ***1,5***	
	**Specify apex point of pyramid or [Ridge/Top]:** ***T***	10. But instead of identifying an apex point, tell AutoCAD to draw a **Top** on the pyramid.
	Specify first corner point for top of pyramid: ***2,2,4*** **Specify second corner point for top of pyramid:** ***2,4,4*** **Specify third corner point for top of pyramid:** ***4,4,4*** **Specify fourth corner point for top of pyramid:** ***4,2,4***	11. Identifying the points for the top is just like identifying the points for the base – except that the top points must be three-dimensional coordinates. Enter these as indicated. Your drawing looks like Figure 4.009 (again, I removed hidden lines for clarity).

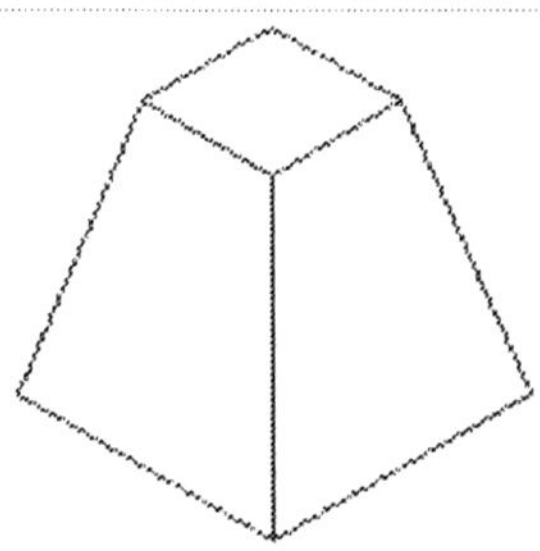

Figure 4.009

I'm not sure that our next pyramid would qualify as a true pyramid – it only has two triangles, but it also has two quadrilaterals and a rectangular base. Still, it fits better with the pyramid command than any other, and an ***AI_Triangles_Quadrilaterals_Rectangle*** command might be too much typing even for a pro! (Of course, the last pyramid didn't have any triangles at all!) Let's draw a ridge.

TOOLS	COMMAND SEQUENCE	STEPS
	**Command:** ***[enter]*** **Enter an option [Box/Cone/DIsh/DOme/Mesh/ Pyramid/Sphere/Torus/Wedge]:** ***p***	12. Repeat the command.
	Specify first corner point for base of pyramid: ***7,1*** **Specify second corner point for base of pyramid:** ***11,1*** **Specify third corner point for base of pyramid:** ***11,5*** **Specify fourth corner point for base of pyramid or [Tetrahedron]:** ***7,5***	13. Draw the base as indicated.

TOOLS	COMMAND SEQUENCE	STEPS
	**Specify apex point of pyramid or [Ridge/Top]:** ***r***	14. But tell AutoCAD to draw a **Ridge** rather than an apex point.
	Specify first ridge end point of pyramid: ***8,3,4*** **Specify second ridge end point of pyramid:** ***10,3,4***	15. Then identify the points on the ridge. Notice that the ridge is parallel to the first base line you drew (Figure 4.010).

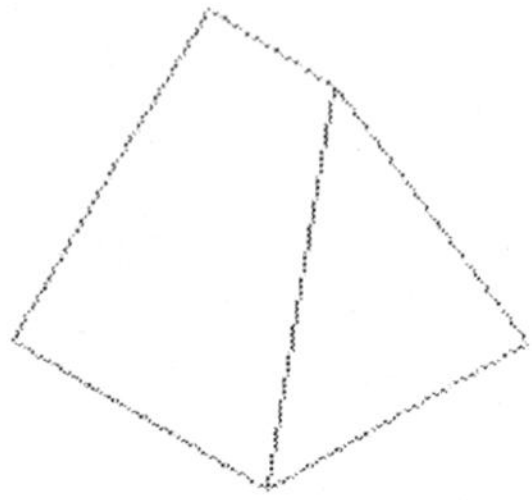

Figure 4.010

TOOLS	COMMAND SEQUENCE	STEPS
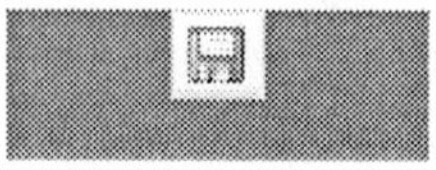	**Command:** ***save***	16. Save the drawing as *MyPyramids* in the C:\Steps3D\Lesson04 folder.

Now let's look at pyramids with round bottoms – let's look at cones.

4.2.4 Cone

Like pyramids, cones are smaller on the top than on the bottom (generally speaking). The top can be pointed like the pyramid's apex point, or it can be flat. But the similarities end there.

> The smaller-top/larger-base definition is simply to help in recognition. It's possible, and at times desirable, to have a pyramid or cone with a larger top than base, or a cone with equal top and base.
>
> You can also use the ***3D*** command's **Cone** option to draw a surface model cylinder. Simply make the top and base radii the same.

The predefined surface models we've examined so far have all had sides that loaned themselves easily to 3D faces. Beginning with cones, we'll look at several predefined shapes that incorporate circles or arcs in their structures. Because these structures are 3D meshes (and will convert to 3D faces when exploded), we must tell AutoCAD how many faces to use when creating the surfaces of the circles or arcs.

The command sequence for cones is:

Command: ***3d***

Enter an option [Box/Cone/DIsh/DOme/Mesh/Pyramid/Sphere/Torus/Wedge]: ***c***

Specify center point for base of cone: ***[identify the center point for the base of the cone]***

Specify radius for base of cone or [Diameter]: ***[identify the radius of the cone or select the diameter option]***

Specify radius for top of cone or [Diameter] <0>: ***[hit enter to accept zero as the top radius – AutoCAD will draw a cone with a point at the top; or identify the radius for the top and AutoCAD will draw an open top]***

Specify height of cone: ***[specify the height of the cone]***

Enter number of segments for surface of cone <16>: ***[tell AutoCAD how many faces to use to create the surface of the cone. Remember that 3D faces are flat, so the more faces you use, the rounder the cone will appear. Remember also that the more faces you use, the larger your drawing becomes. Go for a healthy compromise between appearance and size.]***

We'll draw a couple cones for practice.

Do This: 4.2.4.1	Creating 3D Surfaced Cones

I. Start a new drawing using the *lesson 04 template* file located in the C:\Steps3D\Lesson04 folder.

II. Follow these steps.

TOOLS	COMMAND SEQUENCE	STEPS
Cone Button	**Command:** ***3d*** **Enter an option [Box/Cone/DIsh/DOme/Mesh/ Pyramid/Sphere/Torus/Wedge]:** ***c***	1. Enter the ***3D*** command and select the **Cone** option. Alternately, you can pick the **Cone** button on the Surfaces toolbar.
	Specify center point for base of cone: ***3,3*** **Specify radius for base of cone or [Diameter]:** ***2***	2. Identify the **center point** and the **radius** of the base as indicated.
	Specify radius for top of cone or [Diameter] <0>: ***[enter]***	3. We'll draw a pointed cone first. Accept the default **radius** of zero for the top.
	Specify height of cone: ***4***	4. Specify the **height** of the cone. The height must be a positive number.
	Enter number of segments for surface of cone <16>: ***[enter]***	5. Remember – more surface segments mean a rounder appearance; fewer mean a less pronounced curve. Let's accept the default. Your drawing looks like Figure 4.011.

Figure 4.011

	Command: ***[enter]***	6. Now let's draw a cone with an opening at the top as well as the bottom. Repeat the command.

TOOLS	COMMAND SEQUENCE	STEPS
	Specify center point for base of cone: ***9,3*** **Specify radius for base of cone or [Diameter]:** ***2***	7. Locate the base and specify the radius as you did in Step 2.
	Specify radius for top of cone or [Diameter] <0>: ***1***	8. But this time, identify a radius for the top as well.
	Specify height of cone: ***4***	9. Specify the height as you did in Step 4.
	Enter number of segments for surface of cone <16>: ***32***	10. Let's see what the cone will look like with more surface segments (Figure 4.012).

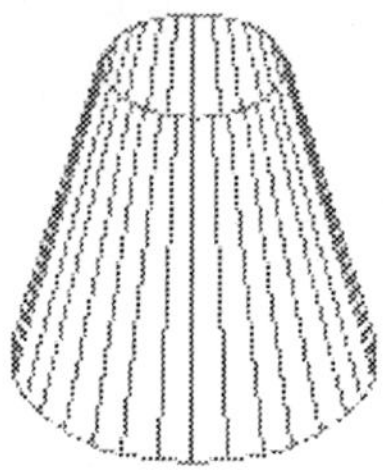

Figure 4.012

	Command: ***save***	11. Save the drawing as *MyCones* in the C:\Steps3D\Lesson04 folder.

4.2.5 Sphere

Spheres also require you to identify the number of faces required to make up the arc. But with spheres, you must identify the number of latitudinal *and longitudinal* faces. In other words, into how many pieces will you divide the sphere from top to bottom (latitude) and side to side (longitude)? Again, it isn't as difficult as it sounds. The prompts look like this:

Command: ***3d***

Enter an option [Box/Cone/DIsh/DOme/Mesh/Pyramid/Sphere/Torus/Wedge]: ***s***

Specify center point of sphere: ***[specify the center point of the sphere; unless you want half the sphere to be underground, use a three-dimensional coordinate to do this]***

Specify radius of sphere or [Diameter]: ***[specify the radius or diameter of the sphere]***

Enter number of longitudinal segments for surface of sphere <16>: ***[how many divisions will you want from side to side?]***

Enter number of latitudinal segments for surface of sphere <16>: ***[how many divisions will you want from top to bottom?]***

Let's draw a sphere.

Do This: 4.2.5.1	Creating a 3D Surfaced Sphere

I. Start a new drawing using the *lesson 04 template* file located in the C:\Steps3D\Lesson04 folder.

II. Follow these steps.

TOOLS	COMMAND SEQUENCE	STEPS
Sphere Button	**Command:** ***3d*** **Enter an option [Box/Cone/DIsh/DOme/Mesh/ Pyramid/Sphere/Torus/Wedge]:** ***s***	1. Enter the ***3D*** command and select the **Sphere** option. Alternately, you can pick the **Sphere** button on the Surfaces toolbar.
	Specify center point of sphere: ***3,3,3***	2. Specify the **center point** of the sphere. Be sure to use a three-dimensional coordinate as indicated.
	Specify radius of sphere or [Diameter]: ***2***	3. Identify the **radius** as indicated.
	Enter number of longitudinal segments for surface of sphere <16>: ***[enter]*** **Enter number of latitudinal segments for surface of sphere <16>:** ***32***	4. Accept the default number of **longitudinal segments** but increase the number of **latitudinal segments** as indicated (this way, you can see the difference). Your drawing looks like Figure 4.013.
	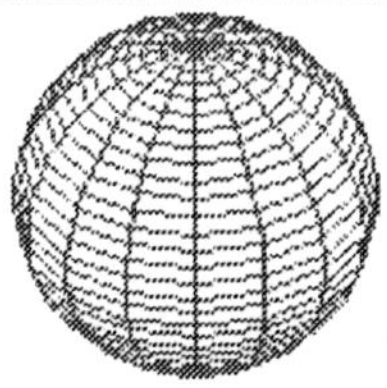 Figure 4.013	
	Command: ***save***	5. Save the drawing as *MySphere* in the C:\Steps3D\Lesson04 folder.

What do you mean you wanted to use the *right half* or *left half* of the sphere?! Some people just have to be difficult!

Well, you can always draw a dome or dish and rotate it using the ***Rotate3d*** command you'll learn in Lesson 6. (Whew!)

4.2.6 Domes and Dishes

So you only need half a sphere, huh? Well, you could draw a sphere, explode it, and then erase what you don't need. But that's too much work; and let's face it, that's not what drafting is all about.

A dome is the upper half of a sphere; a dish is the lower half. The command sequences for both the **Dome** and **Dish** options are identical to the sphere's sequence but result in only half a sphere being drawn.

Command: ***3d***

Enter an option [Box/Cone/DIsh/DOme/Mesh/Pyramid/Sphere/Torus/Wedge]: ***do*** **[or** ***di*****]**

Specify center point of dome: ***[specify the center point of the dome/dish]***

Specify radius of dome or [Diameter]: ***[specify the radius or diameter of the dome/dish]***

Enter number of longitudinal segments for surface of dome <16>: ***[how many divisions will you want from side to side?]***

Enter number of latitudinal segments for surface of dome <8>: ***[how many divisions will you want from top to bottom?]***

Let's draw one of each.

Do This: 4.2.6.1	Creating 3D Surfaced Domes and Dishes

I. Start a new drawing using the *lesson 04 template* file located in the C:\Steps3D\Lesson04 folder.

II. Follow these steps.

TOOLS	COMMAND SEQUENCE	STEPS
 Dome Button	**Command:** ***3d*** **Enter an option [Box/Cone/DIsh/DOme/Mesh/ Pyramid/Sphere/Torus/Wedge]:** ***do***	1. Enter the ***3D*** command and select the **Dome** option. Alternately, you can pick the **Dome** button on the Surfaces toolbar.
	Specify center point of dome: ***3,3,5***	2. Use a three-dimensional coordinate for the **center point** as indicated.
	Specify radius of dome or [Diameter]: ***2***	3. Specify the **radius**.
	Enter number of longitudinal segments for surface of dome <16>: ***[enter]*** **Enter number of latitudinal segments for surface of dome <8>:** ***[enter]***	4. Accept the default number of **longitudinal** and **latitudinal segments**. Your drawing looks like Figure 4.014.

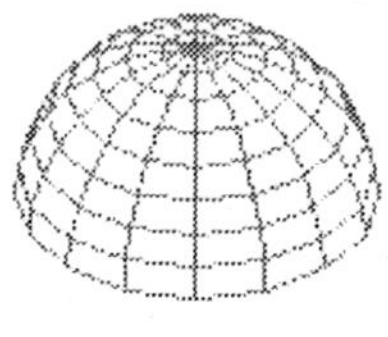

Figure 4.014

TOOLS	COMMAND SEQUENCE	STEPS
 Dish Button	**Command:** ***3d*** **Enter an option [Box/Cone/DIsh/DOme/Mesh/ Pyramid/Sphere/Torus/Wedge]:** ***di***	5. We'll draw the dish below the dome. Enter the ***3D*** command and select the **Dish** option. Alternately, you can pick the **Dish** button on the Surfaces toolbar.
	Specify center point of dish: ***3,3,2*** **Specify radius of dish or [Diameter]:** ***2*** **Enter number of longitudinal**	6. Follow the sequence indicated. Your drawing looks like Figure 4.015.

TOOLS	COMMAND SEQUENCE	STEPS
	segments for surface of dish <16>: ***[enter]*** **Enter number of latitudinal segments for surface of dish <8>:** ***[enter]***	

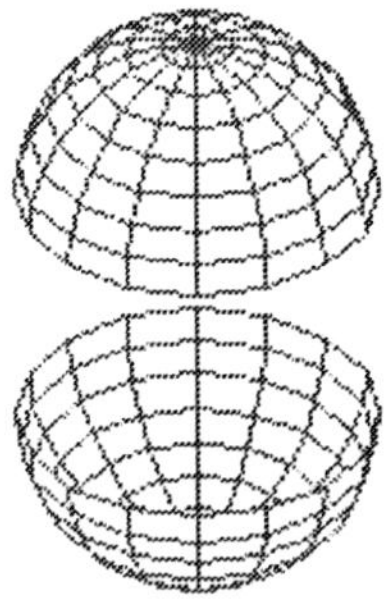

Figure 4.015

	Command: ***save***	7. Save the drawing as *MyDome* in the C:\Steps3D\Lesson04 folder.

4.2.7 Torus

A torus looks like the inner tube you used at the beach or lake when you were a kid. (Okay, some of you are still young enough to enjoy that type of activity without cracking bones that saw the breakup of Pangaea!) Like the sphere and dome/dish objects, you'll be required to identify the number of segments or faces, as well as the radius. But when drawing a torus, you must identify both the radius of the *torus* and the radius of the *tube*.

> Manipulating the various radii of a torus can lead to some startling (and nifty) results as you'll discover when working with 3D solids in Lesson 7.

The command sequence looks like this:

Command: ***3d***

Enter an option [Box/Cone/DIsh/DOme/Mesh/Pyramid/Sphere/Torus/Wedge]: ***t***

Specify center point of torus: ***[specify the center point of the torus; unless you want half of the torus to be underground, use a three-dimensional coordinate and move it upward on the Z-axis]***

Specify radius of torus or [Diameter]: ***[specify the radius or diameter of the torus – this is the distance from the center of your inner tube to the outer edge of the tube (this detail will become important when we look at 3D solid tori)]***

Specify radius of tube or [Diameter]: ***[specify the radius or diameter of the tube – this is how big to make the tube itself (how inflated do you want your inner tube?)]***

Enter number of segments around tube circumference <16>: ***[how many divisions will you want around the circumference of the tube?]***

Enter number of segments around torus circumference <16>: ***[how many divisions will you want along the circumference of the inner tube?]***

Let's give it a try.

Do This: 4.2.7.1	Creating a 3D Surfaced Torus

I. Start a new drawing using the *lesson 04 template* file located in the C:\Steps3D\Lesson04 folder.

II. Follow these steps.

TOOLS	COMMAND SEQUENCE	STEPS
 Torus Button	**Command:** ***3d*** **Enter an option [Box/Cone/DIsh/DOme/Mesh/ Pyramid/Sphere/Torus/Wedge]:** ***do***	1. Enter the ***3D*** command and select the **Torus** option. Alternately, you can pick the **Torus** button on the Surfaces toolbar.
	Specify center point of torus: ***5,5,.5***	2. Locate the torus using a three-dimensional coordinate.
	Specify radius of torus or [Diameter]: ***4*** **Specify radius of tube or [Diameter]:** ***1***	3. Specify the radii of the torus and the tube as indicated.
	Enter number of segments around tube circumference <16>: ***[enter]***	4. Accept the default number of segments around the tube …
	Enter number of segments around torus circumference <16>: ***32***	5. … but increase the number of segments around the torus. This way, you can see the difference. Your drawing looks like Figure 4.016. Notice the number of segments in each direction.

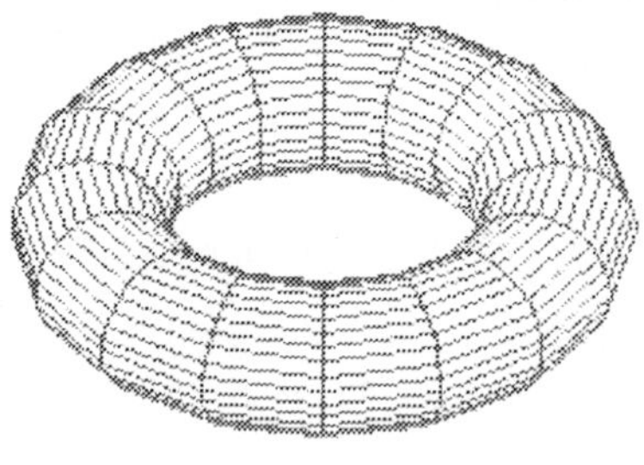

Figure 4.016

	**Command:** ***save***	6. Save the drawing as *MyTorus* in the C:\Steps3D\Lesson04 folder.

4.3 Understanding the Limitations of Predefined Surface Models

By now you may be thinking how wonderful predefined surface models are. And you're right to think so. But remember that you've yet to consider their limitations. Let's look at some of these now.

- The surface models discussed in this lesson are objects. Although you can explode them into 3D faces, neither the original objects nor the 3D faces can be easily modified. You can't trim or extend these objects. Nor can you fillet, chamfer, break, lengthen, or offset them. (You can, however, stretch, mirror, array, rotate, and copy them.)
- Neither 3D faces nor 3D meshes have wall thickness. They are, essentially, two-dimensional objects existing in three-dimensional space.
- While you can use these predefined shapes to build many things, you can't combine them as you can solids. (The ***Union***, ***Subtract***, and ***Intersect*** commands won't work on surface models.)
- Not all OSNAPs will work on predefined surface models.

But take heart, AutoCAD has several other commands (that you'll see in our next lesson) to enable you to draw shapes that aren't predefined. Then in Lesson 6, we'll look at some editing tools that do work!

4.4 Extra Steps

You've probably noticed that there are two toolbars listing similar predefined models – the Surfaces toolbar and the Solids toolbar. We've spent this lesson studying the predefined objects on the Surfaces toolbar. Take a few minutes and compare the two. Are they the same? What are the differences? Do they make the same types of models available?

In a new drawing created with the *Lesson 04 Template*, open both toolbars. Create a box using the surface approach and then create another box using the solid approach. Notice the differences in the command sequences. Repeat this procedure for each of the predefined models.

It's early to be studying Solid Modeling, yet the similarities between solids and surfaces are too tempting to ignore. As you continue your study of Surface Modeling, bear in mind that each procedure probably has a solid modeling equivalent. Then, when we get to Solid Modeling, you'll be a step a head of the game!

4.5 What Have We Learned?

Items covered in this lesson include:

- *AutoCAD's predefined surface models*
  - *Box*
  - *Wedge*
  - *Pyramid*
  - *Cone*
  - *Sphere*
  - *Dome*
  - *Dish*
  - *Torus*

This has been a fun lesson. It's nice to know that they're not all difficult!

We've seen several predefined objects designed by AutoCAD to make three-dimensional drafting move more quickly. But you shouldn't limit these objects to their obvious uses. Whenever you have a complex object to build with surface models, think about these predefined tools as basic – and often elastic – building blocks. For example, when you need a cylinder, consider using a cone with equal radii at both ends. Or when you need a ramp and the wedge is too straight, try using a pyramid with an offset ridge.

Practice the exercises at the end of this lesson for experience. Remember to use the ***UCS*** and **Shademode** commands to assist you. Then try to draw different objects around your desk – how about your mouse or the keyboard?

Our next lesson will cover more complex, user-defined shapes created as surface models. So what you can't draw yet, you'll soon be able to!

4.6 Exercises

1. through 8. Recreate the "su" drawing found in Appendix B using the tools found in this lesson to help you. You'll need to explode many of the primitives to get 3D faces to manipulate, but the drawing should be faster. Use cylinders rather than arcs to line the holes.
9. Create the flying saucer drawing shown in Figure 4.017. The following hints will help:
   - 9.1. Use the *lesson 04 template* to begin.
   - 9.2. Change the layer colors as needed.
   - 9.3. Use these tools: cone, dish, dome, and torus.
   - 9.4. Use clipping planes to create the section.
   - 9.5. Use the direct hatch approach to hatch the section.
   - 9.6. While you're still in Model Space, use the continuous orbit tool to make the saucer fly back and forth across the screen. (Eat your heart out, Marvin the Martian!)
   - 9.7. Save the drawing as *MySaucer* in the C:\Steps3D\Lesson04 folder.

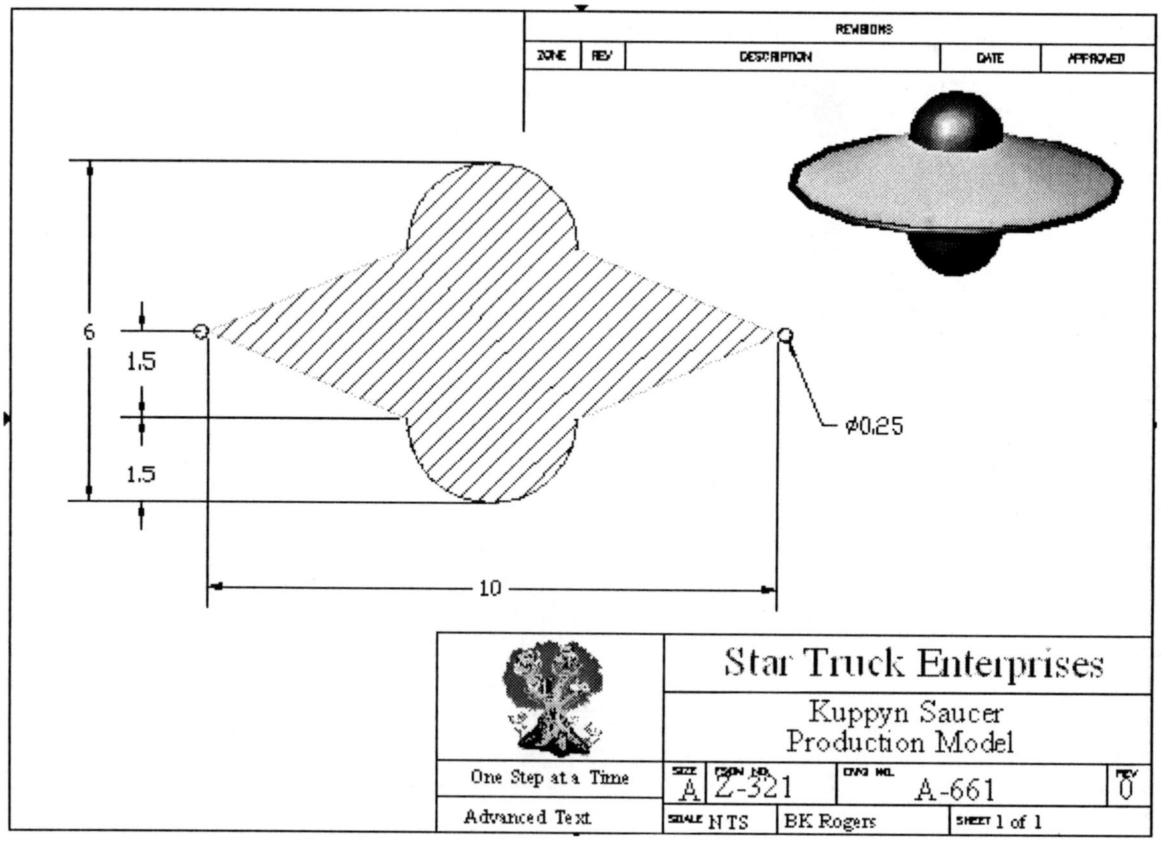

Figure 4.017

10. Create the aquarium aerator drawing shown in Figure 4.018. The following hints will help:
    10.1. Use the *lesson 04 template* to begin.
    10.2. Change the layer colors as needed.
    10.3. Use these tools: box, donut, and dish.
    10.4. Viewport scales are a uniform 1:1.
    10.5. Text uses the Times New Roman font.
    10.6. Save the drawing as *MyAerator* in the C:\Steps3D\Lesson04 folder.

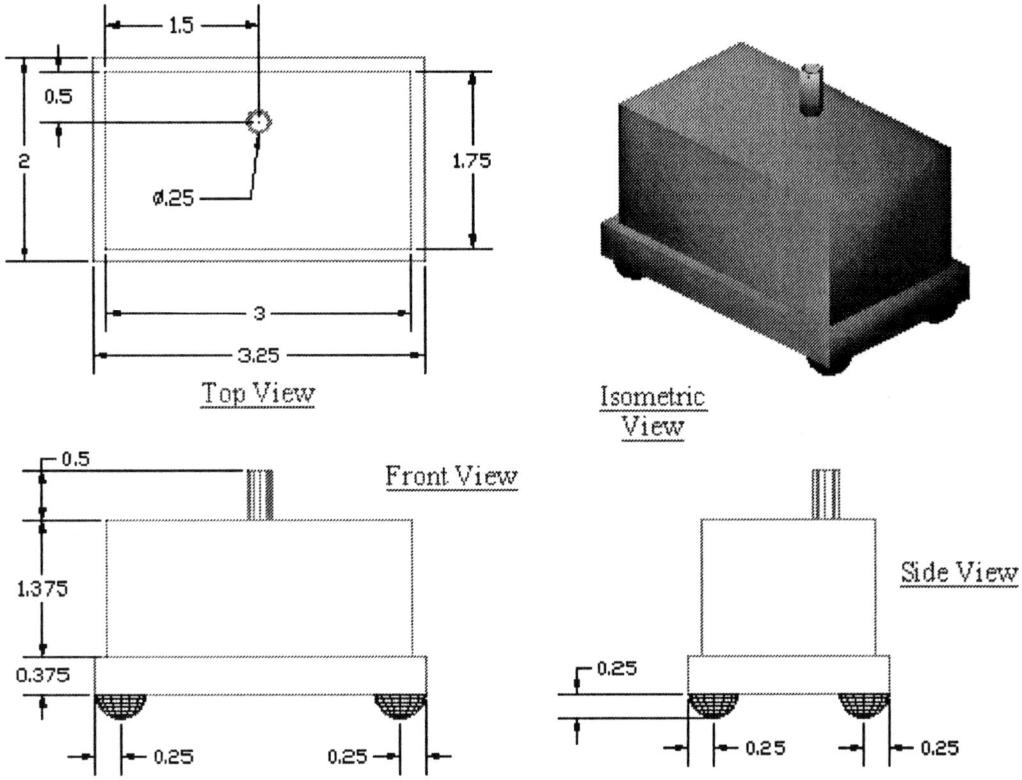

Figure 4.018

11. Create the trailer light drawing shown in Figure 4.019. The following hints will help:
    11.1. Use the *lesson 04 template* to begin.
    11.2. Change the layer colors as needed.
    11.3. Use these tools: box, donut, pyramid, and cone.
    11.4. Viewport scales are a uniform 1:2.
    11.5. Text uses the Times New Roman font.
    11.6. Save the drawing as *MyLight* in the C:\Steps3D\Lesson04 folder.

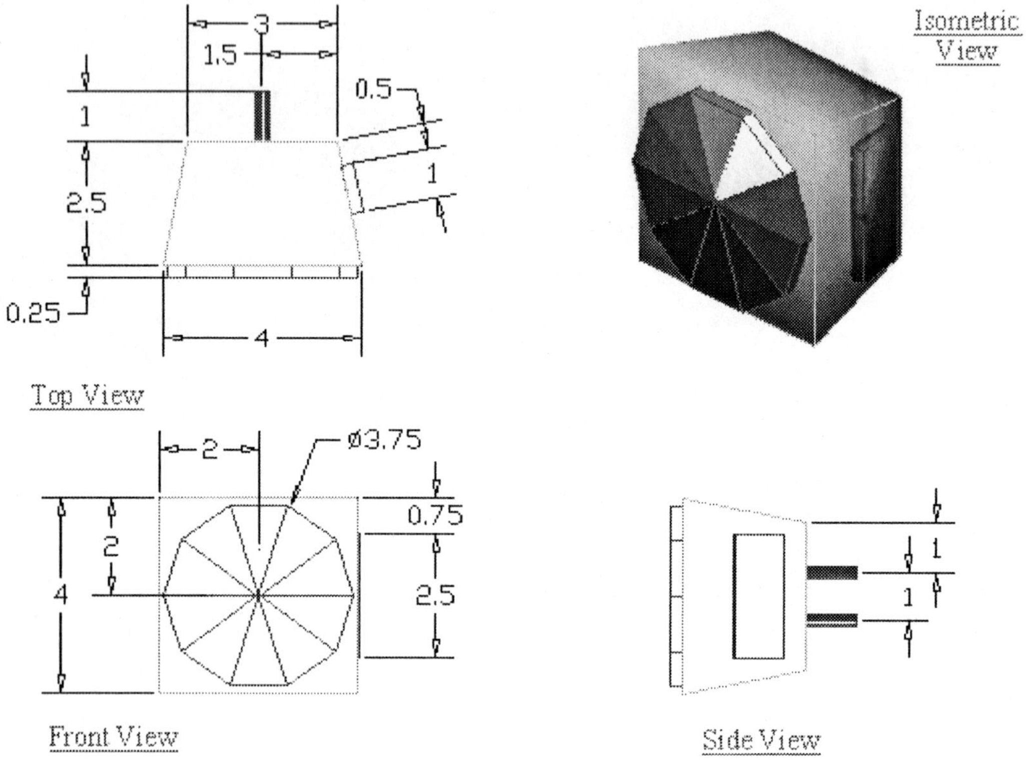

Figure 4.019

12. Create the microphone drawing shown in Figure 4.020. The following hints will help:
    12.1. Use the *lesson 04 template* to begin.
    12.2. Change the layer colors as needed.
    12.3. Use these tools: box, cone, sphere, and wedge.
    12.4. The UCS is tricky on this one. Align it

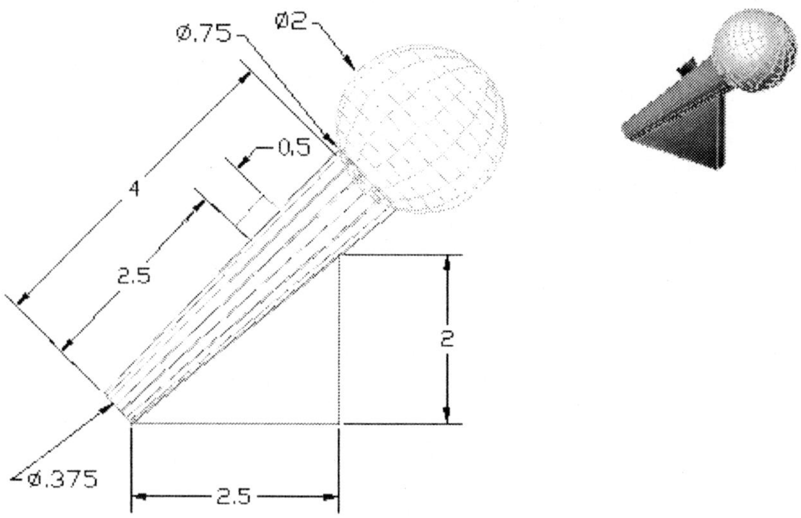

Figure 4.020

with the cone to draw the button. Use the UCS to help you rotate the microphone so that it sits atop the wedge.

12.5. The wedge is ¼" wide.

12.6. Save the drawing as *MyMicrophone* in the C:\Steps3D\Lesson04 folder.

13. Create the coffee table drawing shown in Figure 4.021. The following hints will help:
    13.1. Use these tools: cone, torus, and region.
    13.2. The torus is 36" diameter and the tube is 1" diameter.
    13.3. The legs are 21" long and extend 2" below the bottom shelf..
    13.4. I used the Gouraud Shademode.
    13.5. Save the drawing as *MyCoffeeTable* in the C:\Steps3D\Lesson04 folder.

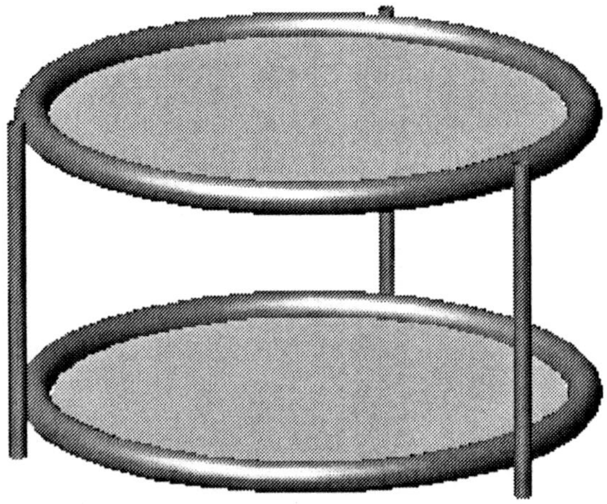

Figure 4.021

14. Create the wagon drawing shown in Figure 4.022. The following hints will help:
    14.1. Use these tools: cone, torus, box, and dome.
    14.2. The wheel tori are 4" diameter with 1" diameter tubes.
    14.3. The axles are 1" diameter x 8" long cones.
    14.4. The axles are 12" apart.
    14.5. The wagon floor is 18" x 6" x ½".
    14.6. The outer boards are 2½" wide and spaced ½" apart.
    14.7. The inner boards are 1½" wide.
    14.8. I used the Gouraud Shademode.
    14.9. Save the drawing as *MyWagon* in the C:\Steps3D\Lesson04 folder.

Figure 4.022

15. Create the pyramid drawing shown in Figure 4.023. The following hints will help:
    15.1. Use these tools: sphere, torus, and pyramid.
    15.2. The sphere is 16" diameter. (You'll need to explode it and erase some of the 3D faces.
    15.3. The pyramid is 5" squared on the base x 5" high.
    15.4. The torus is 10" diameter with a 1" diameter tube.
    15.5. The pyramid is hatched to create the brick pattern.
    15.6. I used the gOuraud+edges Shademode to show both the surfaces and the hatching.
    15.7. Save the drawing as *MyPyrSph* in the C:\Steps3D\Lesson04 folder.

Figure 4.023

16. Create the temple drawing shown in Figure 4.024. The following hints will help:
    16.1. Use these tools: wedge, box, and pyramid.
    16.2. The sphere is the drawing we created in Exercise 15. I inserted it as a block at 1/8 scale. (Be sure the UCS in both drawings is equal to the WCS. We'll discuss more on three-dimensional blocks in Lesson 10.)
    16.3. The ramps are 8" x 2" x 2".
    16.4. The center box is 4" x 4" x 2".
    16.5. The pyramid is 4" x 4" x 2" (with a 2" x 2" top).
    16.6. The torus is 10" diameter with a 1" diameter tube.
    16.7. The pyramid is hatched to create the brick pattern.
    16.8. I used the gOuraud_edges Shademode to show both the surfaces and the hatching.
    16.9. Save the drawing as *MyTemple* in the C:\Steps3D\Lesson04 folder.

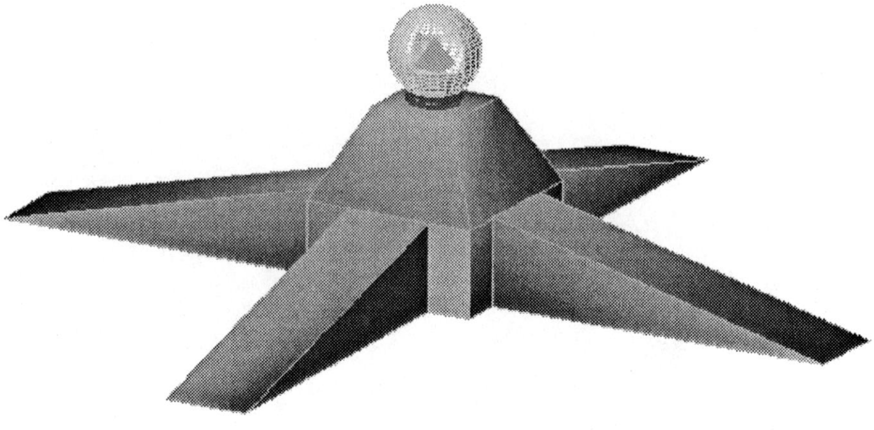

Figure 4.024

17. Create the lamp drawing shown in Figure 4.025. The following hints will help:
    17.1. Use these tools: cone.
    17.2. The center pole is 6' tall x 1" diameter.
    17.3. The base is 12" diameter on the bottom and 1" diameter on the top. It is 2" high.
    17.4. The top is 1" diameter on the bottom and 12" diameter on the top. It is 6" high.
    17.5. The switch is ½" diameter x ½" high. It's located halfway up the center pole.
    17.6. Save the drawing as *MyLamp* in the C:\Steps3D\Lesson04 folder.

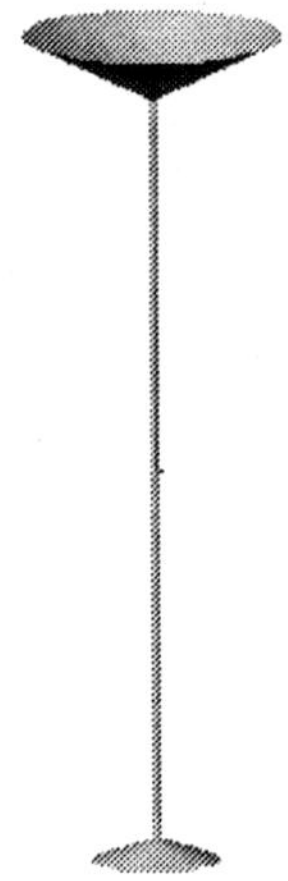

Figure 4.025

4.7	**For Web-Based Review Questions, visit: http://www.uneedcad.com/2006/Files/06R4-3D.pdf**

Lesson 5

Following this lesson, you will:

- ✓ *Know how to build more complex surface models*
  - ***Rulesurf***
  - ***Revsurf***
  - ***Tabsurf***
  - ***Edgesurf***
  - ***3DMesh***
  - ***PFace***
- ✓ *Know how to control the number of surfaces used to draw a surface model*
  - ***Surftab1***
  - ***Surftab2***

Complex Surface Models

In Lesson 4, you discovered some simple tools that you can use to create surface models. But what if you need a model that doesn't easily translate into one of the predefined surface models? For example, suppose you need to draw an I-Beam, a piping elbow, or an ornate lamp. Which predefined model would you use?

The answer of course, is that none of the predefined tools would help. Well then, would you have to draw a wireframe model (a stick figure) and stretch the surfaces over it? Or would you just give up and wait for Autodesk to develop some more predefined shapes? (Sorry, I don't think any more are coming.)

A wireframe model might work for the I-Beam but not for the pipine elbow or the ornate lamp.

The truth is that these tools (wireframe models and predefined surface models) were designed for simple objects, not for complex constructions. But take heart, AutoCAD provides other tools that'll help you deal easily with more intricate designs.

In this lesson, we'll examine tools for creating complex surface models. Here, we'll conclude our study of surface model creation techniques with a look at the procedures needed to put the razzle-dazzle in your three-dimensional drawing.

Let's get started!

5.1 Controlling the Number of Surfaces – Surftab1 and Surftab2

When you drew the sphere, dome, and dish in our last lesson, AutoCAD asked you for the number of segments (or faces) you wanted to use in defining the object. Remember that we defined longitudinal and latitudinal segments differently so that you could see the distinction.

That approach worked well for a predefined shape. But when you draw complex shapes, AutoCAD can't know what you're doing. So, it can't ask you for the number of segments you'll need to define the object. Still, that information will be required for you complex object to take the shape you want.

For this reason, AutoCAD established two system variables to define the number of faces it'll use to create an object. The first – **Surftab1** – defines the number of surfaces AutoCAD will use to create a linear object, or the number of surfaces it will use to create the circumference (axial direction) of a round (or arced) object. The second – **Surftab2** – defines the number of surfaces AutoCAD will use to create latitudinal sections (along the path or rotation) of an object.

Compare the drawings in Figures 5.001, 5.002, and 5.003. Each drawing represents a sphere.

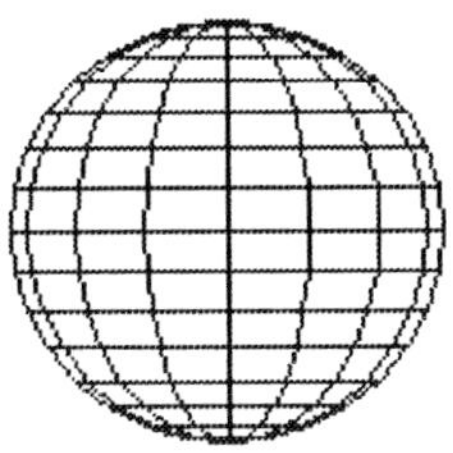

Figure 5.001

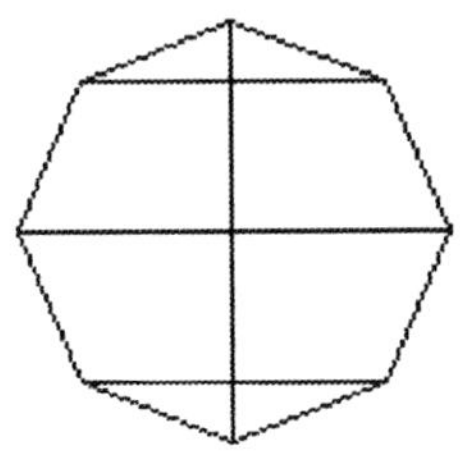

Figure 5.002

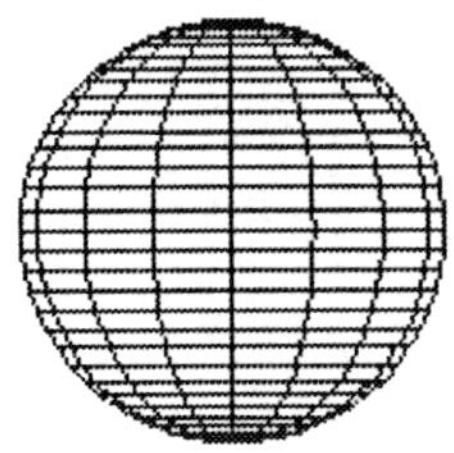

Figure 5.003

- I drew Figure 5.001 using default settings (16 longitudinal and latitudinal segments) and the **Sphere** option of the ***3D*** command.
- I drew Figure 5.002 using the ***Revsurf*** command (Section 5.2.3) with **Surftab1** and **Surftab2** both set to **4**.
- I drew Figure 5.003 using the ***Revsurf*** command with **Surftab1** set to **16** and **Surftab2** set to **32**.

Notice the difference between Figures 5.002 and 5.003. I drew both using the same command with the same size arc and identical axes. The only difference is the **Surftab1** and **Surftab2** settings. In Figure 5.002, AutoCAD used four faces to define the object along the circumference and along the path of revolution. In Figure 5.003, AutoCAD used 16 faces to define the object along the circumference and 32 along the path of revolution. Here you see another important aspect of the **Surftab1** and **Surftab2** system variables – *their settings affect the shape of the object being drawn.*

Make a small copy of Figure 5.003 and write the **Surftab1** and **Surftab2** settings on it. Tape it to your monitor as a guide until you're comfortable with each.

We'll set **Surftab1** and **Surftab2** as needed throughout the exercises in this lesson.

5.2 Different Approaches for Different Goals

AutoCAD provides four basic and two advanced commands to handle the drawing of complex shapes. You'll probably enjoy the basic commands – they're lots of fun! The advanced commands, however, are more challenging and may take some time to master.

The basic commands all have one thing in common – they all use something two dimensional as a guide to create a three-dimensional object. But remember, a two-dimensional object can exist in Z-space. That is, an object can have any two of the three properties required for a three-dimensional object (length, width, and height). The command you use to create the three-dimensional object will provide the third property. This'll become clear once you've used the commands.

We'll discuss the four basic commands in this section and save the advanced commands for the next one. Let's get started.

5.2.1 Follow the Path – The *Tabsurf* Command

Tabsurf is an easy command. You'll need a basic shape – circles, arcs, polylines, and splines make great shapes – and something to indicate a path. AutoCAD will expand the shape into three dimensions along the path you identify.

The command sequence looks like this:

> **Command: *tabsurf***
>
> **Current wire frame density: SURFTAB1=6 *[AutoCAD reports the Surftab1 settings]***
>
> **Select object for path curve: *[select the object that'll give shape to your three-dimensional object]***
>
> **Select object for direction vector: *[select an object that'll tell AutoCAD the direction in which to expand the shape]***

This will become clearer with some practice. (This is a fun project. We'll use the basic commands for drawing complex shapes to create a three-dimensional toy train. So put on your Engineer's hat and let's get started!)

> You can access the ***Tabsurf*** command (and other complex surface commands discussed in this lesson) from the Surfaces toolbar or from the Draw pull-down menu. Follow this path for the pull-down menu commands:
>
> *Draw – Surfaces – Tabulated Surfaces (or other complex surface command)*

Do This: 5.2.1.1	Using Tabsurf to Create a Three-Dimensional Object

I. Open the *train.dwg* file in the C:\Steps3D\Lesson05 folder. The drawing looks like Figure 5.004.

II. Set **Tank** as the current layer.

III. Follow these steps.

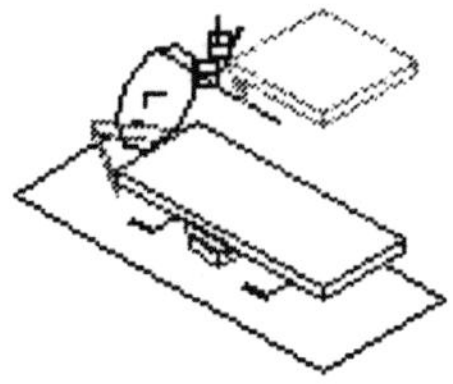

Figure 5.004

Tools	Command Sequence	Steps
No Button Available	**Command: *surftab1*** **Enter new value for SURFTAB1 <6>: *20***	1. Set the **Surftab1** system variable to **20** for a more defined shape.
Tabsurf Button	**Command: *tabsurf***	2. Enter the ***Tabsurf*** command.
	Current wire frame density: SURFTAB1=20 **Select object for path curve: *[select the circle]***	3. (Refer to Figure 5.005 for Steps 3 and 4.) AutoCAD wants you to select the object for path curve – the *shape*. Select the circle indicated.

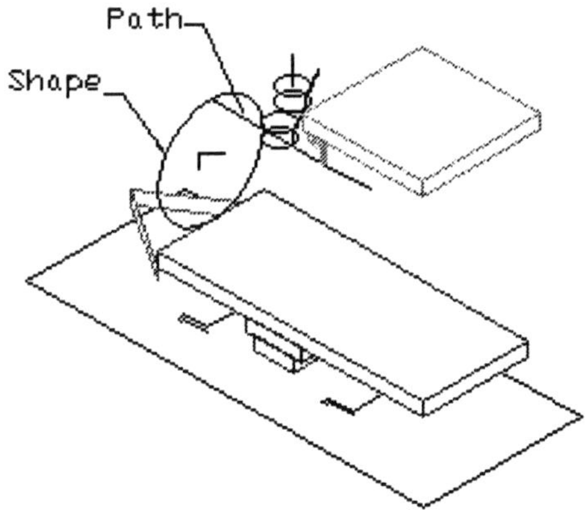

Figure 5.005

Tools	Command Sequence	Steps
	Select object for direction vector: *[select the line]*	4. AutoCAD wants to know in which direction to expand the shape – the *path*. Select the line indicated. (Select next to where the arrow points in Figure 5.005.) Your drawing looks like Figure 5.006. [Note: First AutoCAD determined the path by the selected object. Then it determined the direction for the three dimensional object by where you select on the object for direction vector. Try repeating this step but select the other end of the line. Then undo until your drawing again looks like Figure 5.006.

Tools	**Command Sequence**	**Steps**

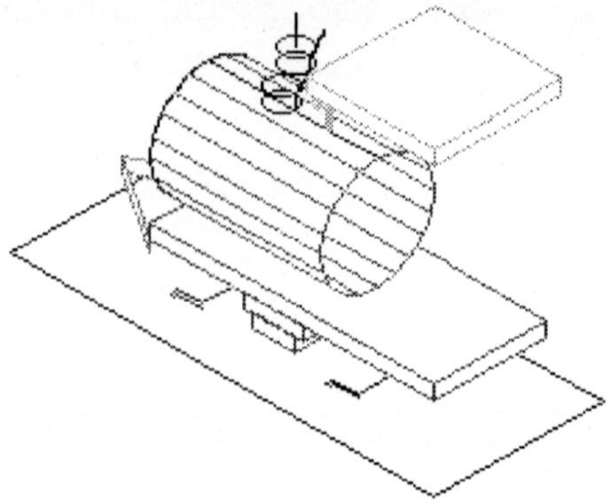

Figure 5.006

	Command: *z*	5. Zoom in around the smokestack (the stacked circles atop the tank) and set the **stack** layer current.
	Command: ***tabsurf***	6. Use the ***Tabsurf*** command to expand the circles along the lines. Notice that the path is independent of the UCS. Your drawing looks like Figure 5.007.

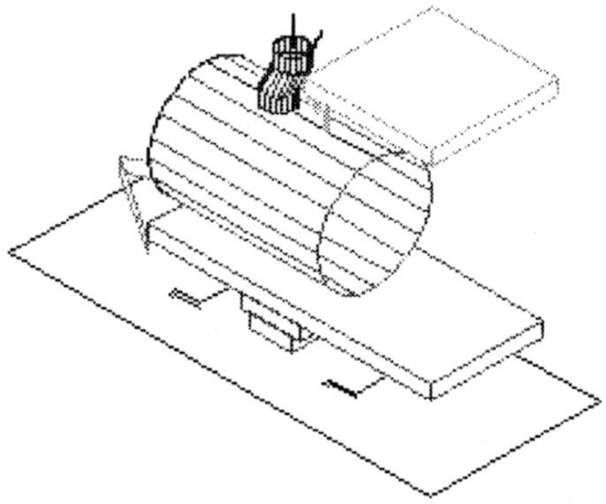

Figure 5.007

	Command: ***qsave***	7. Save the drawing but don't exit.

Some things you may have noticed about the ***Tabsurf*** command and need to remember include:

- The layer of the new object is based on the current layer at the time it is drawn and not on the layer of the original shape or path.
- The part of the path AutoCAD uses to define the new object is the endpoint – *a curved path won't produce a curved object.*
- The original shape and path remain intact after the three-dimensional object is drawn – put them on a separate layer that can be frozen later.
- Where you pick on the path can affect the direction of the expansion.
- The UCS doesn't affect the expansion.

We used circles in our exercise, but **Tabsurf** will work on any predefined shape provided the shape is a single object. Polylines and splines are particularly useful in creating shapes for **Tabsurf** expansion.

Let's take a look at another complex surface modeling command.

5.2.2 Add a Surface Between Objects – The *Rulesurf* Command

Use the ***Rulesurf*** command when you have two existing objects and want to place a surface between them. It's particularly useful when creating surfaces between uneven objects or objects of different size.

The command sequence is:

> **Command: *rulesurf***
>
> **Current wire frame density: SURFTAB1=20 *[AutoCAD reminds you of the current Surftab1 setting; Surftab2 doesn't affect Rulesurf]***
>
> **Select first defining curve: *[select the first edge of the surface to be created]***
>
> **Select second defining curve:*[select the opposite edge of the surface to be created]***

There are some similarities between ***Rulesurf*** and ***Tabsurf***, including:

- The layer of the new object is based on the current layer at the time it's drawn and not on the layer of the original shape or path.
- The original edges remain intact after the three-dimensional object is drawn.
- Where you pick on the path can affect the expansion. You should pick in the same vicinity on both edges (toward the same endpoints).
- The UCS doesn't affect the expansion.

Let's try the ***Rulesurf*** command on our train's cow catcher.

Do This: 5.2.2.1 Using *Rulesurf* to create a Three-Dimensional Object

I. Be sure you're still in the *train.dwg* file in the C:\Steps3D\Lesson05 folder. If not, please open it now.

II. Restore the **–111** view.

III. Set the **tank** layer current.

IV. Follow these steps.

TOOLS	COMMAND SEQUENCE	STEPS
	**Command: *reg*** **Select objects: *[select the circle]*** **Select objects: *[enter]*** **1 loop extracted.** **1 Region created.**	1. Turn the circle in the front of the tank into a region. (This'll improve viewing later.)
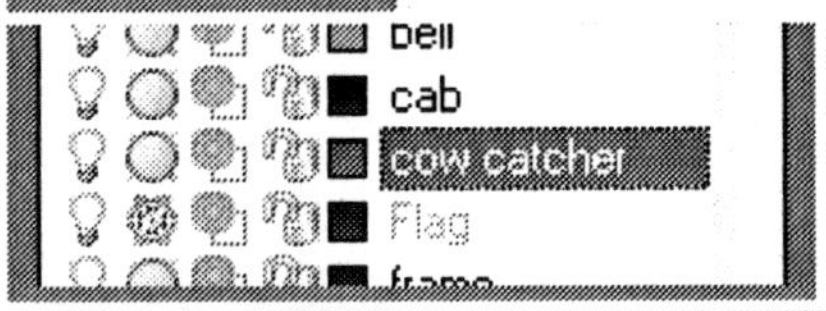		2. Set the **cow catcher** layer current.
 Ruled Surface Button	**Command: *rulesurf***	3. Enter the ***Rulesurf*** command.

Tools	Command Sequence	Steps
	Current wire frame density: SURFTAB1=20 **Select first defining curve:** ***[select the first edge]*** **Select second defining curve:** ***[select the second edge]***	4. (Refer to Figure 5.008.) AutoCAD needs to know where to draw the surface. Select the edges indicated. If the surface appears crossed, erase it and try again. When you select the edges, pick in the same general location of each. (If you have trouble selecting the edges, hold down the control key and select until the edge is found.)

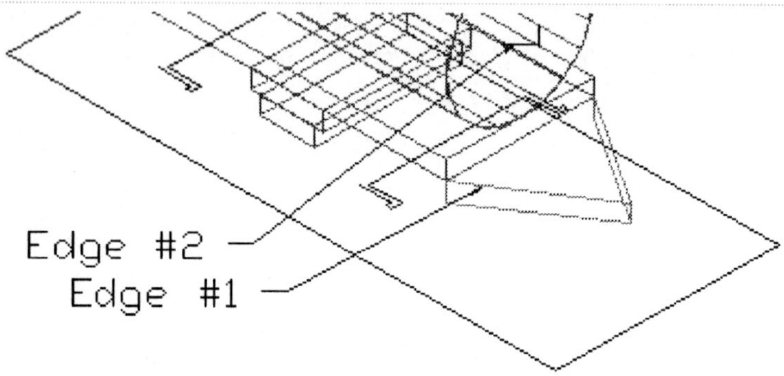

Figure 5.008

Tools	Command Sequence	Steps
	Command: ***rulesurf***	5. Repeat Steps 3 and 4 for the other side of the cow catcher. Your drawing looks like Figure 5.009.

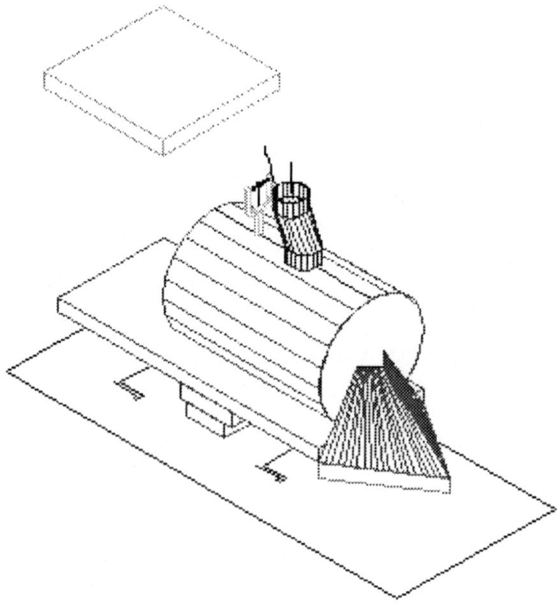

Figure 5.009

Tools	Command Sequence	Steps
	Command: ***qsave***	6. Save the drawing but don't exit.

Like ***Tabsurf***, any type of object will do for a ***Rulesurf*** edge – the fancier the original object, the fancier the results!

Speaking of fancy, let's look at the ***Revsurf*** command!

5.2.3 Creating Circular Surfaces – The *Revsurf* Command

Revsurf is one of the more popular of AutoCAD's surface modeling commands. That could be because ***Revsurf*** is so simple to use; but it's more likely because of the nifty gizmos you can draw with it!

Like the ***Tabsurf*** command, ***Revsurf*** begins with an object that'll define its basic shape and another object that'll define its path (or, in the case of ***Revsurf***, its axis). Devote some time and care when creating the basic shape since the final object will reflect a well-defined shape. Although other objects may occasionally be required (like a circle as a basic shape to define a piping elbow), I'd use polylines or splines almost exclusively. They tend to produce some truly professional results!

The ***Revsurf*** command sequence looks like this:

> **Command: *revsurf***
>
> **Current wire frame density: SURFTAB1=20 SURFTAB2=6 *[AutoCAD reports the current settings for both Surftab1 and Surftab2. Both will be needed for this procedure.]***
>
> **Select object to revolve: *[select the object that defines the basic shape of the object you wish to create]***
>
> **Select object that defines the axis of revolution: *[select an object that defines the axis around which you'll revolve the shape]***
>
> **Specify start angle <0>: *[specify a starting angle]***
>
> **Specify included angle (+=ccw, -=cw) <360>: *[tell AutoCAD if you want a fully or partially revolved shape. The default – 360°– defines a full revolution. For less than a full revolution, enter the degrees that define the arc you wish to fill.]***

Like ***Tabsurf*** and ***Rulesurf***, the UCS doesn't affect ***Revsurf***. However, you might need to adjust it when creating the basic shape you intend to revolve.

Let's take a look at the ***Revsurf*** command.

Do This: 5.2.3.1 Using *Revsurf* to Create a Three-Dimensional Object

I. Be sure you're still in the *train.dwg* file in the C:\Steps3D\Lesson05 folder. If not, please open it now.

II. Restore the **bell** view and set the **Stack** layer current.

III. Follow these steps.

TOOLS	COMMAND SEQUENCE	STEPS
 Revolved Surface Button	**Command: *revsurf***	1. Enter the ***Revsurf*** command.
	Current wire frame density: SURFTAB1=20 SURFTAB2=6 **Select object to revolve:**	2. AutoCAD reports the **Surftab1** and **Surftab2** settings. These suit our purpose, so we'll continue. Select the spline rising from the top of the smokestack as the object to revolve.

Tools	Command Sequence	Steps
	Select object that defines the axis of revolution:	3. Select the line rising from the center of the smokestack as your **axis of revolution**.
	Specify start angle <0>: *[enter]* **Specify included angle (+=ccw, -=cw) <360>:** *[enter]*	4. Accept the **start angle** and **included angle** defaults.
	Command: *revsurf*	5. Repeat Steps 1 to 4 for the bell assembly (be sure to use the **Bell** layer). Your drawing looks like Figure 5.010.

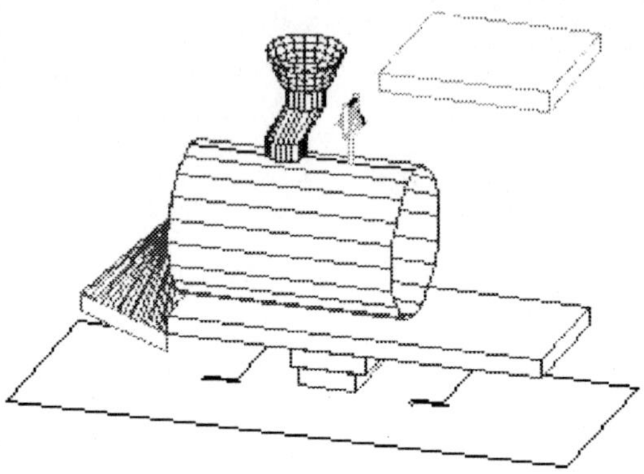

Figure 5.010

Tools	Command Sequence	Steps
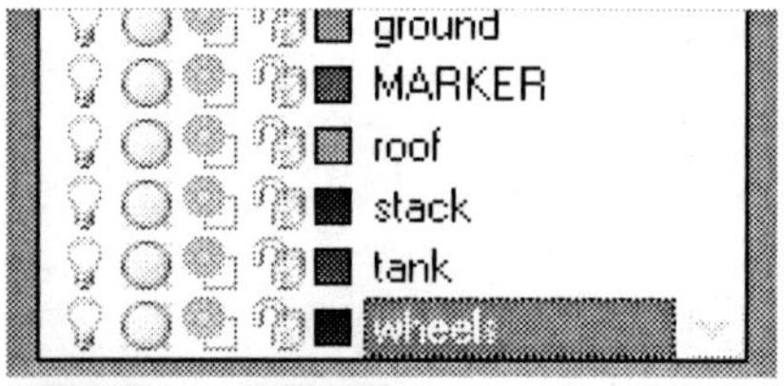		6. Thaw the **MARKER** layer. Notice the nodes, but also notice the lines that appear as axes for the wheel shapes. Set the **wheels** layer current.
	Command: *revsurf*	7. Repeat Steps 1 to 4 for both wheels using the lines on the **MARKER** layer as the axes. (Refreeze the **MARKER** layer when you've finished.) Your drawing looks like Figure 5.011.

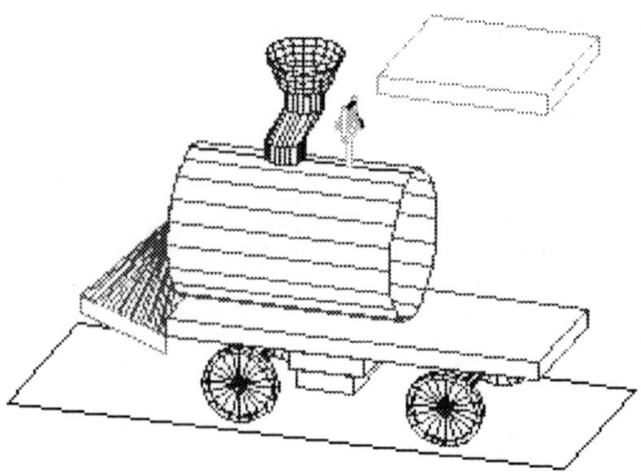

Figure 5.011

Tools	Command Sequence	Steps
	Command: *qsave*	8. Save the drawing but don't exit.

Wasn't that fun?

But remember that I'm providing the basic shapes for these exercises. It's a bit more involved (although not difficult) when you draw them yourself. But just imagine the sense of accomplishment that'll give you!

Our train is starting to take shape (as is our expertise with some cool new commands). Have you tried viewing the model with the **Gouraud Shademode**? It looks very nice (especially if you freeze layer **0**).

But we still have to draw the cab and the ground beneath the wheels. Let's not waste a moment – full *steam* ahead!

5.2.4 Using Edges to Define a Surface Plane – The *Edgesurf* Command

Edgesurf is actually one of the simplest of the complex surface commands. But I've placed it at the end of the basic commands because it makes a good transition to the more advanced commands.

Edgesurf creates a surface plane. That doesn't mean that it creates a surface along the X-, Y-, or Z-planes but rather a plane like an open field. It doesn't have to be flat, although it's not the tool for very complex surfaces.

The ***Edgesurf*** command uses four edges to define a plane. The edges can be parallel to one another or skewed in any direction.

The command sequence simply asks for the four edges:

> **Command: *edgesurf***
>
> **Current wire frame density: SURFTAB1=20 SURFTAB2=6 *[AutoCAD reminds you of the Surftab1 and Surftab2 settings. Like Revsurf, both will be needed here.]***
>
> ***[Use the next four options to defined the edges of the surface.]***
>
> **Select object 1 for surface edge:**
>
> **Select object 2 for surface edge:**
>
> **Select object 3 for surface edge:**
>
> **Select object 4 for surface edge:**

We'll use ***Edgesurf*** to create the ground beneath our train's wheels.

Do This: 5.2.4.1 — Using *Edgesurf* to Create a Three-Dimensional Object

I. Be sure you're still in the *train.dwg* file in the C:\Steps3D\Lesson05 folder. If not, please open it now.

II. Remain in the **bell** view but zoom out so you can see the ground.

III. Set the **ground** layer current, and freeze the **MARKER** layer.

IV. Follow these steps.

TOOLS	COMMAND SEQUENCE	STEPS
 Edge Surface Button	**Command: *edgesurf***	1. Enter the ***Edgesurf*** command.

TOOLS	COMMAND SEQUENCE	STEPS
	Current wire frame density: SURFTAB1=20 SURFTAB2=6 **Select object 1 for surface edge:** **Select object 2 for surface edge:** **Select object 3 for surface edge:** **Select object 4 for surface edge:**	2. AutoCAD reports the current **Surftab1** and **Surftab2** settings and then asks you to select the edges of your object. Select the four lines that form the boundary of the ground (pick one of the shorter lines first). Your drawing looks like Figure 5.012.

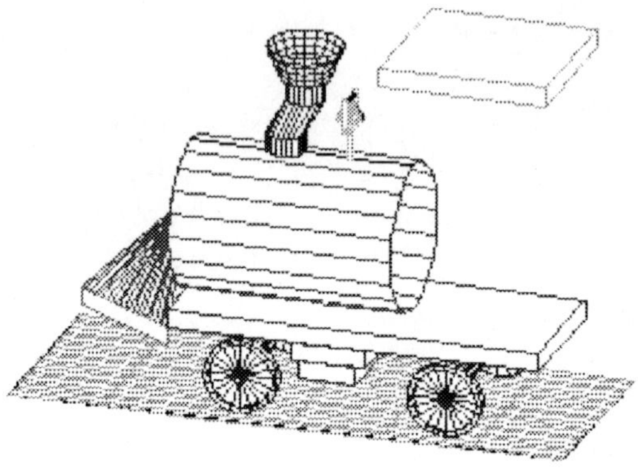

Figure 5.012

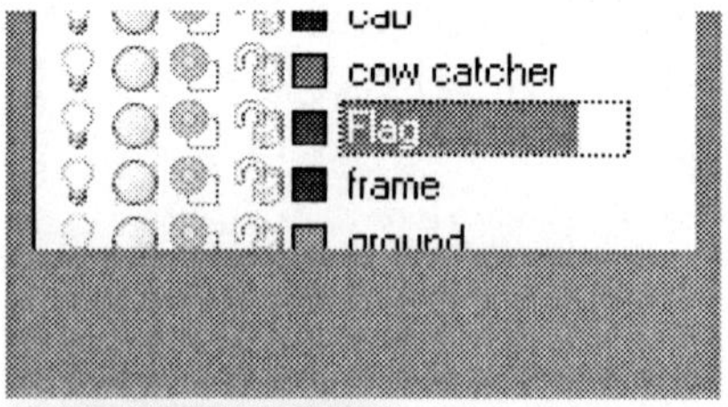		3. You've seen how ***Edgesurf*** works. But let me show you what it can do with a little imagination. Thaw the **Flag** layer and set it current. Notice the outline of the flag atop the roof (refer to Figure 5.013).
	Command: ***surftab2*** **Enter new value for SURFTAB2 <6>:** ***20***	4. Set **Surftab2** to **20**. This will enhance our resolution.
	Command: ***edgesurf***	5. Now do an ***Edgesurf*** using the four sides (two lines and two splines) of the flag as your edges. Your drawing looks like Figure 5.013.

Tools	Command Sequence	Steps
	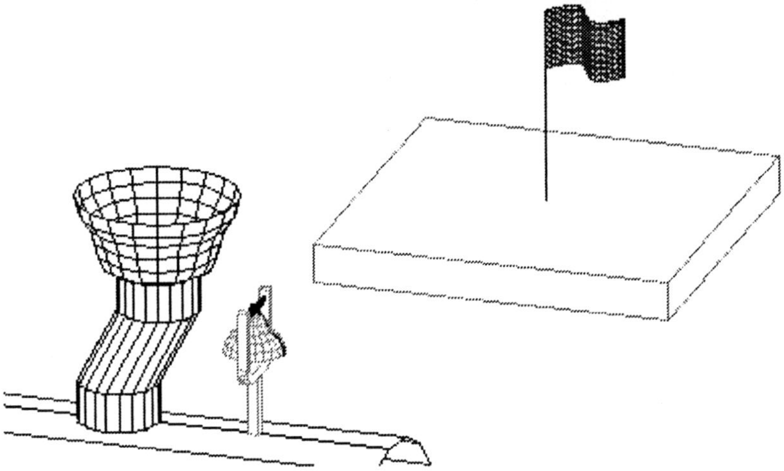 Figure 5.013	
	Command: ***qsave***	6. Save the drawing but don't exit.

How's that for a bit of razzle-dazzle?!

This concludes the basic Surface Modeling commands. But there's one part of our train we have yet to draw – the cab. We could create it using simple 3D face commands. However, that would give us several objects with which to deal later. There is a way to create the cab as a single object just as the sphere, box, torus, and other predefined surface models are single objects. That involves the advanced surface model commands, which we'll look at next.

5.3 More Complex Surfaces

There are actually two advanced surface model commands – ***3DMesh*** and ***PFace***. I don't recommend either for the faint of heart!

AutoCAD invented these commands for you to create objects similar to the predefined surface models so you can manipulate many surfaces as a single object (as in a single sphere rather than dozens of faces). But be forewarned: creating an object with either the ***3DMesh*** or ***PFace*** command means manually specifying *every vertex on the object*! Besides that, you must identify the vertices *in a specific order*!

Does that sound like a lot of work? It is! But in AutoCAD's defense, I must add that these commands are actually better suited for lisp routines or other third-party programs. You might say that you're on the border where CAD operation ends and CAD programming begins.

> A third-party program is one that's designed to use AutoCAD as a base. In other words, AutoCAD becomes something like a CAD operating system (as Windows is your computer's operating system). The third-party program builds on AutoCAD by providing shortcuts toward a specific end. Popular third-party programs include *Mechanical Desktop*, *Architectural Desktop*, *Propipe*, *ProISO*, and many others.

We'll start with the ***3DMesh*** command.

5.3.1 Creating Meshes with the *3DMesh* Command

The ***3DMesh*** command is similar to the ***Edgesurf*** command. (The similarity is akin to that between a sculptor and a whittler – both will give you a carving. But what the sculptor produces will embarrass the whittler.) Both work well in creating that open field look. But where four edges define the ***Edgesurf*** object, you have no limit to the number of defined vertices with the ***3DMesh*** command. As I've mentioned, however, it requires tedious effort to identify each vertex involved.

The command sequence looks like this:

> **Command:** ***3dmesh***
>
> **Enter size of mesh in M direction:** ***[tell AutoCAD how many lines are required to define the columns of faces you'll need]***
>
> **Enter size of mesh in N direction:** ***[tell AutoCAD how many lines are required to define the rows of faces you'll need]***
>
> **Specify location for vertex (0, 0):** ***[this prompt will repeat for each vertex (intersection) of lines defining the rows and columns]***

Notice that AutoCAD defines the rows and columns in terms of **M** and **N**. This helps avoid any confusion with the X-, Y-, and Z-axes. ***3DMesh*** works independently of the UCS.

There's an interesting point to remember when defining the size of the mesh in terms of rows and columns. Notice that AutoCAD doesn't ask for the number of rows and columns, but rather, it asks for the number of lines required to define the rows and columns. *You'll be working with the vertices that create the surfaces, not the spaces between the vertices.* The easiest way to determine the number of lines required is simply to add one to the number of rows and columns you want.

This'll become clearer in our next exercise. Let's draw a hill in front of our train (we'll call it *the little engine that could*).

Do This: 5.3.1.1 Using *3DMesh* to Create a Three-Dimensional Object

I. Be sure you're still in the *train.dwg* file in the C:\Steps3D\Lesson05 folder. If not, please open it now.

II. Thaw the **MARKER** layer and set the **ground** layer current.

III. Begin in the **Bell** view, but pan and orbit so you can see the nodes at the front of the train.

IV. Set your running OSNAPs to **Node**; clear all other settings.

V. Follow these steps.

TOOLS	COMMAND SEQUENCE	STEPS
 3D Mesh Button	**Command:** ***3dmesh***	1. Enter the ***3DMesh*** command.
	Enter size of mesh in M direction: ***6*** **Enter size of mesh in N direction:** ***6***	2. (Refer to Figure 5.014.) Tell AutoCAD you want 5 rows and 5 columns. (Remember that you need 6 lines to define 5 rows or columns.)

TOOLS	COMMAND SEQUENCE	STEPS

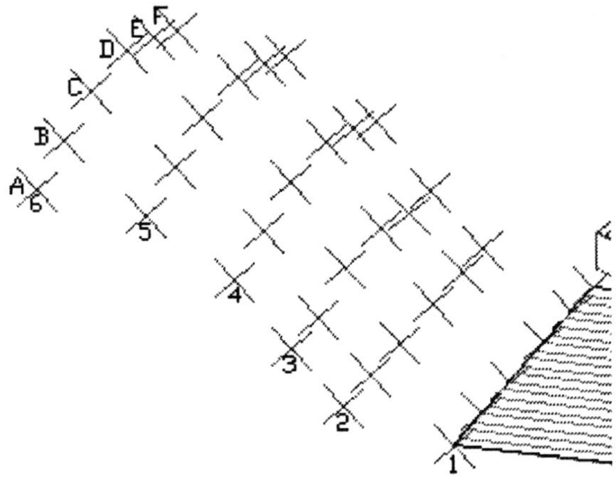

Figure 5.014

	Specify location for vertex (0, 0): **Specify location for vertex (0, 1):** **Specify location for vertex (0, 2):** **Specify location for vertex (0, 3):** **Specify location for vertex (0, 4):** **Specify location for vertex (0, 5):**	3. Now we'll identify the vertices. Select nodes A-1 through A-6 sequentially. (Hint: Use running OSNAPs.)
	Specify location for vertex (1, 0):	4. Follow Step 3 for rows B through F. (Be sure to select the nodes sequentially.)
		5. Freeze the **MARKER** layer and remove the hidden lines. Your drawing looks like Figure 5.015.

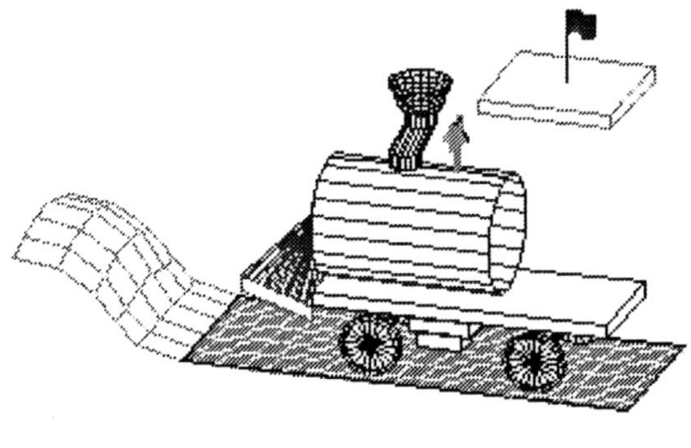

Figure 5.015

	Command: ***qsave***	6. Save the drawing but don't exit.

Note that, once the basic shape of our hill has been defined by using the ***3DMesh*** command, you can add or remove faces using the ***PEdit*** command. We'll look at that in more detail in Lesson 6.

Did you find it tedious having to select a node to define each of the 36 vertices? Imagine what it's like defining a large area! Bear in mind that I provided the nodes as a guide for this exercise.

Normally, you'll identify the location and place the nodes as well (or enter coordinates at the **Specify location** prompt). Do you see now why I described this command as the edge between CAD operating and CAD programming? You'll often create a 3D mesh using a lisp routine or a Visual Basic program in conjunction with coordinates defined in a database or spread sheet. This is really the only way to create large surfaces (like mountains and valleys).

But wait! The ***PFace*** command has even more options!

5.3.2 Creating Meshes with the *PFace* Command

We still haven't created our cab. Why didn't we use the ***3DMesh*** command? Simply because it doesn't lend itself easily to creating surfaces with holes (or, in the case of our cab, with windows).

In some instances, the ***PFace*** command actually appears easier to use than the ***3DMesh*** command. You don't have to enter coordinates sequentially – at least not as you did with the ***3DMesh*** command. In fact, you'll identify all the vertices making up the object and then tell AutoCAD which vertex to use for which face. All you have to remember is which vertex is which. AutoCAD will do the rest. But with the ***PFace*** command, you'll also have the opportunity to change the color and/or layer of each face. This might come in handy for construction walk-throughs.

The command sequence for the ***PFace*** command is:

> **Command:** ***pface***
>
> **Specify location for vertex 1:** ***[this prompt repeats for each vertex; specify the location of all the vertices]***
>
> **Specify location for vertex X or <define faces>:** ***[once you've identified all the vertices, hit enter to define the faces]***
>
> **Face 1, vertex 1:**
>
> **Enter a vertex number or [Color/Layer]:** ***[here you'll tell AutoCAD which vertex to use for each face]***
>
> **Face 1, vertex 2:**
>
> **Enter a vertex number or [Color/Layer] <next face>:** ***[hit enter to identify the vertices on the next face; the prompt will repeat until you've identified each face. Type* C *or* L *to change the color or layer of the face]***

This command sequence is a bit tricky, but it'll be clearer when you complete the next exercise. Trickier still will be the identification of the vertex locations. This may involve some math and a comfortable understanding of the Cartesian Coordinate System. I'll provide the vertex locations for our cab so that you can concentrate on the workings of the ***PFace*** command.

Let's begin. (The cab will have a front and two sides drawn as three separate objects for simplicity. Each will consist of four faces framing an opening for a window.)

Do This: 5.3.2.1 Using *PFace* to Create a Three-Dimensional Object

I. Be sure you're still in the *train.dwg* file in the C:\Steps3D\Lesson05 folder. If not, please open it now.

II. Set the **Cab** layer current and return to the **bell** view.

III. Clear all running OSNAPs.

IV. Set the UCS to World and turn off dynamic input.

V. Follow these steps.

TOOLS	COMMAND SEQUENCE	STEPS
No Button Available	**Command:** ***pface***	1. Enter the ***PFace*** command.
	Specify location for vertex 1: ***7,1,1.5*** **Specify location for vertex 2 or <define faces>:** ***7,5,1.5*** **Specify location for vertex 3 or <define faces>:** ***7,5,5.75*** **Specify location for vertex 4 or <define faces>:** ***7,1,5.75*** **Specify location for vertex 5 or <define faces>:** ***7,1.75,5.75*** **Specify location for vertex 6 or <define faces>:** ***7,1.75,7.75*** **Specify location for vertex 7 or <define faces>:** ***7,1,7.75*** **Specify location for vertex 8 or <define faces>:** ***7,1,8.5*** **Specify location for vertex 9 or <define faces>:** ***7,5,8.5*** **Specify location for vertex 10 or <define faces>:** ***7,5,7.75*** **Specify location for vertex 11 or <define faces>:** ***7,4.25,7.75*** **Specify location for vertex 12 or <define faces>:** ***7,4.25,5.75*** **Specify location for vertex 13 or <define faces>:** ***7,5,5.75***	2. We'll draw the front of the cab first. Enter the coordinates indicated at the prompts.
	Specify location for vertex 14 or <define faces>: ***[enter]***	3. Hit ***enter*** to define the faces.
	Face 1, vertex 1: **Enter a vertex number or [Color/Layer]:** ***1*** **Face 1, vertex 2:** **Enter a vertex number or [Color/Layer] <next face>:** ***2*** **Face 1, vertex 3:** **Enter a vertex number or [Color/Layer] <next face>:** ***3*** **Face 1, vertex 4:** **Enter a vertex number or [Color/Layer] <next face>:** ***4***	4. Identify the bottom face (**Face 1**) as indicated.

Tools	Command Sequence	Steps
	Face 1, vertex 5: **Enter a vertex number or [Color/Layer] <next face>:** *[enter]*	5. Hit ***enter*** to move to the next face (**Face 2**).
	Face 2, vertex 1: **Enter a vertex number or [Color/Layer]:** *4* **Face 2, vertex 2:** **Enter a vertex number or [Color/Layer] <next face>:** *5* **Face 2, vertex 3:** **Enter a vertex number or [Color/Layer] <next face>:** *6* **Face 2, vertex 4:** **Enter a vertex number or [Color/Layer] <next face>:** *7*	6. Identify the vertices for the second face as indicated.
	Face 2, vertex 5: **Enter a vertex number or [Color/Layer] <next face>:** *[enter]*	7. Hit ***enter*** to move to the next face (**Face 3**).
	Face 3, vertex 1: **Enter a vertex number or [Color/Layer]:** *7* **Face 3, vertex 2:** **Enter a vertex number or [Color/Layer] <next face>:** *8* **Face 3, vertex 3:** **Enter a vertex number or [Color/Layer] <next face>:** *9* **Face 3, vertex 4:** **Enter a vertex number or [Color/Layer] <next face>:** *10*	8. Identify the vertices for the third face as indicated.
	Face 3, vertex 5: **Enter a vertex number or [Color/Layer] <next face>:** *[enter]*	9. Hit ***enter*** to move to the next face (**Face 4**).
	Face 4, vertex 1: **Enter a vertex number or [Color/Layer]:** *10* **Face 4, vertex 2:** **Enter a vertex number or [Color/Layer] <next face>:** *11* **Face 4, vertex 3:** **Enter a vertex number or**	10. Identify the vertices for the fourth face as indicated.

Tools	Command Sequence	Steps
	[Color/Layer] <next face>: ***12*** **Face 4, vertex 4:** **Enter a vertex number or [Color/Layer] <next face>:** ***13***	
	Face 4, vertex 5: **Enter a vertex number or [Color/Layer] <next face>:** ***[enter]***	11. Hit ***enter*** to complete this face.
	Face 5, vertex 5: **Enter a vertex number or [Color/Layer] <next face>:** ***[enter]***	12. Hit ***enter*** again to complete the command. Your drawing looks like Figure 5.016.

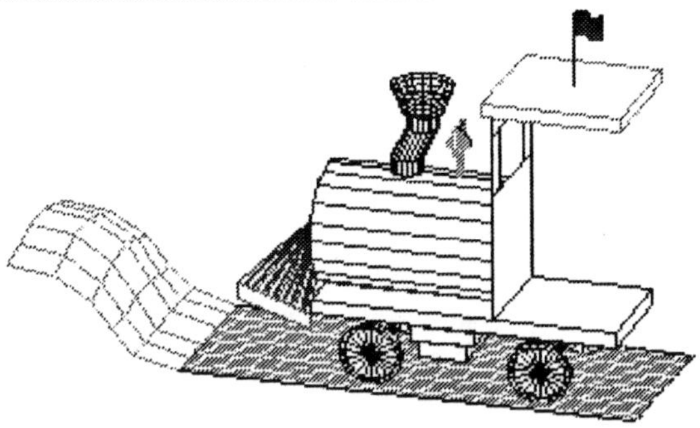

Figure 5.016.

	**Command:** ***qsave***	13. Save this drawing but don't exit. (This exercise continues.)

Now I suppose you're saying to yourself, "I drew it ... I see it ... but I don't know how I did it!" That's an understandable reaction to this exercise. Let me explain what you did.

- Look closely at the new mesh. You'll see four faces (you may need to orbit slightly to see the top one). The 13 coordinates you entered in Step 2 are the coordinates of each corner of the faces (some coordinates are used twice). As you entered the coordinates, AutoCAD identified it with a number (**Specify location for vertex 6**, and so forth).
- In Steps 4, 6, 8, and 10, you told AutoCAD which coordinates to use to identify a specific corner of a specific face (**Face 1, vertex 2**, and so forth).

So you see, the complexity lies not in the command itself but in the amount of data required (and in calculating that data).

Let's continue.

Tools	Command Sequence	Steps
		14. Repeat Steps 1 through 13 using the following coordinates to draw the first side wall.

Tools	Command Sequence	Steps

Number	Coordinate	Number	Coordinate	Number	Coordinate
1	11,1,1.5	6	10.25,1,7.75	11	7.75,1,7.75
2	7,1,1.5	7	11,1,7.75	12	7.75,1,5.75
3	7,1,5.75	8	11,1,8.5	13	7,1,5.75
4	11,1,5.75	9	7,1,8.5		
5	10.25,1,5.75	10	7,1,7.75		

Tools	Command Sequence	Steps
	**Command:** ***co***	15. Copy the new wall to the other side of the cab.
	Command: ***vpoint*** **Current view direction: VIEWDIR=7.0000,-22.0000,9.0000** **Specify a view point or [Rotate] <display compass and tripod>: -*30,21,6***	16. Change the viewpoint as indicated, set the **Shademode** to **Gouraud**, and freeze layer **0**. Your drawing looks like Figure 5.017.

Figure 5.017 (It looks really cool in color!)

Tools	Command Sequence	Steps
	**Command:** ***qsave***	17. Save the drawing.

Be proud of yourself! You've accomplished AutoCAD's 3D Surface commands (*not* a small feat)!

5.4 Extra Steps

Try to incorporate all you've learned thus far into the train drawing.

- Adjust the views so you can see it from all sides.
- Freeze the **ground** layer and use the Continuous Orbiter to make the train revolve about the screen.
- Set up the train drawing for plotting:
    - Show it in three views and an isometric.
    - Dimension it.
    - Put it on a title block with your school/business name.

5.5 What Have We Learned?

Items covered in this lesson include:

- *Controlling the number of surfaces on a surface model*
- *Basic and Advanced Surface Modeling Commands*
  - ***Surftab1***
  - ***Surftab2***
  - ***Rulesurf***
  - ***Tabsurf***
  - ***Revsurf***
  - ***Edgesurf***
  - ***3DMesh***
  - ***PFace***

Congratulations! You've finished AutoCAD's Wireframe and Surface Modeling commands. You've really come a long way in five lessons!

The decisions you'll now face concern which procedure or method you'll need to create the objects that you want to create. The best help you can get for that is *practice*! So work through the problems at the end of this lesson until you're comfortable with your new abilities.

When you've finished with the exercises, go on to Lesson 6. There we'll discuss the three-dimensional aspects of several editing tools you already know … and some new ones that are specific to three-dimensional drawings. Then, at least where Wireframe and Surface Modeling is concerned, your training will be complete. After that, we'll start a whole new ballgame – we'll learn how to create solid models!

So do the problems, pat yourself on the back for having come this far, and then move onward … ever onward!

5.6 Exercises

1. Create the Window Guide drawing in Figure 5.018. The following information will help.
    1.1. Use a title block of your choice.
    1.2. Use the Times New Roman font – 3/16" and 1/8".
    1.3. Adjust the dimstyle as needed.
    1.4. Create layers as needed.
    1.5. The object is a tabulated surface model created from a polyline shape.
    1.6. Use either a region or a 3D face to close the ends.
    1.7. Use **Flat+Edges** Shademode for the isometric figure.
    1.8. Save the drawing as *MyWinGuide* in the C:\Steps3D\Lesson05 folder.

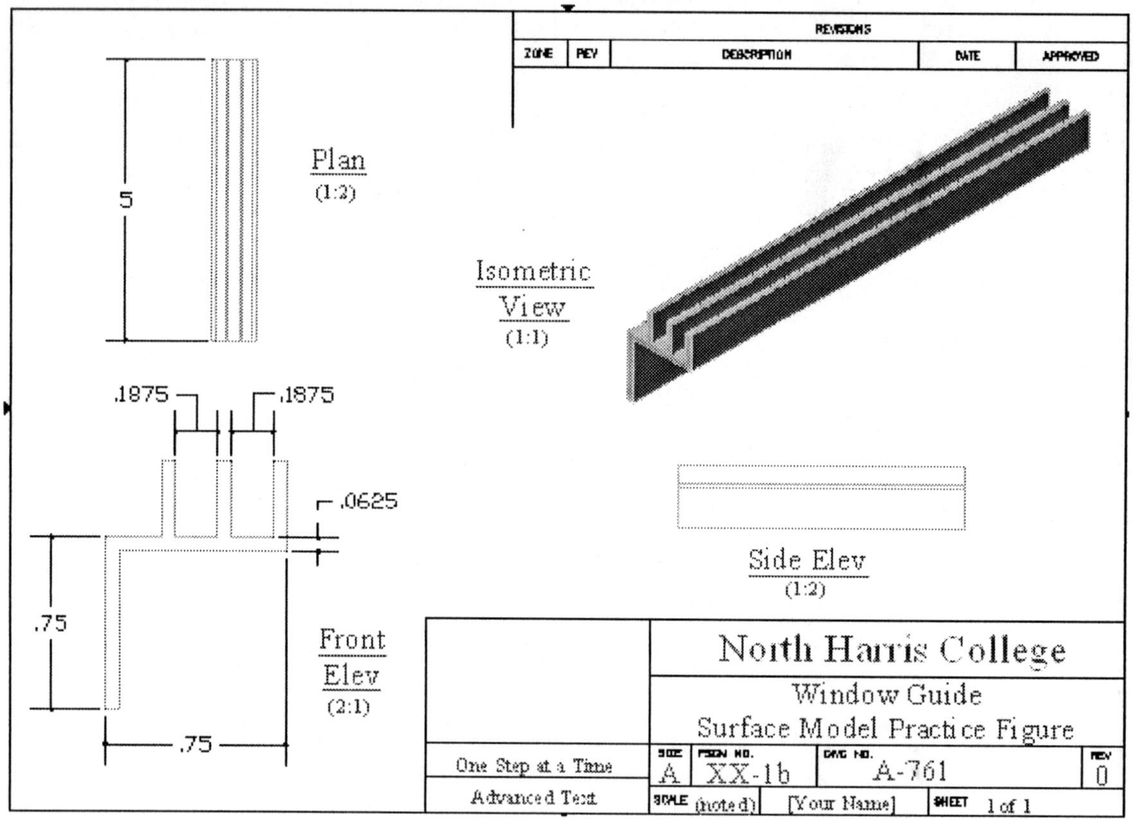

Figure 5.018

2. Create the Light Fixture drawing in Figure 5.019. The following information will help.
   2.1. Use a title block of your choice.
   2.2. Use the Times New Roman font – 3/16" and 1/8".
   2.3. Adjust the dimstyle as needed.
   2.4. Create layers as needed.
   2.5. I used a **Surftab1** setting of **32** and a **Surftab2** setting of **6**.
   2.6. The object is a simple revolved surface model. I created the shape on one layer and the revolved surface model on another. I did the cross section by adjusting the viewport and freezing unnecessary layers.
   2.7. Save the drawing as *MyFixture* in the C:\Steps3D\Lesson05 folder.

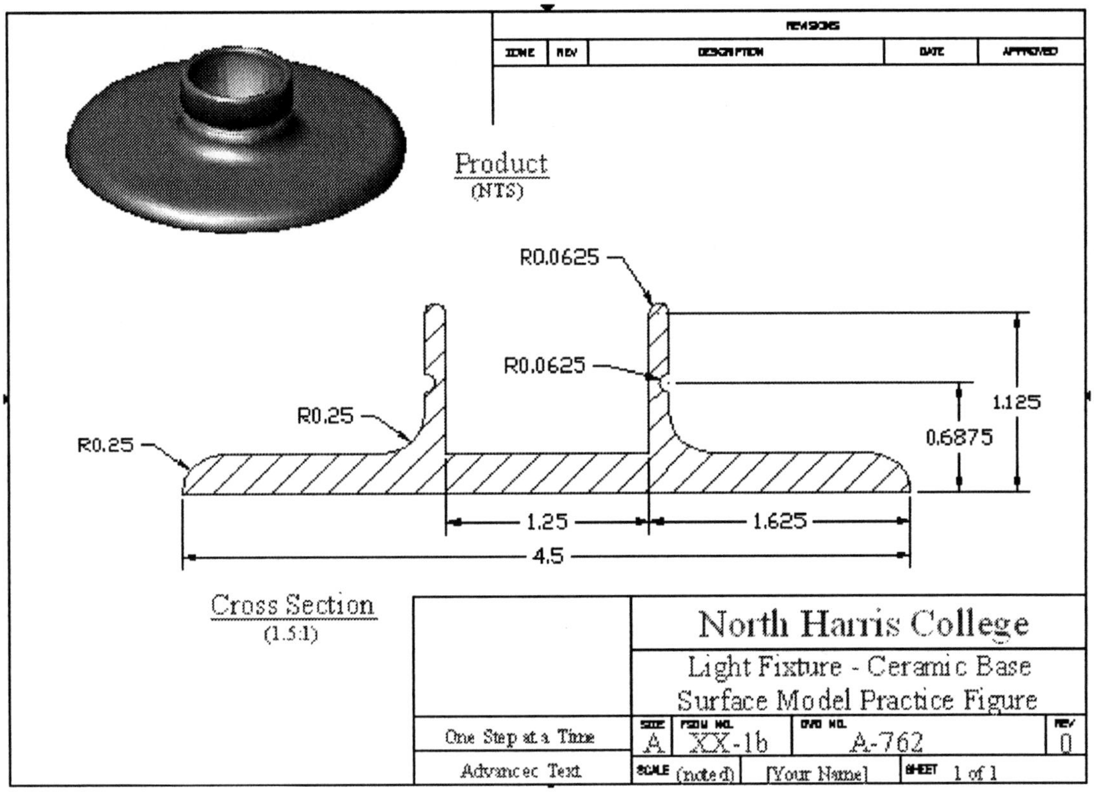

Figure 5.019

3. Create the Alan Wrench drawing in Figure 5.020. The following information will help.
   3.1. Use a title block of your choice.
   3.2. Use the Times New Roman font – 3/16" and 1/8".
   3.3. Adjust the dimstyle as needed.
   3.4. Create layers as needed.
   3.5. I used the default settings for both surftabs.
   3.6. This object is made up of two tabulated surfaces and a revolved surface. I used six-sided polygons as my basic shapes.
   3.7. Save the drawing as *MyWrench* in the C:\Steps3D\Lesson05 folder.

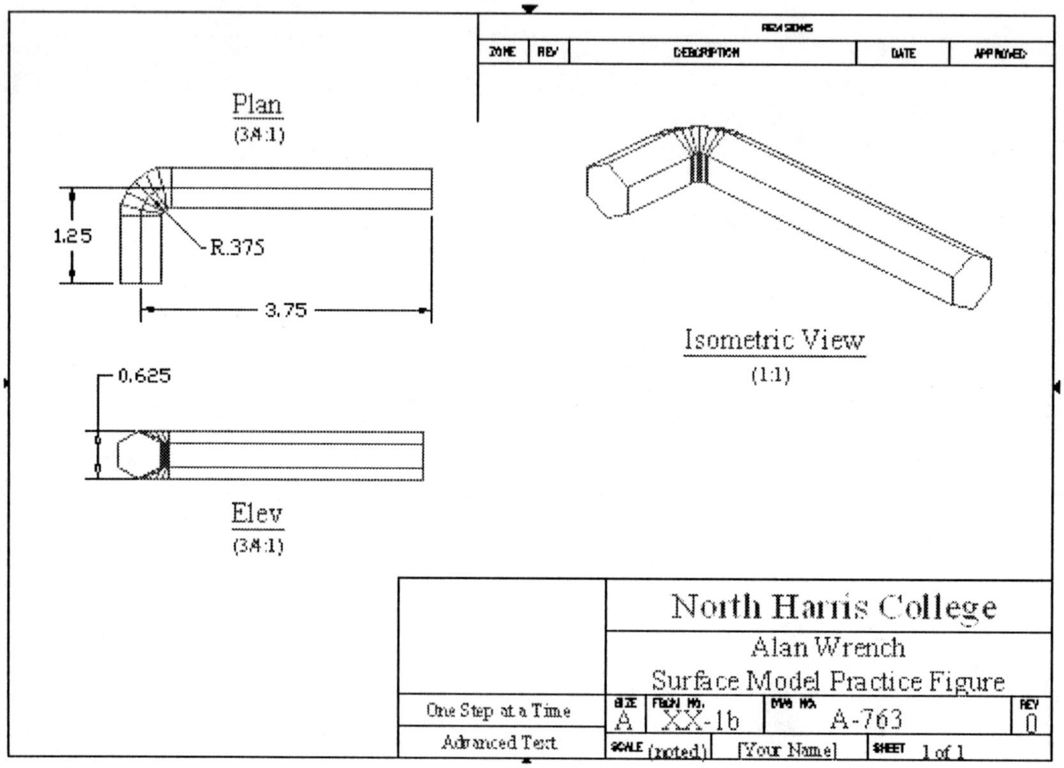

Figure 5.020

4. Create the caster drawing in Figure 5.021. The following information will help.
   - 4.1. Use the Times New Roman font – 3/16" and 1/8".
   - 4.2. Adjust the dimstyle as needed.
   - 4.3. Create layers as needed (you may need more than you think).
   - 4.4. I used 16 as my setting for both surftabs.
   - 4.5. The spindle is a revolved surface.
   - 4.6. The upper place is an edged surface.
   - 4.7. The frame is a single polygon mesh (use the PFace command).
   - 4.8. The axle is a revolved surface.
   - 4.9. The wheel is a revolved surface with a hole in it for the axle.
   - 4.10. The ball bearings are 1/8" diameter spheres.
   - 4.11. The Product view uses the Gouraud Shademode.
   - 4.12. To create the sections, use the same technique you used in Exercise 2.
   - 4.13. Save the drawing as *MyCaster* in the C:\Steps3D\Lesson05 folder.

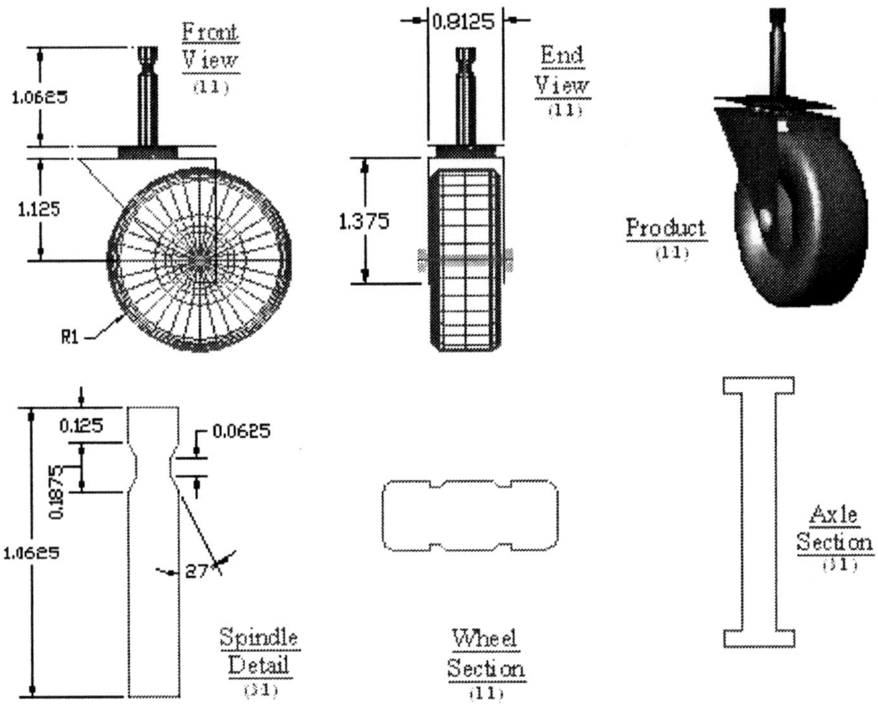

Figure 5.021

5. Create the Toothpaste Tube drawing in Figure 5.022. The following information will help.
   5.1. The main tube is 7¼" long x 2" wide on the flat end. The round end has an additional ¼" bubble.
   5.2. The round end is 1¼" diameter.
   5.3. The cap is ½" long, 5/8" diameter at the base and ½" diameter at the top.
   5.4. Use ***Edgesurf*** to create the body of the tube.
   5.5. Use ***Revsurf*** to create the cap and the bubble-end of the tube.
   5.6. Create layers as needed.
   5.7. Save the drawing as *MyToothpaste* in the C:\Steps3D\Lesson05 folder.

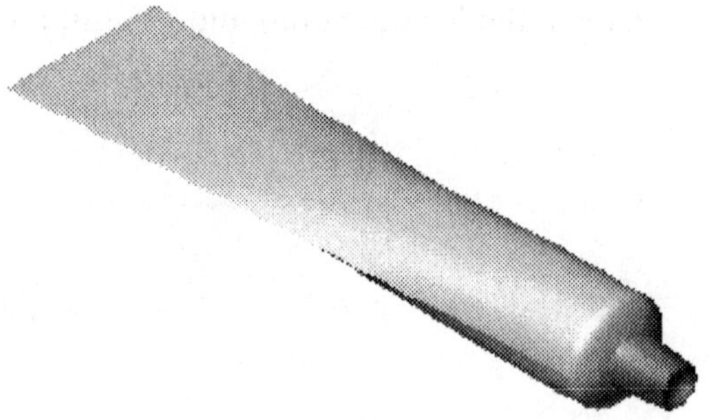

Figure 5.022

6. Create the Lighter drawing in Figure 5.023. The following information will help.
   6.1. The main bottle is 2¼" tall and is based on an ellipse that is 1" x ½".
   6.2. The top is based on half the bottle's ellipse and is ½" tall.
   6.3. The button is also based on half the bottle's ellipse, is 1/16" thick, and is at an angle of 15°.
   6.4. The striker wheel is 5/16" wide and ¼" diameter.
   6.5. Use ***Revsurf*** to create the striker wheel.
   6.6. Use ***Edgesurf*** to create the bottle and the top.
   6.7. Use regions where needed.
   6.8. Save the drawing as *MyLighter* in the C:\Steps3D\Lesson05 folder.

Figure 5.023

7. Create the Remote drawing in Figure 5.024. The following information will help.
   7.1. The main face is 5¾" long x 2" wide.
   7.2. There's an additional ½" molded arc on the sides and ends.
   7.3. The thickness of the instrument is ½"
   7.4. Round buttons are 3/8" diameter and ¼" diameter domes.
   7.5. Ellipse buttons are ½" x ¼".
   7.6. Other buttons are either pyramids or tabsurfed polylines with regions closing the tops. These are 1/8" high.
   7.7. Use the Times New Roman font at a 1/8" text height.
   7.8. Save the drawing as *MyRemote* in the C:\Steps3D\Lesson05 folder.

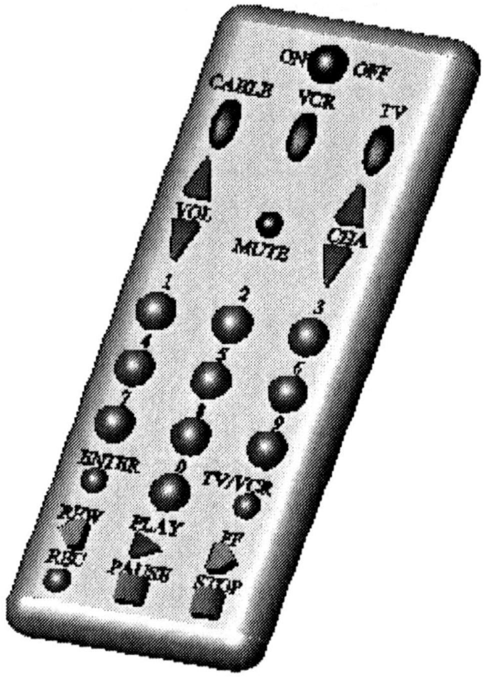

Figure 5.024

8. Create the racer drawing in Figure 5.026. The ½" grid in Figure 5.025 will help. Save the drawing as *MyRacer* in the C:\Steps3D\Lesson05 folder.

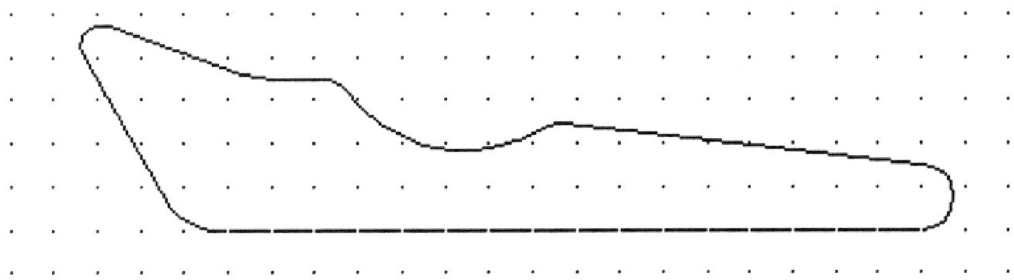

Figure 5.025

Figure 5.026

5.7	**For Web-Based Review Questions, visit: http://www.uneedcad.com/2006/Files/06R5-3D.pdf**

Lesson

6

Following this lesson, you will:

- ✓ *Know how to use AutoCAD's basic editing tools in Z-Space*
  - ***PEdit***
  - ***Properties***
  - ***Grips***
  - ***Trim and Extend***
  - ***Align***
- ✓ *Know how to use basic Z-Space-Specific editing tools*
  - ***Rotate3d***
  - ***Mirror3D***
  - ***3DArray***

Z-Space Editing

It was just about here in our basic text that we began to look at editing tools. So it's appropriate, I suppose, that we stop now to look at how those editing tools work both on three-dimensional objects and on two-dimensional objects drawn in Z-Space. You'll find little difference in the way some tools and procedures work, but the differences in others may unsettle you. You'll find still others to be brand new (although strangely familiar).

Regardless of their differences or newness, however, you'll find each tool we discuss in this lesson to be invaluable in your three-dimensional efforts. As they did in the two-dimensional world, editing tools will enhance your speed and drawing ability in Z-Space.

To keep it simple (KISS), I'll divide the tools into two categories of study – Familiar (two-dimensional tools with which you're already familiar), and New (three-dimensional tools).

Let's start with the Familiar.

6.1 Three-Dimensional Uses for Familiar (Two-Dimensional) Tools

As we studied editing tools in our basic text, we occasionally came across a prompt that I said would appear in the 3D text. AutoCAD designed these prompts to allow you to continue using some of the more common (and useful) tools when you made the transition into Z-Space. The tools – ***Trim***, ***Extend***, and ***Align*** – weren't difficult to learn in the basic book. And now that you're familiar with them, covering their three-dimensional functions will be a snap!

Once we've looked at those, we'll look at the Properties palette and some of the things it can do for a surface model.

Lastly, we'll take a new look at the ***PEdit*** command. You might be surprised (if not thrilled) at what you find!

Let's start at the beginning.

6.1.1 Trimming and Extending in Z-Space

Trim and ***Extend*** in Z-Space begin very much like ***Trim*** and ***Extend*** in two-dimensional space. In fact, the commands are the same, so all you must learn is the option required to control what gets trimmed/extended in a three-dimensional view.

Remember how the **Edgemode** system variable controlled the **Edge** option in both ***Trim*** and ***Extend*** commands? It still holds true in Z-Space, but there's an additional system variable to consider now – **Projmode**. Like **Edgemode**, **Projmode** affects both the ***Trim*** and ***Extend*** commands. But where **Edgemode** controls your ability to trim/extend to an imaginary extension of the selected cutting edge/boundary, **Projmode** controls how the ***Trim*** and ***Extend*** commands behave in three-dimensional space.

There are two ways to set the **Projmode** system variable – by accessing the **Project** option at the [**Project/Edge/Undo**] prompt of either command, or by entering ***Projmode*** at the command prompt. The command prompt requires that you enter a number code for the option you wish to use; the **Project** option of the ***Trim/Extend*** command presents the available settings. The number codes and their corresponding settings follow.

CODE	SETTING	FUNCTION
0	None	This is the *True 3D* setting. It requires that both the cutting edge/boundary and the object to trim/extend be in the same plane. That is, they must actually intersect or, using the **Edgemode** system variable, intersect at an imaginary

		extension.
1	UCS	(Default Setting) This setting projects the cutting edge/boundary and object to trim/extend onto the XY plane of the current UCS and then performs the task as though the objects exist in two-dimensional space.
2	View	This setting causes AutoCAD to project the cutting edge/boundary and object to trim/extend onto the current view as though your monitor's screen is the XY plane. It then performs the task as though the objects exist in two-dimensional space.

Let's see how these settings work in a couple exercises.

Do This: 6.1.1.1	Trimming in Z-Space

I. Open the *trim.dwg* file in the C:\Steps3D\Lesson06 folder. The drawing looks like Figure 6.001.

II. Be sure the **Edgemode** system variable is set to 1.

III. Follow these steps.

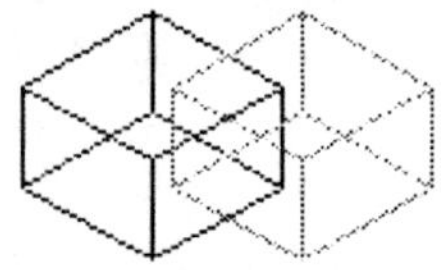

Figure 6.001

TOOLS	COMMAND SEQUENCE	STEPS
No Button Available	**Command: *Projmode*** **Enter new value for PROJMODE <1>: *0***	1. Let's begin by setting the **Projmode** system variable to zero (the *True 3D* setting).
	Command: *tr* **Current settings: Projection=None, Edge=Extend** **Select cutting edges ...** **Select objects or <select all>:** **Select objects: *[enter]***	2. Now begin the ***Trim*** command and select the line indicated in Figure 6.002 as your cutting edge.

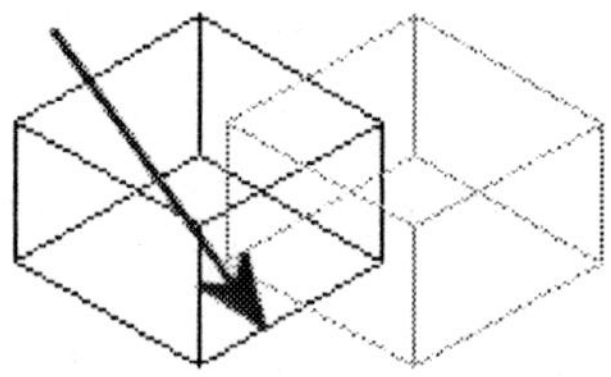

Figure 6.002

	Select object to trim or shift-select to extend or [Fence/ Crossing/Project/Edge/ eRase/Undo]:	3. Now select the lines indicated in Figure 6.003 to trim. Select the bottom line first. Notice that the lower line trims, but the upper line doesn't. With the **Projmode** system variable set to zero, the lines must be on the same plane (as I explained in the chart). The upper line doesn't intersect the cutting edge, so it didn't trim.

TOOLS	COMMAND SEQUENCE	STEPS

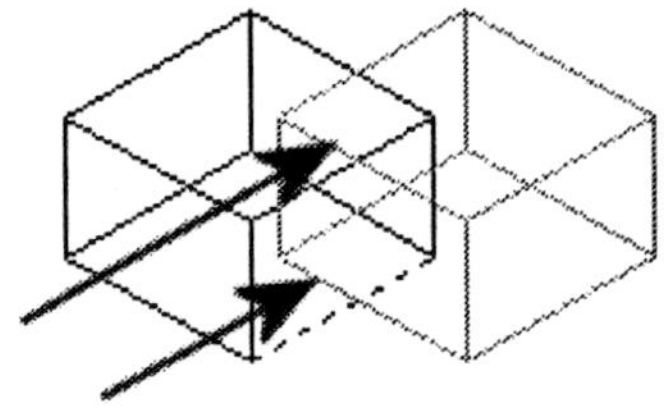

Figure 6.003

TOOLS	COMMAND SEQUENCE	STEPS
	**Select object to trim or shift-select to extend or [Fence/ Crossing/Project/Edge/ eRase/Undo]:** ***p*** **Enter a projection option [None/Ucs/View] <None>:** ***u***	4. Without leaving the command, change **Projmode** to the **UCS** setting as shown.
	Select object to trim or shift-select to extend or [Fence/ Crossing/Project/Edge/ eRase/Undo]:	5. Now trim the line that wouldn't trim previously. It trims now because AutoCAD projects the cutting edge and object to trim against the XY plane of the current UCS (as if both lines were drawn in 2D space).
	Select object to trim or shift-select to extend or [Fence/ Crossing/Project/Edge/ eRase/Undo]: ***[enter]***	6. Complete the command. Your drawing looks like Figure 6.004.

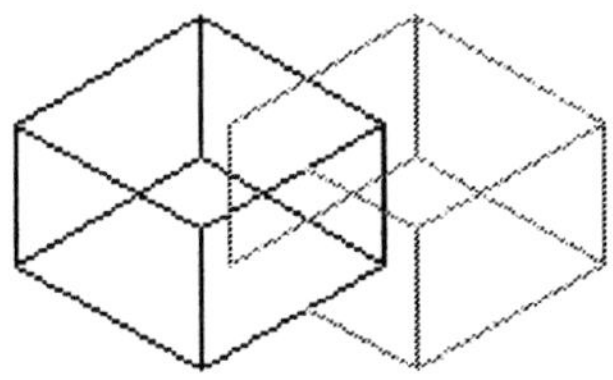

Figure 6.004

TOOLS	COMMAND SEQUENCE	STEPS
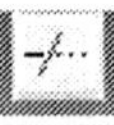	**Command:** ***[enter]***	7. Repeat Step 2.
	Select object to trim or shift-select to extend or [Fence/ Crossing/Project/Edge/ eRase/Undo]:	8. Select the line indicated in Figure 6.005 to trim. Notice that the line won't trim. The **Projmode** system variable setting of **1** (**UCS**) means that the lines must intersect in a two-dimensional projection of the current UCS. The line doesn't intersect the cutting edge, so it didn't trim.

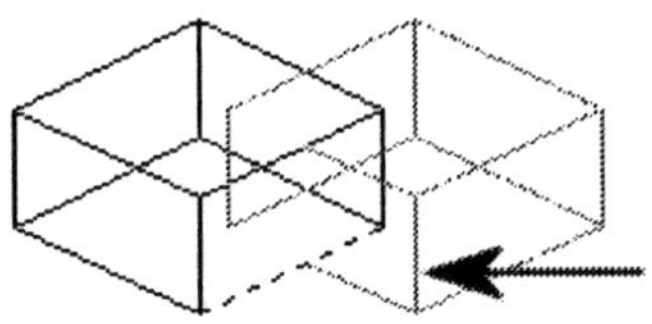

Figure 6.005

Tools	Command Sequence	Steps
	**Select object to trim or shift-select to extend or [Fence/ Crossing/Project/Edge/ eRase/Undo]:** ***p*** **Enter a projection option [None/Ucs/View] <Ucs>:** ***v***	9. Change the **Projmode** to the **View** setting as indicated. (Note: you can use cursor or dynamic input menus if you prefer.)
	Select object to trim or shift-select to extend or [Fence/ Crossing/Project/Edge/ eRase/Undo]:	10. Now trim the line that wouldn't trim previously. It trims now for two reasons: (1) AutoCAD has projected the **cutting edge** and **object to trim** against your screen (the current view) as if your screen defined the XY plane, and (2) AutoCAD has extended the cutting edge according to the **Edgemode** setting.
	Select object to trim or shift-select to extend or [Fence/ Crossing/Project/Edge/ eRase/Undo]: ***[enter]***	11. Complete the command. Your drawing looks like Figure 6.006.

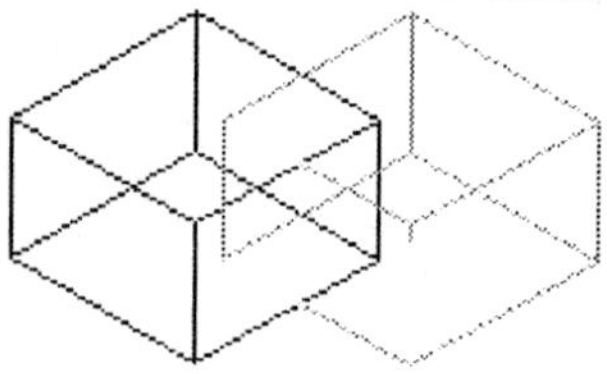

Figure 6.006

	Command: ***quit***	12. Exit the drawing without saving.

Now let's look at the ***Extend*** command.

Do This: 6.1.1.2	Extending in Z-Space

I. Open the *ext.dwg* file in the C:\Steps3D\Lesson06 folder. The drawing looks like Figure 6.007.

II. Be sure the **Edgemode** system variable is set to **1** and the **Projmode** system variable is set to **0**.

III. Follow these steps.

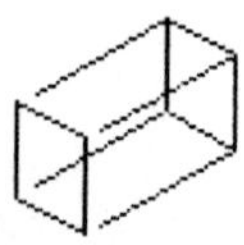

Figure 6.007

Tools	Command Sequence	Steps
	**Command:** ***ex*** **Current settings: Projection=None, Edge=Extend** **Select boundary edges ...** **Select objects or <select**	1. Begin the ***Extend*** command and select the line indicated in Figure 6.008 as your boundary edge.

Tools	Command Sequence	Steps
	all>: **Select objects:** ***[enter]***	

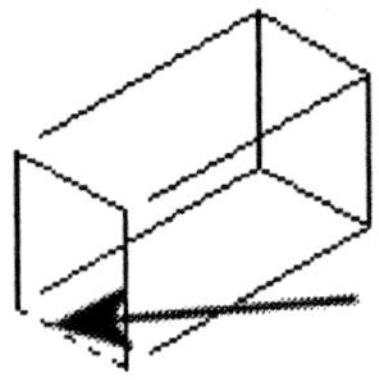

Figure 6.008

Tools	Command Sequence	Steps
	Select object to extend or shift-select to trim or [Fence/Crossing/Project/ Edge/Undo]:	2. Select the four open-ended lines to extend. Select the bottom lines first. Notice that the lower lines extend but not the upper lines. As in the ***Trim*** command, with the **Projmode** set to **0**, the lines must be in the same XY plane. The upper lines don't share the boundary edge's plane, so they didn't extend.
	Select object to extend or shift-select to trim or [Fence/Crossing/Project/ Edge/Undo]: ***p*** **Enter a projection option [None/Ucs/View] <None>:** ***u***	3. Change the **Projmode** to the **UCS** setting.
	Select object to extend or shift-select to trim or [Fence/Crossing/Project/ Edge/Undo]:	4. Now extend the lines that wouldn't extend previously. They extend now because AutoCAD has projected them against the XY plane of the current UCS.
	Select object to extend or shift-select to trim or [Fence/Crossing/Project/ Edge/Undo]: ***[enter]***	5. Complete the command. Your drawing looks like Figure 6.009.

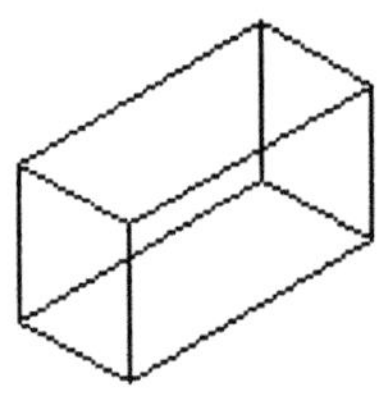

Figure 6.009

Tools	Command Sequence	Steps
	**Command:** ***ex***	6. Repeat Step 1, but select the boundary edge indicated in Figure 6.010.

Tools	Command Sequence	Steps

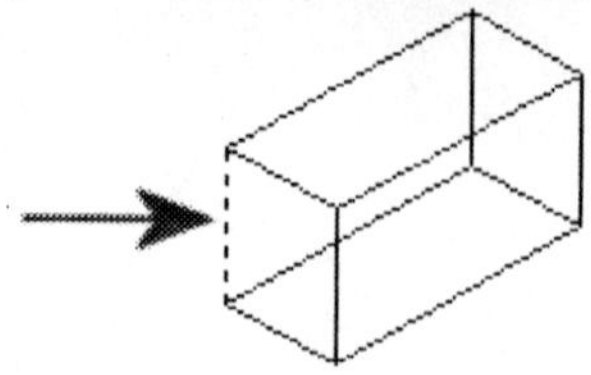

Figure 6.010

Tools	Command Sequence	Steps
	Select object to extend or shift-select to trim or [Fence/Crossing/Project/ Edge/Undo]:	7. Select the line indicated in Figure 6.011 to extend. Notice that the line won't extend. With **Projmode** set to **1**, the lines must intersect in a two-dimensional projection of the current UCS. The lines don't, so the selected **object to extend** didn't extend.

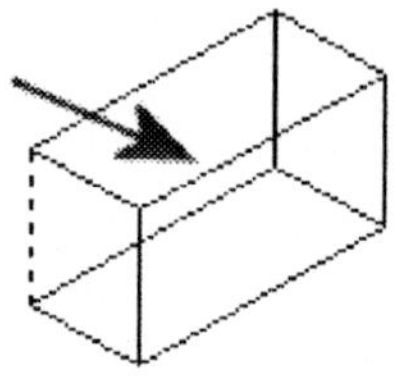

Figure 6.011

Tools	Command Sequence	Steps
	Select object to extend or shift-select to trim or [Fence/Crossing/Project/ Edge/Undo]: ***p*** **Enter a projection option [None/Ucs/View] <Ucs>:** ***v***	8. Change the **Projmode** to the **View** settings.
	Select object to extend or shift-select to trim or [Fence/Crossing/Project/ Edge/Undo]:	9. Now extend the line that wouldn't extend previously. As with the ***Trim*** command, it extends now because AutoCAD is looking at the lines projected against your screen (the current view).
	Select object to extend or shift-select to trim or [Fence/Crossing/Project/ Edge/Undo]: ***[enter]***	10. Complete the command. Your drawing looks like Figure 6.012.

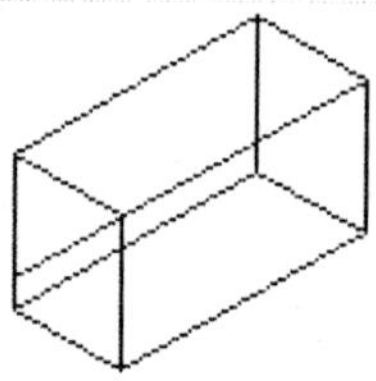

Figure 6.012

Tools	Command Sequence	Steps
	Command: ***quit***	11. Exit the drawing without saving.

You've seen that, although still fairly simple to use, the ***Trim*** and ***Extend*** commands have some different options with which you must become familiar if you're to use them to full advantage in Z-Space.

But there are some additional things that I should highlight before continuing.

- These commands work on the same objects for which you used them in 2D space – you can't trim/extend a 3D face, 3D mesh, region or solid.
- You can't use a 3D face, 3D mesh, or solid as a cutting edge or boundary edge when trimming/extending.
- You *can*, however, use a region as a cutting edge or boundary edge when trimming or extending.

Let's look at how another 2D command works with three-dimensional objects. Let's look at the ***Align*** command.

6.1.2 Aligning Three-Dimensional Objects

When you aligned objects in the basic text, you used two source points and two destination points. The main difference between that two-dimensional exercise and aligning three-dimensional objects is that AutoCAD requires three points of alignment for the three-dimensional object.

Another difference lies in your ability to scale the aligned objects as you did in the 2D exercise. You won't have that option with three-dimensional objects. But you can always use the ***Scale*** command once the objects are aligned.

Let's perform a three-dimensional alignment using AutoCAD's ***Align*** command.

Do This: 6.1.2.1 Aligning Three-Dimensional Objects

I. Open the *align.dwg* file in the C:\Steps3D\Lesson06 folder. The drawing looks like Figure 6.013. We'll align the eastern face of the ridged pyramid with the top of the box.

II. Follow these steps.

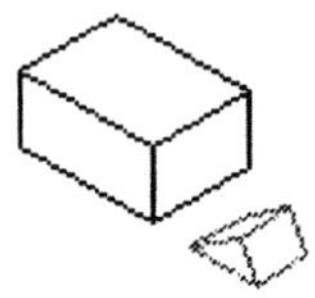

Figure 6.013

Tools	Command Sequence	Steps
No Button Available	**Command: *al***	1. Enter the ***Align*** command.
	Select objects: **Select objects: *[enter]***	2. Select the wedge.
	Specify first source point: *[Point 1a]* **Specify first destination point: *Point 1b]*** **Specify second source point: *[Point 2a]*** **Specify second destination point: *[Point 2b]*** **Specify third source point or**	3. (Refer to Figure 6.014.) Select the alignment points as indicated. Your drawing looks like Figure 6.015.

Tools	**Command Sequence**	**Steps**
	<continue>: ***[Point 3a]*** **Specify third destination point:** ***[Point 3b]***	

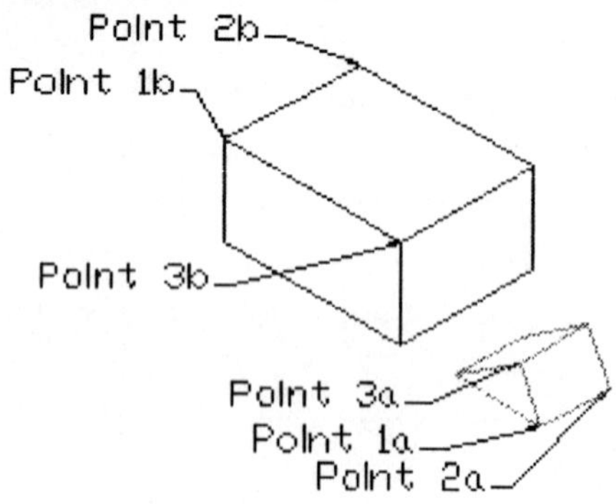

Figure 6.014

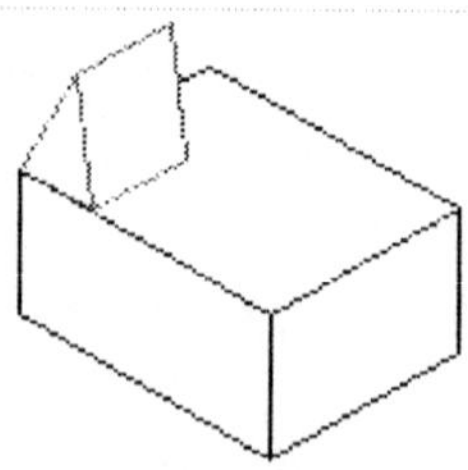

Figure 6.015

	Command: ***quit***	4. Exit the drawing without saving it.

I wish they were all that easy!

The really bright side to the ***Align*** command is that it doesn't care what types of objects are being aligned. You can align the objects you aligned in 2D space or you can align 3D faces, 3D meshes, regions, or solids!

6.1.3 Three-Dimensional Object Properties

When you began creating surface models, you opened the door to a wide variety of new objects – 3D faces, 3D meshes, regions, and solids – that you may need to modify at one time or another.

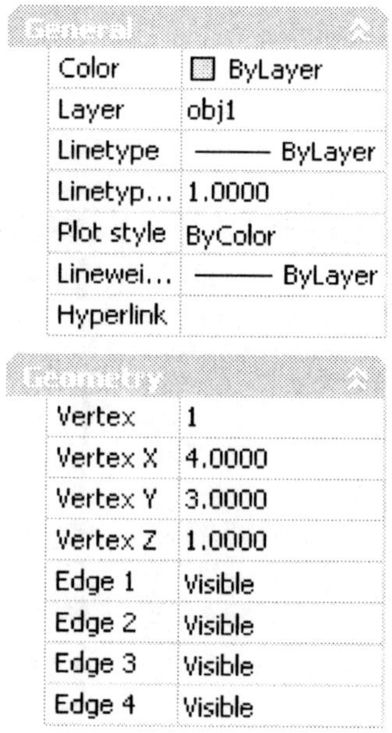

Figure 6.016

You can immediately halve your list, however, because regions and three-dimensional solids have no editable properties! Additionally, 3D faces have very few (and very elementary) editable properties, so you can relax in your approach to three-dimensional object properties.

On the other hand, 3D meshes can make you long for the simplicity of a two-dimensional world!

We'll spend some time with the different approaches to editing 3D meshes later, but for now, let's begin by examining the 3D face properties AutoCAD makes available in the Properties palette. Refer to Figure 6.016 as we consider the possibilities.

- The **General** section contains property information with which you're already familiar from your basic studies.
- The **Geometry** section contains editable information for the specific 3D face that you've selected.
    - The **Vertex** row identifies a specific vertex (corner) on the 3D face. AutoCAD identifies the vertex on the drawing with an "X" – similar to the way it identifies a vertex when

editing with the ***PEdit*** command's **Edit Vertex** option. When you pick in the **Vertex** row, a set of directional arrows appears in the value column. Use these to scroll through the vertices or type in the number of the vertex you wish to modify.

- The **Vertex X**, **Vertex Y**, and **Vertex Z** rows present the corresponding coordinate value for that vertex. The coordinate values reflect the current UCS. When you pick one of these rows, AutoCAD presents a **Pick Point** button. You may use this button to pick a point on the screen or you can enter a typed number in the value column to change the location of the vertex.
- The **Edge** rows provide toggles that you can use to change the visibility of specific 3D face edges.

Let's see how the Properties palette interacts with three-dimensional objects.

Do This: 6.1.3.1	Using the Properties Palette with Three-Dimensional Objects

I. Open the *3Dfaces.dwg* file in the C:\Steps3D\Lesson06 folder. The drawing looks like Figure 6.017. (We made the open box using 3D faces; the rectangle is a region, and the other box is a solid.)

II. Follow these steps.

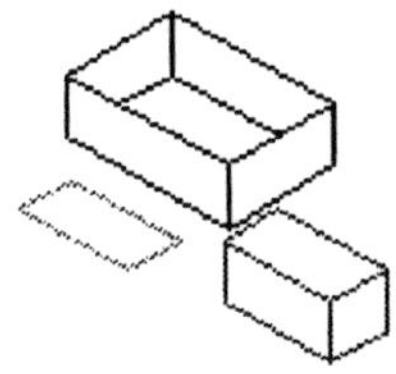

Figure 6.017

TOOLS	COMMAND SEQUENCE	STEPS
	Command: ***props***	1. Open the Properties palette. Move it to one side and adjust the display so you can see the objects.
		2. Select the region (the flat rectangle). Notice the **Geometry** section of the palette (Figure 6.018. It displays two properties for reference, but you can't access either for modification.

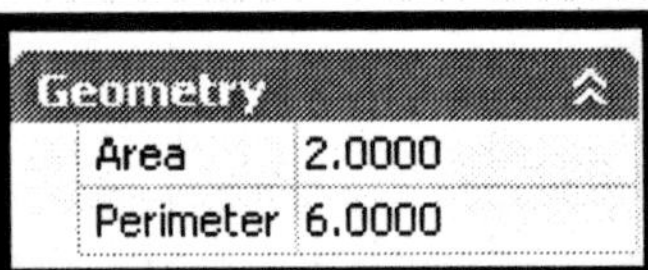

Figure 6.018

TOOLS	COMMAND SEQUENCE	STEPS
Esc		3. Deselect the region.
		4. Repeat Step 2, but this time select the solid box. Notice that AutoCAD makes no geometry available for editing or reference.
Esc		5. Repeat Step 3.

TOOLS	COMMAND SEQUENCE	STEPS
		6. Select the front 3D face on the open box. Notice the Geometry section of the Properties palette (see Figure 6.016).
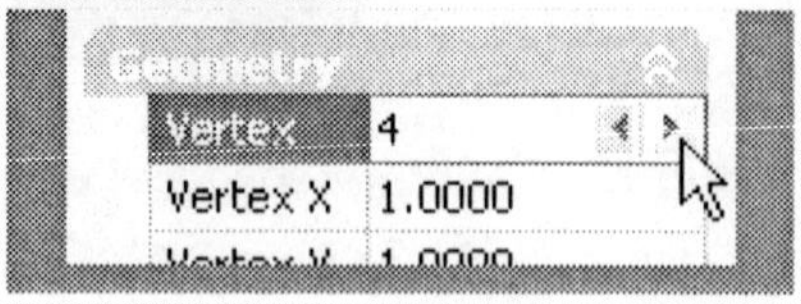		7. Let's perform some modifications. Change the Vertex that you'll edit to ***4***. Notice an "X" identifies the vertex on the 3D face (Figure 6.019).

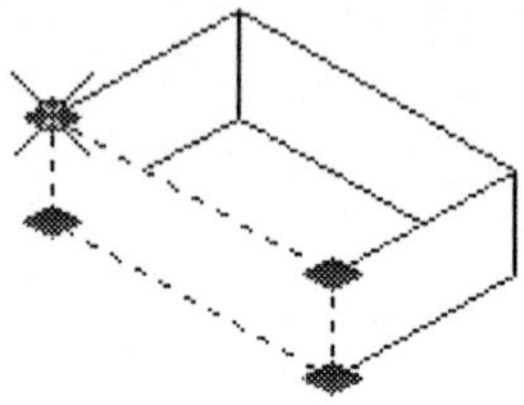

Figure 6.019

TOOLS	COMMAND SEQUENCE	STEPS
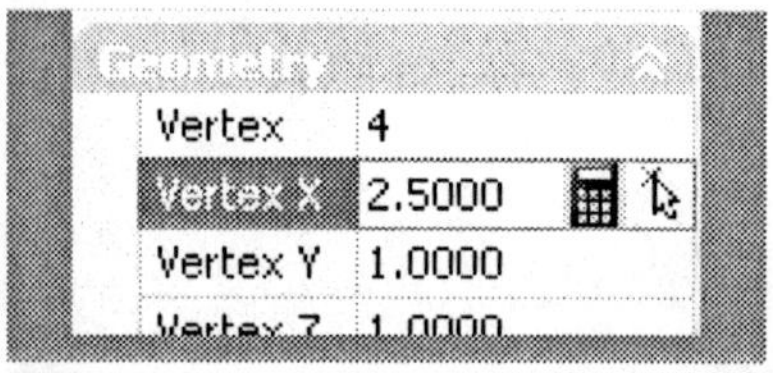		8. Change the value of **Vertex X** to **2.5**. Notice the change on the 3D face (Figure 6.020). (Note: You must hit ***enter*** after changing the value for AutoCAD to accept the change.)
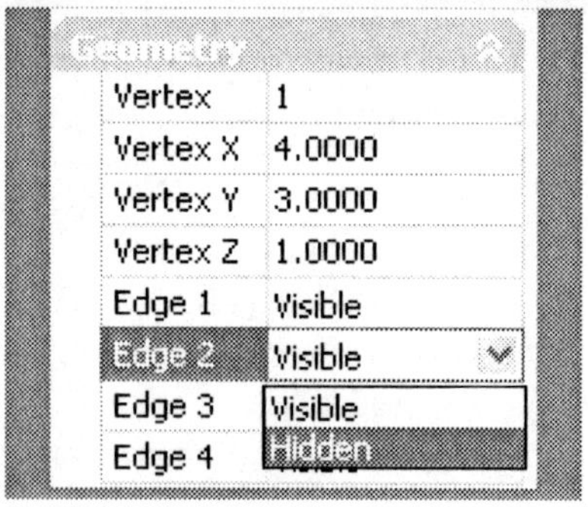		9. Now make **Edge 2** of the back face invisible, as shown. Notice the change in the 3D face (Figure 6.020).

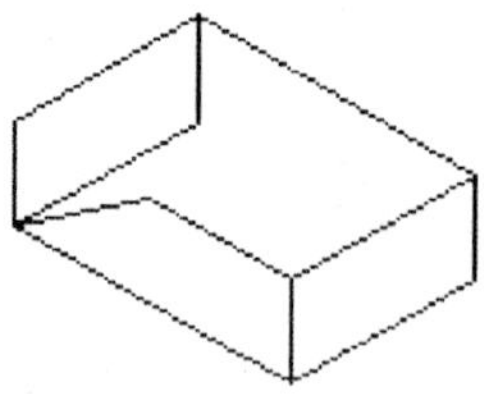

Figure 6.020

TOOLS	COMMAND SEQUENCE	STEPS
		10. Exit the drawing without saving the changes.

As you can see, although you can't modify a region or solid, modifying a 3D face with the Properties palette is quite easy.

6.1.4	Modifying a 3D Mesh

3D mesh modification is considerably more complicated than 3D face modification primarily because it's a more complex object. When you modify a 3D face, you have only the one object with which to work. That object has three or four vertices and edges. You have nothing else with which to work that might complicate matters.

A 3D mesh, on the other hand, has any number of vertices, several mesh values, and even a polyline fit/smooth value.

Luckily, AutoCAD has provided three methods for modifying 3D meshes – the ***PEdit*** command, the Properties palette, and grips. Each has its place, and knowing when to use each procedure will go a long way toward preserving your sanity when facing a 3D mesh with dozens (or hundreds) or vertices.

Let's start with the ***PEdit*** command.

6.1.4.1	Modifying a 3D Mesh

Remember how much fun you had with the ***PEdit*** command in the basic text? I told you then that you'd probably never need the **Edit vertex** tier of options – at least not in the two-dimensional world. Well, you're not in Kansas anymore. But this is where you get the payoff for struggling through the exercise that covered the **Edit vertex** options.

The ***PEdit*** command looks slightly different when you select a 3D mesh instead of a polyline. This is the sequence:

Enter an option [Edit vertex/Smooth surface/Desmooth/Mclose/Nclose/Undo]:

- **Mclose/Mopen** and **Nclose/Nopen** serve the same function as the **Open/Close** option of the 2D ***PEdit*** command. But on the 3D mesh, AutoCAD draws a closing line between first and last points of the M-columns or N-rows.
- As always, **Undo** undoes the last modification. (Remember; don't confuse the **Undo** option with the ***Undo*** command, which undoes the last command.)
- **Smooth surface** creates a curved shape from the 3D mesh. This handy tool is really quite useful for images that'll be rendered (more on rendering in Lesson 11).

  AutoCAD provides three types of smooth surface (Figure 6.021) – Quadratic B-Spline, Cubic B-Spline, and Bezier. The type of surface created with the **Smooth surface** option depends on the current setting of the **Surftype** system variable (refer to Figure 6.021). Two other system variables control the number of M-columns and N-rows on the smoothed 3D mesh. These are **SurfU** (to control the number of M-columns) and **SurfV** (to control the number of N-rows).

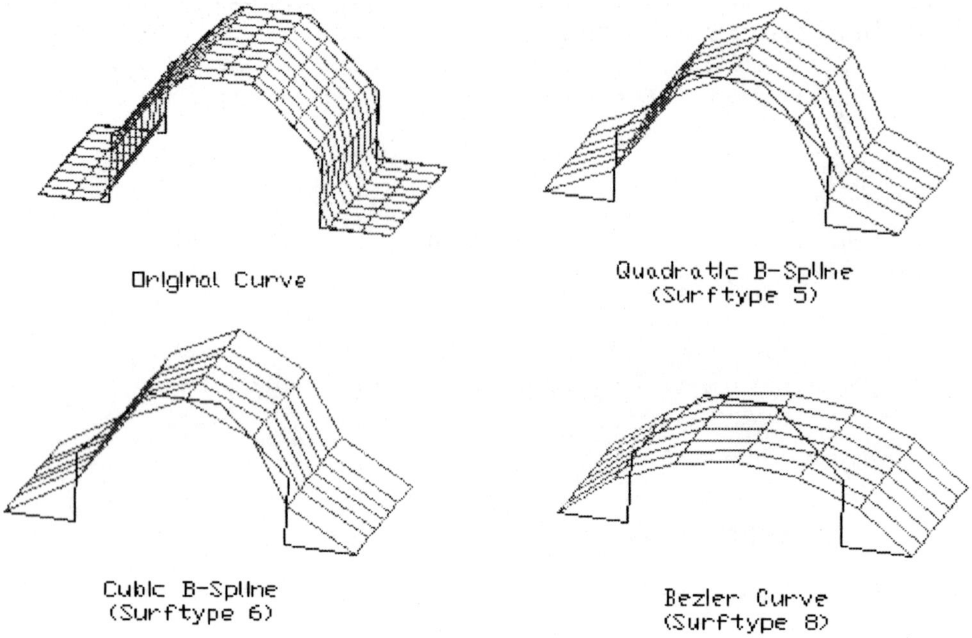

Figure 6.021

- Of course, the **Desmooth** option removes any changes made with the **Smooth surface** option.
- The **Edit vertex** options resemble the same options you received when you selected a two-dimensional polyline. But it has some additional tools. It looks like this:

  Current vertex (0,0).

  Enter an option [Next/Previous/Left/Right/Up/Down/ Move/REgen/eXit] <N>:

  - The first thing you'll notice is that AutoCAD identifies the vertex both with an "X" (as it did on the polyline) and by column and row (**M,N**) coordinate at the command prompt.
  - The **Next/Previous** tools work just as they did for a polyline – to maneuver along the mesh from vertex to vertex.
  - The **Left/Right/Up/Down** tools supplement the **Next/Previous** tools, making it easier to maneuver to a specific vertex without having to pass through countless vertices to get there.
  - The **Move** option, of course, allows you to move the currently selected vertex to a new point.
  - **REgen** and **eXit** work the same on 2D and 3D objects. (**REgen** regenerates the mesh/polyline, and **eXit** exits this tier of options.)

Let's experiment with the ***PEdit*** command and 3D meshes.

Do This: 6.1.4.1.1	Using *PEdit* on 3D Meshes

I. Open the *3DMesh.dwg* file in the C:\Steps3D\Lesson06 folder. The drawing looks like Figure 6.022.

II. Close the Properties palette.

III. Follow these steps.

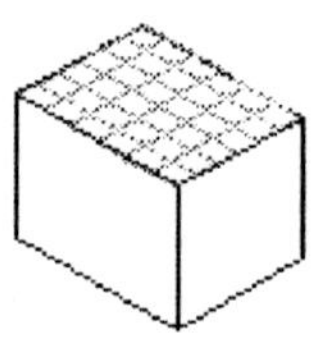

Figure 6.022

TOOLS	COMMAND SEQUENCE	STEPS
	**Command:** ***pe***	1. Enter the ***PEdit*** command.
	Select polyline or [Multiple]:	2. Select the 3D mesh atop the figure.
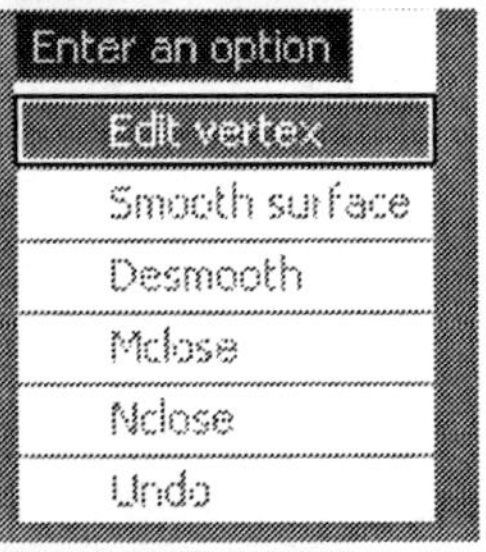	**Enter an option [Edit vertex/Smooth surface/ Desmooth/Mclose/Nclose/ Undo]:** ***e***	3. Tell AutoCAD to use the **Edit vertex** option.
	Current vertex (0,0). **Enter an option [Next/ Previous/Left/Right/Up/Down/ Move/REgen/eXit] <N>:** ***[enter]*** **Current vertex (0,1).** **Enter an option [Next/ Previous/Left/Right/Up/Down/ Move/REgen/eXit] <N>:** ***[enter]*** **Current vertex (0,2).** **Enter an option [Next/ Previous/Left/Right/Up/Down/ Move/REgen/eXit] <N>:** ***[enter]***	4. Hit ***enter*** three times (accepting the **Next** option) to move the locator to the middle of the east end of the 3D mesh (Figure 6.023). (The **Current vertex** will be **0,3**.)

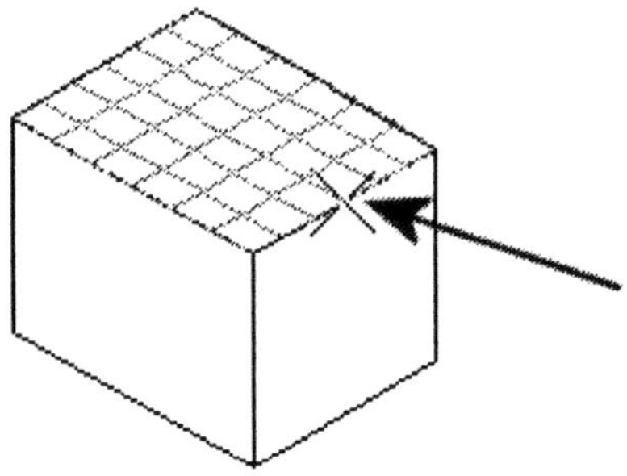

Figure 6.023

	**Enter an option [Next/Previous/ Left/Right/Up/Down/Move/ REgen/eXit] <N>:** ***m*** **Specify new location for marked vertex:** ***@0,0,1.5***	5. Use the **Move** option to move the vertex upward 1.5 units as shown. Your drawing looks like Figure 6.024.

TOOLS	COMMAND SEQUENCE	STEPS

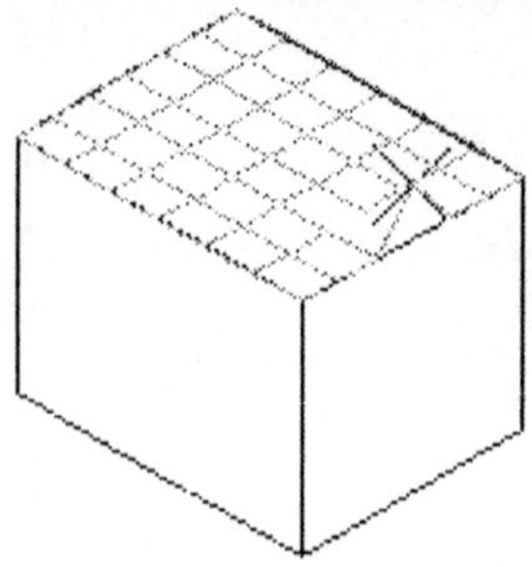

Figure 6.024

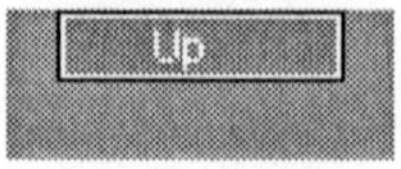		6. Use the **Up** option to move the locator to vertex **1,3** (Figure 6.025).

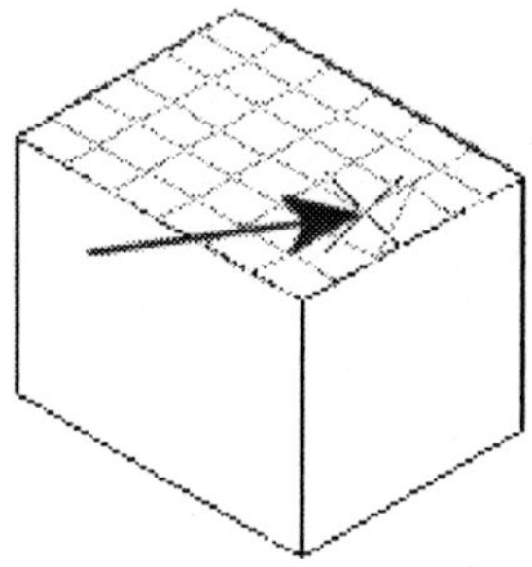

Figure 6.025

	Enter an option [Next/Previous/Left/Right/Up/Down/Move/REgen/eXit] <U>: *m* **Specify new location for marked vertex:** *@0,0,1.5*	7. Repeat Step 5.
		8. Repeat Steps 6 and 7 until the entire column has been raised. Your drawing looks like Figure 6.026.

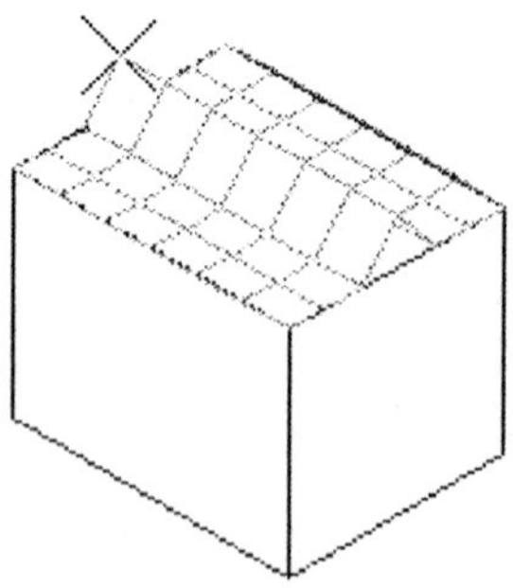

Figure 6.026

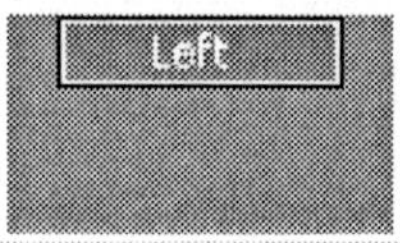	**Enter an option [Next/Previous/Left/Right/Up/Down/Move/REgen/eXit] <U>:** *l*	9. Use the **Left** option to move the locator to vertex **6,2** (Figure 6.027).

Tools	Command Sequence	Steps

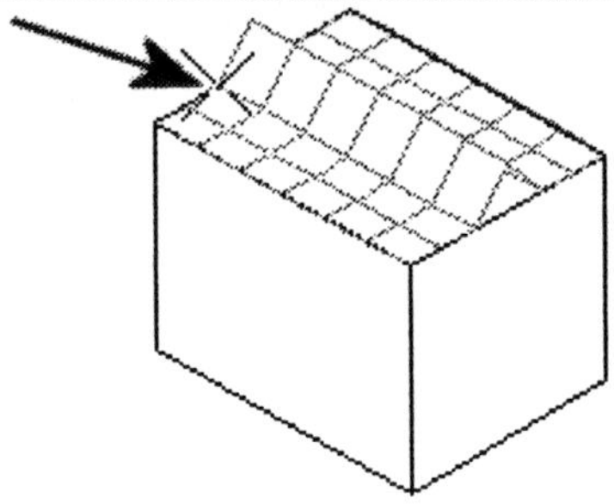

Figure 6.027

Tools	Command Sequence	Steps
Move	**Enter an option [Next/Previous/ Left/Right/Up/Down/Move/ REgen/eXit] <L>:** ***m*** **Specify new location for marked vertex:** ***@0,0,.75***	10. Move the vertex upward three-fourths of a unit as shown.
Down		11. Move the locator **Down** to the next vertex and repeat Step 10. Repeat this procedure until the roof looks like Figure 6.028. (You'll use the **Right** option to get to the other side of the ridge.)

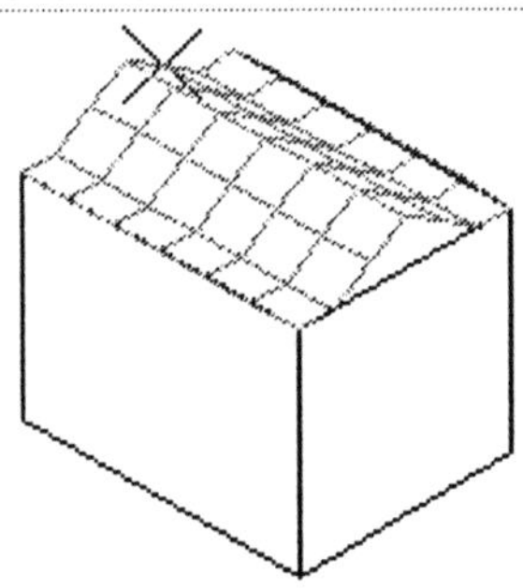

Figure 6.028

Tools	Command Sequence	Steps
eXit	**Enter an option [Next/Previous/ Left/Right/Up/Down/Move/ REgen/eXit] <U>:** ***x*** **Enter an option [Edit vertex/ Smooth surface/Desmooth/Mclose/ Nclose/Undo]:** ***[enter]***	12. Exit the command.
	Command: ***surftype*** **Enter new value for SURFTYPE <6>:** ***5***	13. Now we'll experiment with different roof shapes. Set the **Surftype** to **5** for a Quadratic B-Spline.

Tools	Command Sequence	Steps
	Command: ***surfu*** **Enter new value for SURFU <6>:** ***12*** **Command:** ***surfv*** **Enter new value for SURFV <6>:** ***18***	14. Set the **SurfU** and **SurfV** system variables as indicated for a more rounded roof.
	Command: ***pe*** **Select polyline or [Multiple]:**	15. Repeat the ***PEdit*** command and select the same 3D mesh.
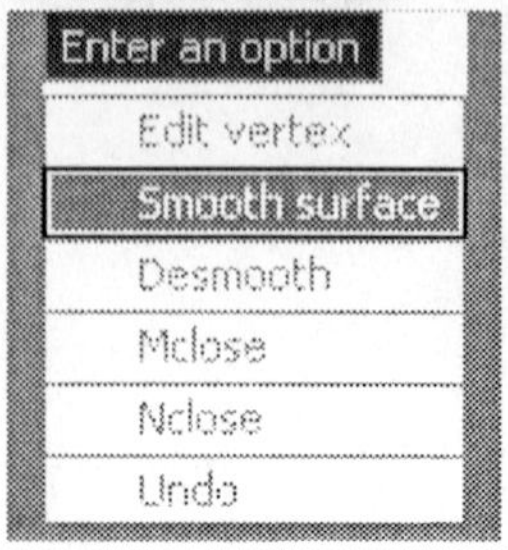	**Enter an option [Edit vertex/ Smooth surface/Desmooth/ Mclose/Nclose/Undo]:** ***s***	16. Smooth the surface.
	Enter an option [Edit vertex/ Smooth surface/Desmooth/ Mclose/Nclose/Undo]:	17. Exit the command. Your drawing looks like Figure 6.029.

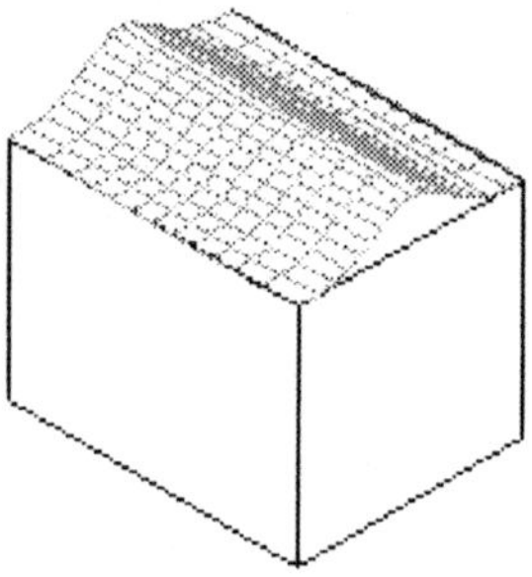

Figure 6.029

Tools	Command Sequence	Steps
		18. Repeat Steps 15 through 17 with the **Surftype** set to **6** (Cubic B-Spline). It'll be difficult to see the difference, but if you look closely, you'll notice that the Cubic B-Spline is slightly more curved (Figure 6.030).

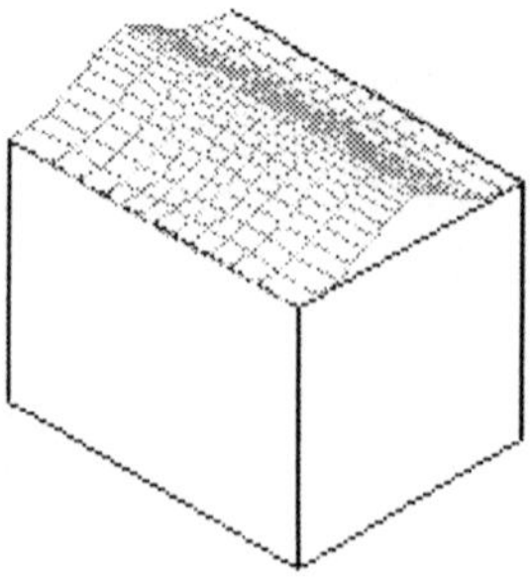

Figure 6.030

Tools	Command Sequence	Steps
	Command: ***qsave***	19. Save the drawing, but don't exit.

Experiment with the Shademode for each of the different types of roofs. Which do you like best? (Return the Shademode to **Hidden**.)

Let's look next at using the Properties palette to edit the 3D mesh.

6.1.4.2	Modifying a 3D Mesh

The Properties palette offers the same options as the ***PEdit*** command but makes it easier to select a specific vertex if you know its M,N coordinate. [Refer to Figure 6.1.4.2a for this discussion.]

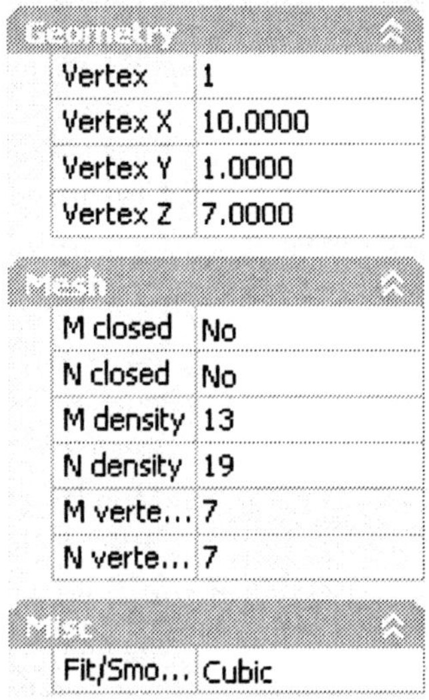

Figure 6.031

- Vertices work the same as they did when you modified the 3D face in Exercise 6.031. The only difference is the possible number of vertices with which to work.
- Mesh properties include four that you can change and two for reference.
  - **M closed** and **N closed** are toggles. They work like the **Close** option of the ***PEdit*** command. Remember that the **M** value controls columns of faces and the **N** value controls rows. I generally rely on AutoCAD to set these values when I create a surface model – I haven't found the situation where it fails to set them properly at design time.
  - **M** and **N density** control the density of columns and rows on the 3D mesh. AutoCAD bases these on the values of the **SurfU** and **SurfV** system variables, except that they reflect the number of lines defining the rows/columns rather than the number of rows/columns.
  - You can't change the **M vertex count** or **N vertex count**. AutoCAD gives these numbers as a reference only.
- The **Fit/Smooth** option works like the **Smooth vertex** option of the ***PEdit*** command.

Let's experiment with 3D meshes and the Properties palette.

Do This: 6.1.4.2.1	Modifying a 3D Mesh with the Properties Palette

I. Be sure you're still in the *3DMesh.dwg* file in the C:\Steps3D\Lesson06 folder. If not, please open it now.

II. Open the Properties palette.

III. Follow these steps.

TOOLS	COMMAND SEQUENCE	STEPS
		1. Select the roof. Notice that the Properties palette changes to reflect the properties of the 3D mesh (Figure 6.031).

TOOLS	COMMAND SEQUENCE	STEPS

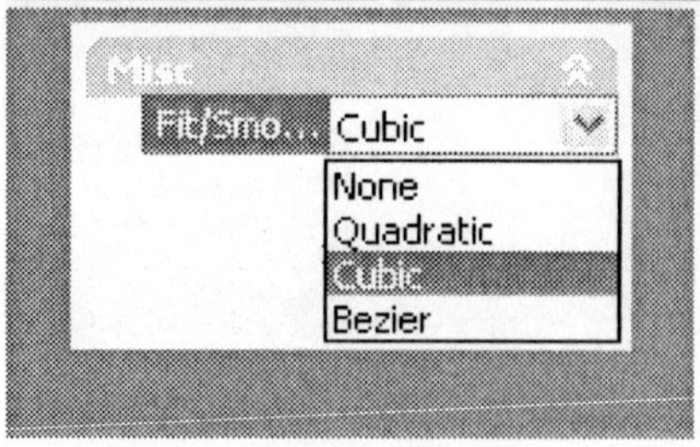

2. Change the type of surface to **Bezier** as indicated. (There's a scroll bar on the left side of the Properties palette. Use it to scroll down until you see the **Misc** section.) Notice how much softer the curve is than the other two you've seen (Figure 6.032).

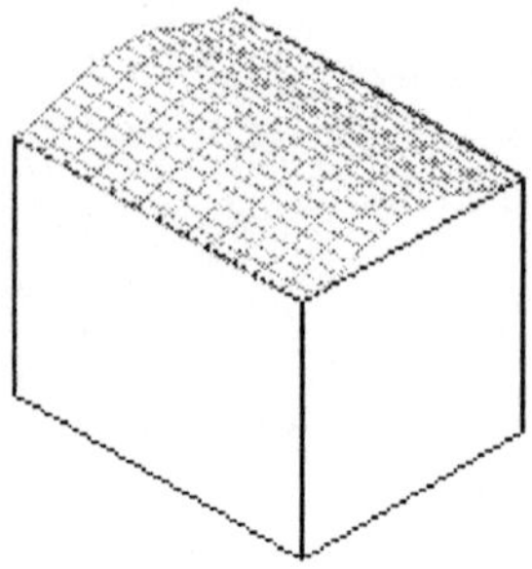

Figure 6.032

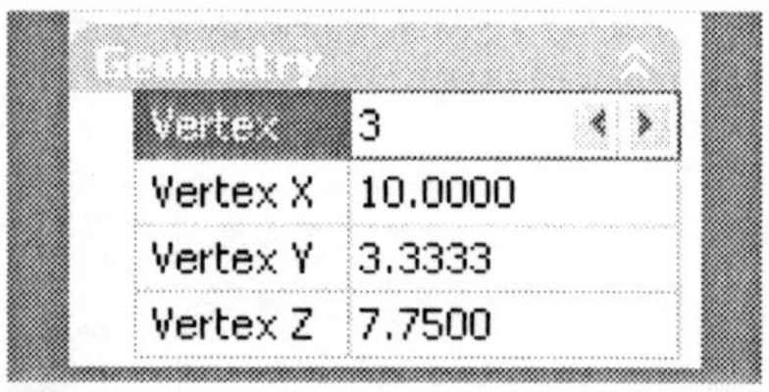

3. We'll use the Properties palette to seal both ends of the roof against the wall. Make **Vertex 3** active. (Select the **Vertex** row and use the directional arrows to change the value to **3**).

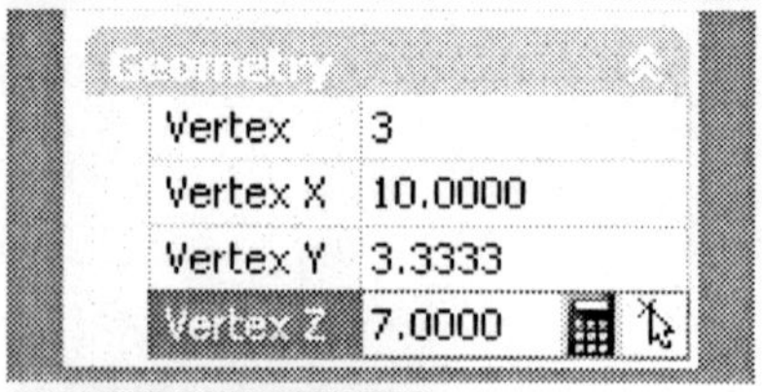

4. Change the value of **Vertex Z** to **7**.

Notice the change on the drawing.

5. Repeat Steps 3 and 4 for vertices 4, 5, 45, 46, and 47. (Normally, you'd use the arrow keys in the **Vertex** row while watching the locator "X", but I've already determined the vertices you'll need to change. I list them here to save time.)

Your drawing looks like Figure 6.033.

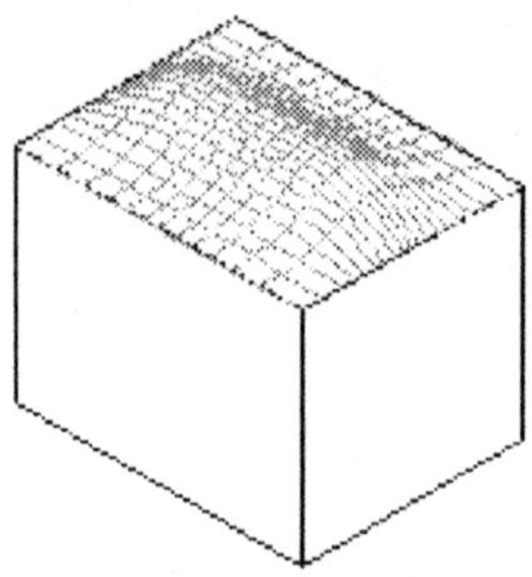

Figure 6.033

TOOLS	COMMAND SEQUENCE	STEPS
	Command: ***qsave***	6. Save the drawing, but don't exit.

Again, experiment with the Shademode settings. Orbit the view to see the building from all sides and then return to this view.

Which procedure do you like best so far? Certainly the Properties palette is easiest, but let's take a look at grips next.

6.1.4.3 Using Grips to Modify a 3D Mesh

There's very little to add to what you've already learned about grips (Lesson 16 of *AutoCAD 2006: One Step at a Time – Part I*). But I want to show you how easy it is to modify a 3D mesh using these marvelous tools.

Let's get right to it.

Do This: 6.1.4.3.1 Using Grips to Modify a 3D Mesh

I. Be sure you're still in the *3DMesh.dwg* file in the C:\Steps3D\Lesson06 folder. If not, please open it now.

II. Set the **Fit/Smooth** option to **None**.

III. Set the running OSNAP to **Node** and clear all other settings. Thaw the **Marker** layer (notice the nodes above the roof).

IV. Follow these steps.

TOOLS	COMMAND SEQUENCE	STEPS
		1. Select the roof. Notice the grips.
		2. Pick the center grip at the top of the roof (use the coordinate display to the left on the status bar – the grip is at coordinate 5.5, 4.5, 8.25).
		3. Stretch the 3D mesh to the center node above the roof.
		4. Stretch the 3D mesh using the grip at point 7, 4.5, 8.25 to the eastmost node and the grip at point 4, 4.5, 8.25 to the westmost node.
		5. Freeze the **Marker** layer, set the **Fit/Smooth** option to **Quadratic**, and set Shademode to **Gouraud**. Your drawing looks like Figure 6.034.

TOOLS	COMMAND SEQUENCE	STEPS

Figure 6.034

	**Command:** ***qsave***	6. Save the drawing.

You can see how much easier it is to move a vertex using grips than any other method. The only requirement is that you know the destination point!

6.2 Editing Tools Designed for Z-Space

You've seen how AutoCAD has adapted several modification tools you already knew to help you in Z-Space and with three-dimensional objects. It's good to maintain some familiarity between the 2D and 3D worlds.

In this section, we'll look at some new tools designed specifically for working in Z-Space, but as their names imply (***Rotate3d***, ***Mirror3d***, and ***3DArray***), they serve familiar functions. The major difference between these and their two-dimensional counterparts involves the use of axes rather than base or rotation points.

Let's look at each.

6.2.1 Rotating About an Axis – The *Rotate3d* Command

Rotating about a base point was easy – you simply selected what to rotate and a base point. Then you told AutoCAD what angle you wanted.

Rotating about an axis is slightly more complex. But if you ever need to rotate an object in Z-Space, you'll find the ***Rotate3d*** command irreplaceable. It works like this:

> **Command:** ***rotate3d***
>
> **Current positive angle: ANGDIR=counterclockwise ANGBASE=0** ***[AutoCAD reminds you how the drawing was set up]***
>
> **Select objects:** ***[select the object(s) you want to rotate]***
>
> **Select objects:** ***[confirm completion of the selection set]***
>
> **Specify first point on axis or define axis by**
>
> **[Object/Last/View/Xaxis/Yaxis/Zaxis/2points]:** ***[select a point on the axis about which you wish to rotate the objects]***
>
> **Specify second point on axis:** ***[select a second point to identify the axis]***
>
> **Specify rotation angle or [Reference]:** ***[tell AutoCAD how much to rotate the object(s)]***

It might look frightening compared with the ***Rotate*** command, but once you've used it, you'll find it fairly simple and straightforward. Let's consider each of the axis-defining options.

- The default option is to specify **2points** on the axis. AutoCAD needs you to define the axis of rotation by picking any two points on it. Once you've done that, AutoCAD will prompt you to

  Specify rotation angle or [Reference]:

  Then tell AutoCAD how much to rotate the selected objects.
- The **Object** option is probably the easiest. If you have an object drawn that can serve as an axis, all you have to do is select it. When you choose this option, AutoCAD prompts

  Select a line, circle, arc, or 2D-polyline segment:

  - If you select a line, AutoCAD uses the line as your axis of rotation.
  - If you select a circle or arc, AutoCAD rotates the objects parallel to the plane of the circle or arc and about an imaginary axis drawn through the center of it.
  - AutoCAD treats a straight 2D-polyline segment as a line, and a 2D-polyline arc as an arc.
- The **Last** option refers to the last axis you used in the ***Rotate3d*** command.
- When you use the **View** option, AutoCAD rotates the objects about an imaginary axis drawn perpendicular to your monitor's screen.
- The **Xaxis/Yaxis/Zaxis** options align the axis of rotation with the X-, Y-, or Z-axis that runs through a selected point. AutoCAD prompts:

  Specify a point on the X [or Y or Z] axis <0,0,0>:

  Enter the point's coordinates or pick it (with an OSNAP) on the screen.

Let's try it!

> The ***Rotate3d*** command (and the other 3D commands in this section) can also be found in the Modify pull-down menu. Follow this path:
>
> *Modify – 3D Operation – Rotate 3D (or Mirror 3D or 3D Array)*

Do This: 6.2.1.1	**Rotating Objects in Z-Space**

I. Open the *ro3d.dwg* file in the C:\Steps3D\Lesson06 folder. The drawing looks like Figure 6.035.

II. Follow these steps.

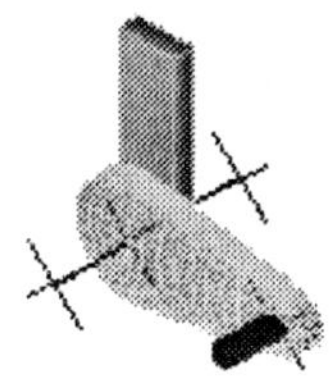

Figure 6.035

TOOLS	COMMAND SEQUENCE	STEPS
No Button Available	**Command: *rotate3d***	1. Enter the ***Rotate3d*** command.
	Select objects: **Select objects: *[enter]***	2. Select the handle.
	Specify first point on axis or define axis by [Object/Last/View/Xaxis/Yaxis/Zaxis/2points]:	3. We'll begin by selecting two points to define the axis of rotation (the default). Select the node at the end of the line in front of the handle ...

Tools	Command Sequence	Steps
	Specify second point on axis:	4. … and then select the node at the other end of the line.
	Specify rotation angle or [Reference]: *45*	5. Rotate the handle 45°. The drawing looks like Figure 6.036. Notice that the handle rotated downward. When we selected points on the axis, AutoCAD assumed the direction we defined (from the first node to the second) to be the positive Z-direction of our rotation. It then rotated the object counterclockwise. (Use the right-hand rule to verify this for yourself.)

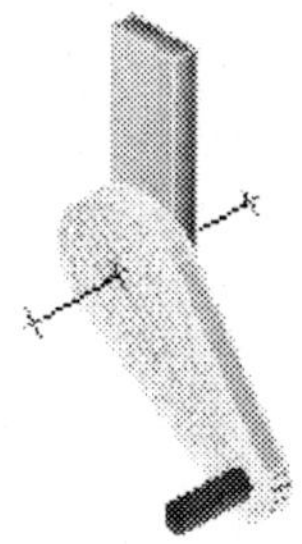

Figure 6.036

Tools	Command Sequence	Steps
	Command: *[enter]*	6. We'll use an object to define our axis of rotation this time. Repeat the command.
	Select objects: **Select objects:** *[enter]*	7. Select the handle again.
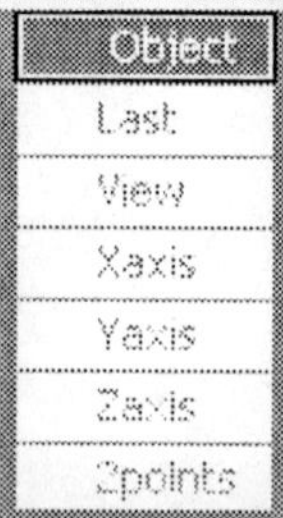	**Specify first point on axis or define axis by [Object/Last/View/Xaxis/ Yaxis/Zaxis/2points]:** *o*	8. Now choose the **Object** option.
	Select a line, circle, arc, or 2D-polyline segment:	9. Select the line between the two nodes. (Note: I drew the line from the front node to the rear node – thus defining the positive Z-direction for the object).
	Specify rotation angle or [Reference]: *-45*	10. Rotate the handle -45°. It returns to its original position.
	Command: *vp*	11. Next we'll try the **View** option. Reset the viewpoint to 0,-1,0 for a front view of the objects. (Caution: Do *not* use the **Front View** button on the View toolbar, as this will also change the UCS.)

Tools	Command Sequence	Steps
		12. Repeat Steps 1 and 2.
View	**Specify first point on axis or define axis by [Object/Last/View/Xaxis/ Yaxis/Zaxis/2points]:** *v*	13. Select the **View** option.
	Specify a point on the view direction axis <0,0,0>:	14. Select the node in the center of the large end of the handle.
	Specify rotation angle or [Reference]: ***135***	15. Rotate the handle 135°. This time, AutoCAD assumes the view represents a plan view of the drawing (with +Z rising outward from the monitor).
	Command: *z*	16. Restore the previous view. Your drawing looks like Figure 6.037.

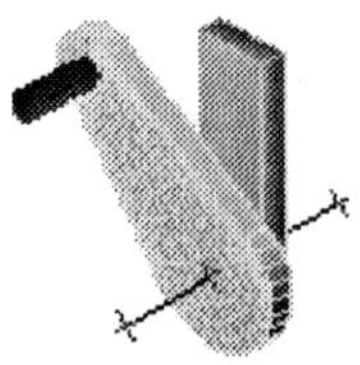

Figure 6037

Tools	Command Sequence	Steps
	Command: *ucsicon* **Command:** *rotate3d*	17. Next, we'll rotate the handle about the Y-axis (the axis along which the line is drawn). (Turn on the UCS icon to make this clear.) Repeat Steps 1 and 2.
Yaxis	**Specify first point on axis or define axis by[Object/Last/View/Xaxis/ Yaxis/Zaxis/2points]: y**	18. Choose the **Yaxis** option.
	Specify a point on the Y axis <0,0,0>:	19. Select one of the nodes on the line …
	Specify rotation angle or [Reference]: ***135***	20. … and tell AutoCAD to rotate the handle 135°. Then handle returns to its original position.
	Command: ***quit***	21. Exit the drawing without saving it.

As you can see, the only real difficulty in three-dimensional rotations is deciding which option to use (and remembering in which direction the positive Z-axis runs)!

6.2.2 Mirroring Three-Dimensional Objects – The *Mirror3d* Command

The differences between the ***Rotate3d*** and ***Rotate*** commands are really quite similar to the differences between the ***Mirror3d*** and ***Mirror*** commands. Rather than selecting a point around which to rotate an object in 2D space, you had to pick two points on an axis to satisfy the ***Rotate3d*** command. Rather than picking two points on a mirror *line* as you did in 2D space, you must pick three points to identify a mirror *plane* (the actual face of the mirror) when you use the ***Mirror3d*** command.

The options offered by the ***Mirror3d*** command are also very similar to those presented by the ***Rotate3d*** command.

Command: ***mirror3d***

Select objects: ***[select the object(s) you want to mirror]***

Select objects: ***[confirm the selection set]***

Specify first point of mirror plane (3 points) or

[Object/Last/Zaxis/View/XY/YZ/ZX/3points] <3points>: ***[use these three options to identify the mirror plane]***

Specify second point on mirror plane:

Specify third point on mirror plane:

Delete source objects? [Yes/No] <N>: ***[this option is the same as the 2D* Mirror *command – hit enter to keep the source objects or enter* Y *to remove them]***

Let's get right to an exercise.

Do This: 6.2.2.1	Rotating Objects in Z-Space

I. Open the *Star.dwg* file in the C:\Steps3D\Lesson06 folder. The drawing looks like Figure 6.038.

II. Set the **Endpoint** running OSNAP.

III. Follow these steps.

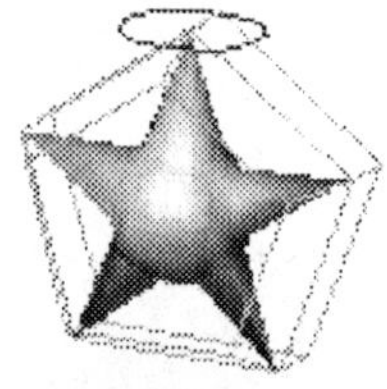

Figure 6.038

TOOLS	COMMAND SEQUENCE	STEPS
No Button Available	**Command:** ***mirror3d***	1. Enter the ***Mirror3d*** command.
	Select objects: **Select objects:** ***[enter]***	2. Select the star.
	[Object/Last/Zaxis/View/XY/YZ/ZX/3points] <3points>: **Specify second point on mirror plane:** **Specify third point on mirror plane:**	3. We'll use the default **3points** approach first. Pick the points indicated in Figure 6.039.

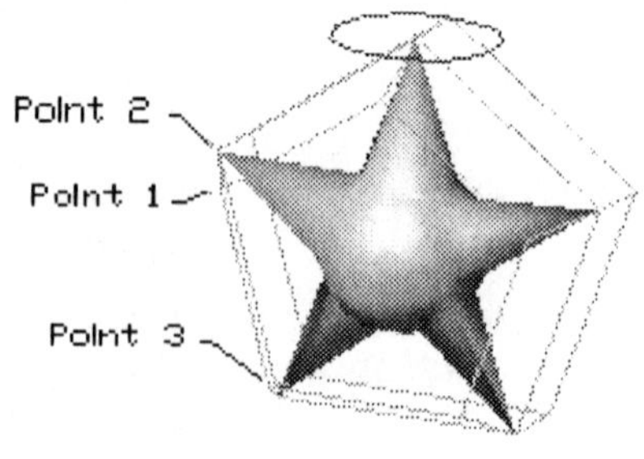

Figure 6.039

TOOLS	COMMAND SEQUENCE	STEPS
	Delete source objects? [Yes/No] <N>: ***n***	4. Don't delete the source objects. Your drawing looks like Figure 6.040. The star has been mirrored along the plan you identified (look at it in plan view –Figure 6.041 – for a better understanding of the angles).

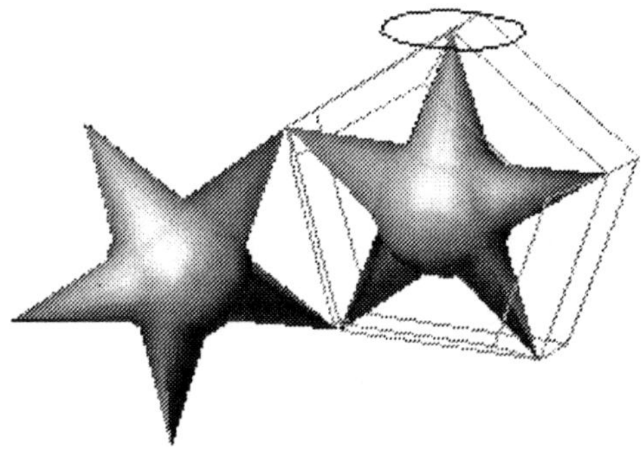

Figure 6.040

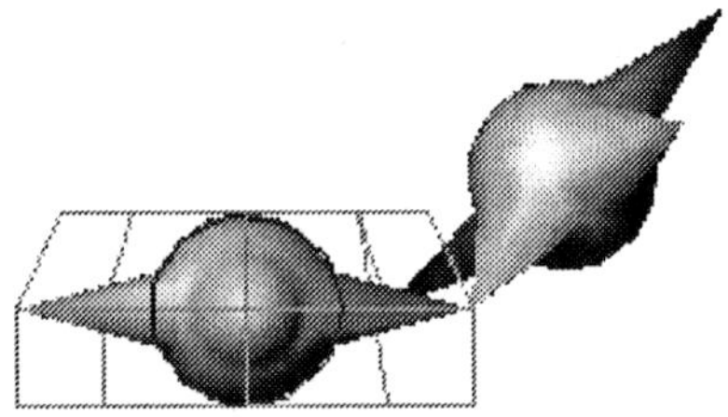

Figure 6.041

TOOLS	COMMAND SEQUENCE	STEPS
	**Command:** ***e***	5. Erase the new star.
	Command: ***mirror3d***	6. Let's use the **Object** option to stand the star on its head. Repeat Steps 1 and 2.
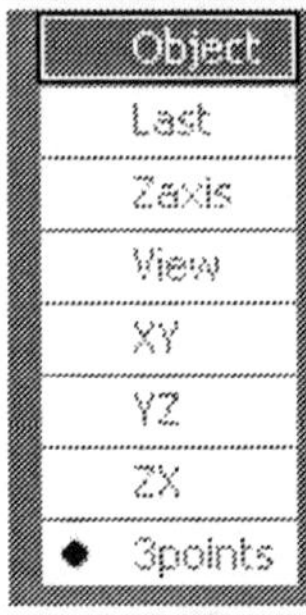	**Specify first point of mirror plane (3 points) or [Object/Last/Zaxis/View/ XY/YZ/ZX/3points] <3points>:** ***o***	7. Choose the **Object** option.
	Select a circle, arc, or 2D-polyline segment:	8. Select the star's halo (the circle).
	Delete source objects? [Yes/No] <N>: y	9. This time, delete the source objects. Your drawing looks like Figure 6.042. The star has been mirrored using the plane in which the circle was drawn.

TOOLS	COMMAND SEQUENCE	STEPS

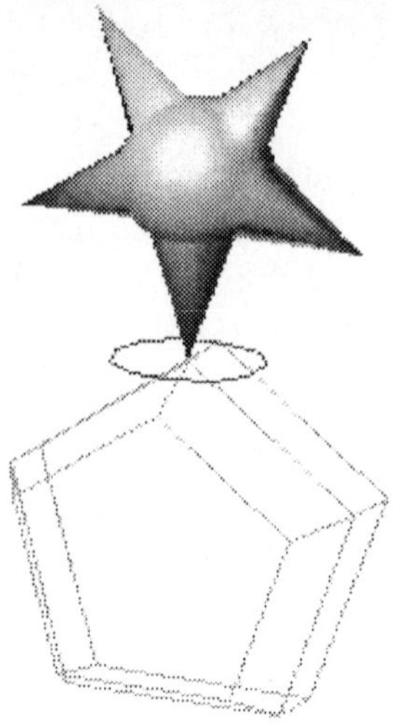

Figure 6.042

TOOLS	COMMAND SEQUENCE	STEPS
	Command: ***[enter]***	10. Now we'll mirror the star using the YZ plane. Repeat Steps 1 and 2. (Turn on the UCS icon to identify the YZ plane.)
	**Specify first point of mirror plane (3 points) or [Object/Last/Zaxis/View/ XY/YZ/ZX/3points] <3points>:** *yz* **Specify point on YZ plane <0,0,0>:**	11. Choose the **YZ** option and select the leftmost point of the star.
	Delete source objects? [Yes/No] <N>:	12. Don't delete the source objects. Your drawing looks like Figure 6.043.

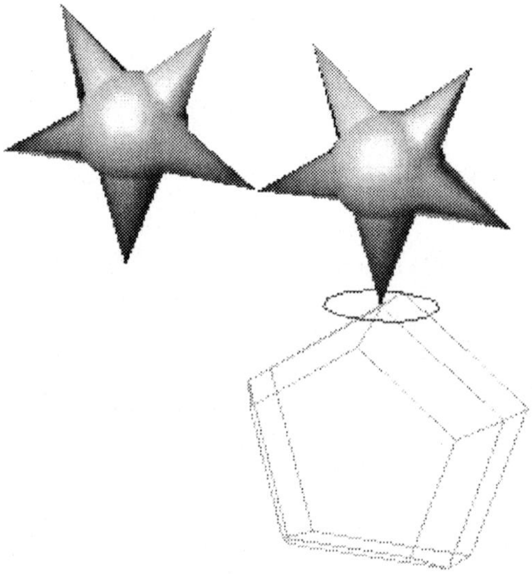

Figure 6.043

TOOLS	COMMAND SEQUENCE	STEPS
	Command: ***quit***	13. Exit the drawing without saving your changes.

If there were only one suggestion I could make about both the ***Rotate3d*** and ***Mirror3d*** commands, it would be to always check your image from more than one viewpoint (preferably three or four). Remember that, in Z-Space, object positions seen from one angle are not necessarily true three-dimensional positions.

6.2.3 Arrayed Copies in Three Dimensions – The *3DArray* Command

Of the three modification commands in this section, the ***3DArray*** command most closely resembles its two-dimensional counterpart. In fact, the most important difference between the two-dimensional rectangular array and the three-dimensional rectangular array is the addition of prompts for number and spacing of levels. The most important difference between the two-dimensional polar array and the three-dimensional polar array is that, rather than selecting a center point of the array, you must identify two points on an axis.

I should mention another important difference between the ***Array*** and ***3DArray*** commands. ***3DArray*** has no dialog box with which to work. But if you're comfortable with the Array dialog box, the command line prompts and options will be familiar to you.

Let's array some objects in Z-Space.

> A notable difference between ***3DArray*** and the other modification commands in this section is that ***3DArray*** has a hotkey – ***3a***.

Do This: 6.2.3.1 Arraying Objects in Z-Space – Rectangular Arrays

I. We'll begin this exercise by creating a three-dimensional piperack. Open the *3darray-rec.dwg* file in the C:\Steps3D\Lesson06 folder. The drawing looks like Figure 6.044.

Figure 6.044

II. Follow these steps.

<table>
<tr><th>TOOLS</th><th>COMMAND SEQUENCE</th><th>STEPS</th></tr>
<tr><td>No Button Available</td><td>Command: 3a</td><td>1. Enter the 3DArray command.</td></tr>
<tr><td></td><td>Select objects:
Select objects: [enter]</td><td>2. Select the vertical 10' I-Beam.</td></tr>
<tr><td></td><td>Enter the type of array [Rectangular/Polar] <R>: [enter]</td><td>3. Accept the default Rectangular type of array.</td></tr>
<tr><td></td><td>Enter the number of rows (---) <1>: 2
Enter the number of columns (|||) <1>: 3
Enter the number of levels (...) <1>: 2</td><td>4. Tell AutoCAD you want two rows, three columns, and two levels.</td></tr>
</table>

<table>
<tr><th>Tools</th><th>Command Sequence</th><th>Steps</th></tr>
<tr><td></td><td>Specify the distance between rows (---): 9'
Specify the distance between columns (|||): 15'
Specify the distance between levels (...): 11'</td><td>5. Specify the distances as shown.
Your drawing looks like Figure 6.045. (Adjust your view as required to see the entire drawing.)</td></tr>
</table>

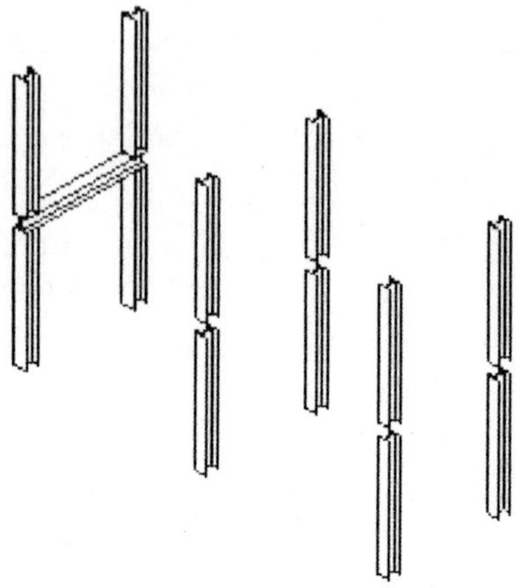

Figure 6.045

<table>
<tr><td></td><td>Command: [enter]</td><td>6. Now we'll array the horizontal support. Repeat the 3DArray command.</td></tr>
<tr><td></td><td>Select objects:
Select objects: [enter]</td><td>7. Select the horizontal support.</td></tr>
<tr><td></td><td>Enter the type of array [Rectangular/Polar] <R>: [enter]</td><td>8. Accept the default Rectangular type of array.</td></tr>
<tr><td></td><td>Enter the number of rows (---) <1>: [enter]
Enter the number of columns (|||) <1>: 3
Enter the number of levels (...) <1>: 2</td><td>9. You'll want to create one row, three columns, and two levels …</td></tr>
<tr><td></td><td>Specify the distance between columns (|||): 15'
Specify the distance between levels (...): 11'</td><td>10. …at the spacing indicated.
Your drawing looks like Figure 6.046.</td></tr>
</table>

TOOLS	COMMAND SEQUENCE	STEPS

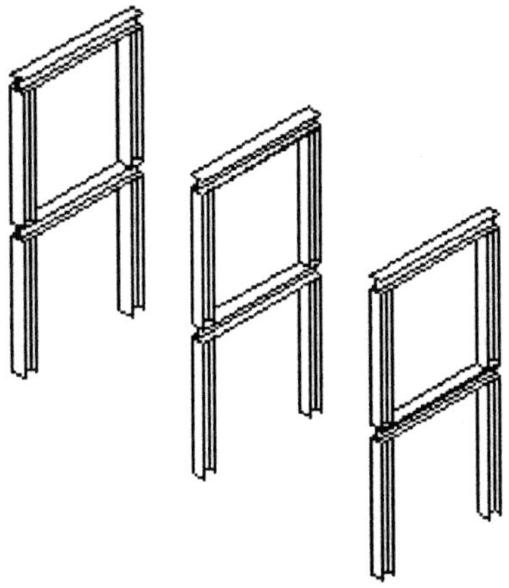

Figure 6.046

	Command: ***saveas***	11. Save the drawing as *MyPiperack.dwg* in the C:\Steps3D\Lesson08 folder, and then exit.

Do This: 6.2.3.2	Arraying Objects in Z-Space – Polar Arrays

I. Now we'll use the **Polar** option of the ***3DArray*** command. Open the *3darray-polar.dwg* file in the C:\Steps3D\Lesson06 folder. The drawing looks like Figure 6.047.

II. Set the **Intersection** and **Endpoint** running OSNAPs. Clear all other settings.

III. Freeze the **obj2** layer to temporarily remove the sphere.

IV. Follow these steps.

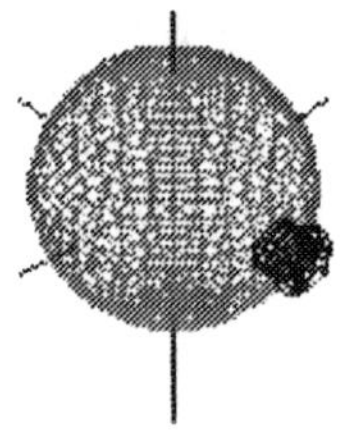

Figure 6.047

TOOLS	COMMAND SEQUENCE	STEPS
No Button Available	**Command:** ***3a***	1. Enter the ***3DArray*** command.
	Select objects: **Select objects:** ***[enter]***	2. Select the nozzle.
• Rectangular Polar	**Enter the type of array [Rectangular/Polar] <R>:** ***p***	3. Tell AutoCAD you wish to create a **Polar** array.
	Enter the number of items in the array: ***4*** **Specify the angle to fill (+=ccw, -=cw) <360>:** ***[enter]***	4. We'll create four copies of the nozzle and fill a full circle.
	Rotate arrayed objects? [Yes/No] <Y>: ***[enter]***	5. We do want to rotate the nozzles as they're copied.
	Specify center point of array: **_int of**	6. Select the intersection of the guidelines as the **center point of array**.

Tools	Command Sequence	Steps
	Specify second point on axis of rotation:	7. Pick the rightmost endpoint of the north-south horizontal line (the one running from lower left to upper right). Your drawing looks like Figure 6.048.

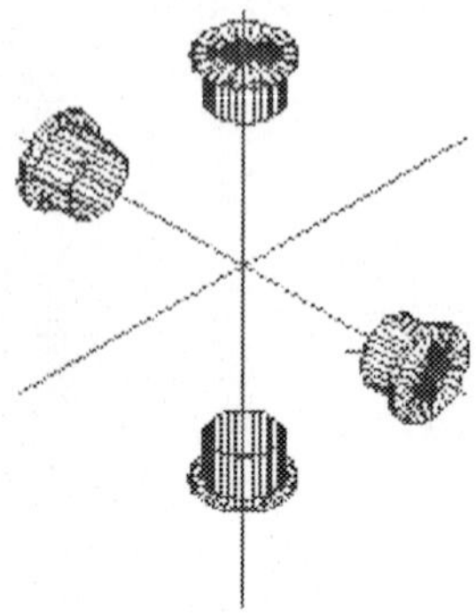

Figure 6.048

Tools	Command Sequence	Steps
		8. Repeat Steps 2 through 8, but this time select an endpoint on the vertical line in Step 8. Your drawing looks like Figure 6.049.

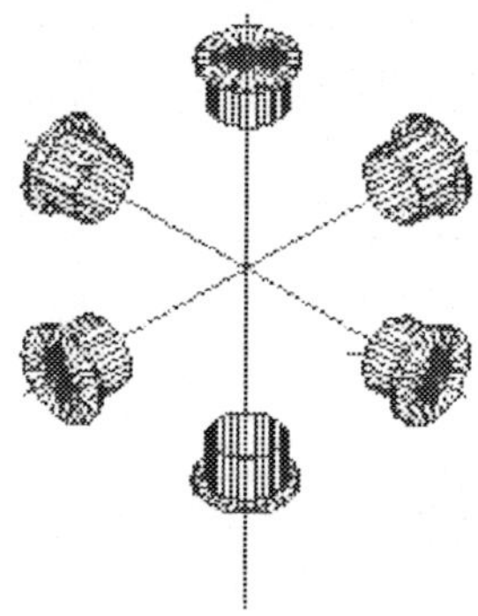

Figure 6.049

Tools	Command Sequence	Steps
		9. Thaw the **obj2** layer, freeze the **marker** layer, and set the Shademode to **Gouraud**. Your drawing looks like Figure 6.050.

Figure 6.050

Tools	Command Sequence	Steps
	Command: *saveas*	10. Save the drawing as *Weird Vessel.dwg* in the C:\Steps3D\Lesson06 folder, and then exit.

6.3 Extra Steps

Create several 3D meshes – similar to those shown in Figure 6.021. Use different shapes to begin – hat or stair shapes are good as starters, but don't limit yourself. Use the ***Edgesurf*** command to help you.

Once you have four or five meshes, experiment with the different **Smooth surface** options of the ***PEdit*** command.

It's an important step in your training to combine the different tools you've learned.

- Try editing each mesh before and after you've smoothed it. Note the differences in outcome.
- Try doing the same editing chores using different Shademode settings. Which setting is easier? Which editing tool (***PEdit***, Properties palette, or grips) is easiest for each of the settings?

6.4 What Have We Learned?

Items covered in this lesson include:

- *Two-dimensional modification tools used on three-dimensional objects*
    - ***PEdit***
    - ***Trim***
    - ***Extend***
    - ***Align***
    - *Grips*
    - *The Properties palette*
- *Tools designed specifically for Z-Space*
    - ***Surftype***
    - ***SurfV** and **SurfU***
    - ***Rotate3d***
    - ***Mirror3d***
    - ***3DArray***

This has been a busy (and full) lesson, but you've learned so much!

When combined with your knowledge of wireframe and surface modeling (and some practice), these tools will enable you to create almost any structure you wish to draw. With some creative use of the Shademode system variable, you can produce professional-quality, colorful drawings of almost anything for any industry!

But what must you have that I can't provide?

PRACTICE ... PRACTICE ... PRACTICE!

Remember: Only through practice does training become experience. And it's experience that creates successful, efficient, economical, and sound designs; and it's experience that earns top dollar!

So repeat any lesson as needed for the proper training, and then work through the exercises at the end of the lesson for experience.

Our next lesson begins the wonderful world of Solid Modeling. There, you'll see things that are guaranteed to amaze and confound, bemuse and befuddle. But above all, you'll see why Solid Modeling is the future of CAD.

6.5 Exercises

1. Using the *Star-Root.dwg* file in the C:\Steps3D\Lesson06 folder, create the star drawing we used in Exercise 6.2.2.1. (Hint: Grips make this exercise much easier.)
2. Create the conveyor belt drawing shown in Figure 6.051. Follow these guidelines:
   2.1. Draw only one shaft and one roller (use the ***Revsurf*** command for best results).
   2.2. Use the ***3DArray***, ***Rotate3d***, and ***Mirror3d*** commands to arrange the guides and the rollers on the guides.
   2.3. Use splines and the ***Rulesurf*** command to create the belt.
   2.4. I used a **Surftab1** setting of **18** or the shaft and roller, and **Surftab1** setting of **100** and **Surftab2** setting of **36** when I created the belt.
   2.5. Save the drawing as *MyBelt.dwg* in the C:\Steps3D\Lesson06 folder.

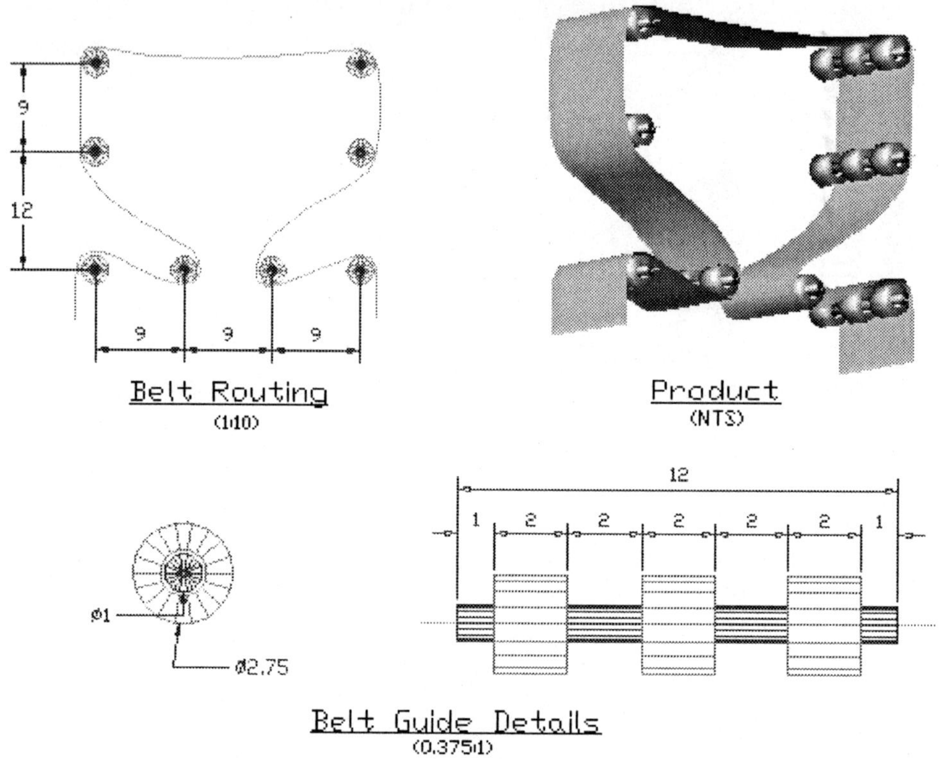

Figure 6.051

3. Starting with the *MyPiperack.dwg* file you created in Exercise 6.2.3.1 (or the *Piperack.dwg* file if that one isn't available), create the piping drawing shown in Figure 6.052. Follow these guidelines:
    3.1. The tank has a 15' diameter and a 10' height. The top has a 3' pointed cone.
    3.2. Pipe is 12" diameter (12.75" ID, or outer diameter).
    3.3. Elbows are 18" from open face to centerline of bend.
    3.4. There's a 1/8" gasket between the flange and the nozzle at the tank.
    3.5. **Surftab1** and **Surftab2** values are 16.
    3.6. The dike wall around the tank is 2' high. The top of the wall is one mesh grid wide.
    3.7. I used ***Revsurf*** to create the elbows and ***Rulesurf*** to create the pipe.
    3.8. Save the drawing as *MyPipingPlan.dwg* in the C:\Steps3D\Lesson06 folder.

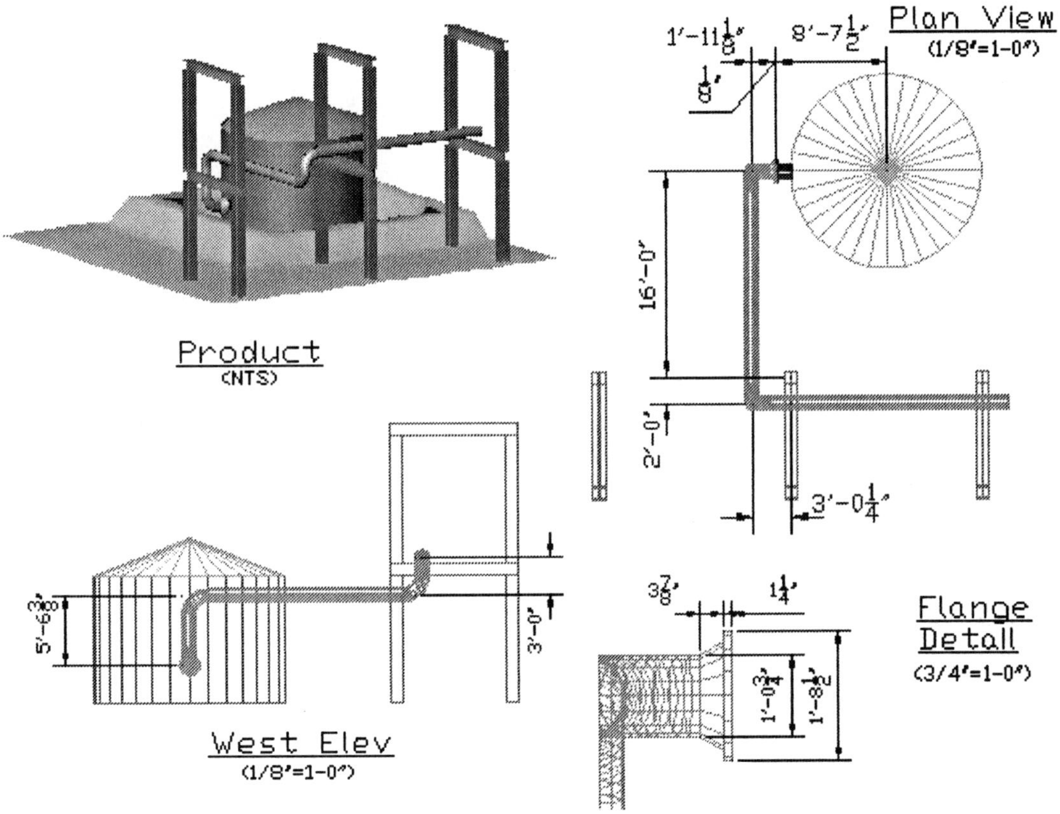

Figure 6.052

4. Create the propeller drawing shown in Figure 6.053. Follow these guidelines:
   4.1. The blade is a three-dimensional curve – use a spline as the arc and rise to the end of the upper line as shown in the *top blade detail*. Use as many vertices a you need – but I wouldn't use less than five.
   4.2. I used a **Surftab1** setting of **8** to create the hub, and a **Surftab1** setting of **16** and **Surftab2** setting of **18** to create the blade.
   4.3. Once you've drawn the blade, turn it into a block. Insert the block into its proper place on the hub, but then explode it.
   4.4. Use the ***Rotate3d*** command to rotate the blade 105° on the hub.
   4.5. Use the 3D mesh editing tools you learned in this lesson to attach the ends of the blade to the hub.
   4.6. (Hint: The ***Stretch*** command works as well in Z-Space as it did in 3D space.)
   4.7. Save the drawing as *MyProp.dwg* in the C:\Steps3D\Lesson06 folder.

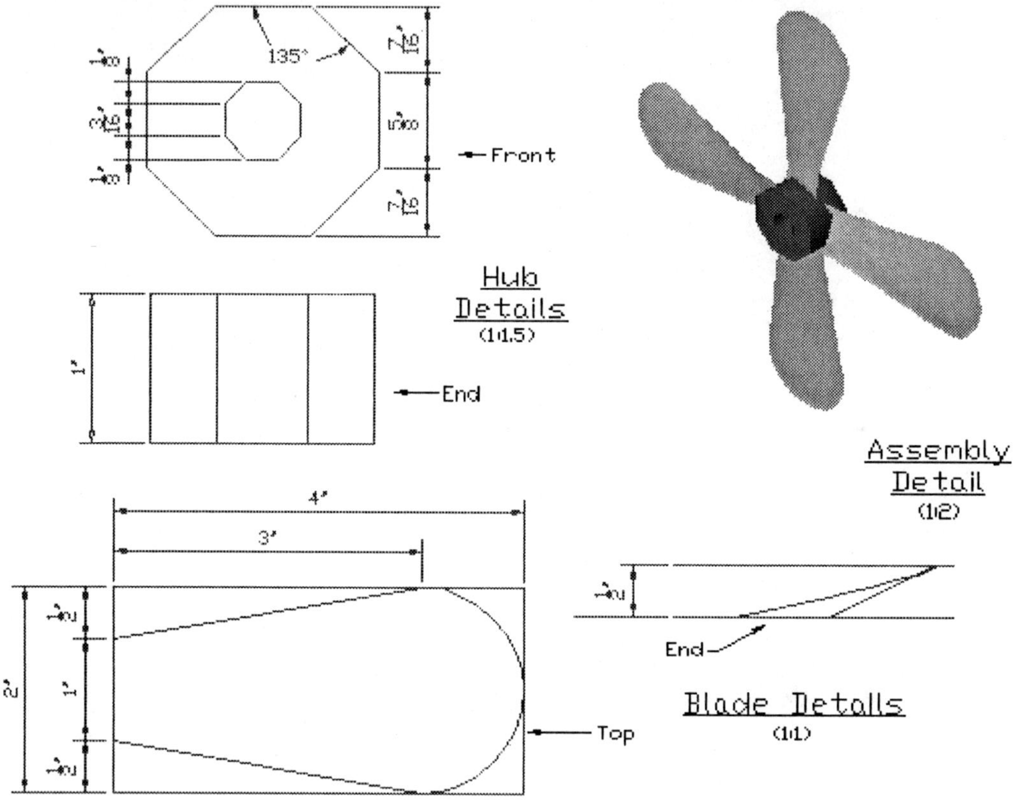

Figure 6.053

5. Create the three-dimensional chess drawing shown in Figure 6.054. Follow these guidelines:
   5.1. Each square is 1½".
   5.2. The boards are rotated at 15° increments.
   5.3. The post is 1" diameter.
   5.4. The frames are ½" wide x ¾" deep.
   5.5. The boards are 8" apart.
   5.6. Save the drawing as *My3DChess.dwg* in the C:\Steps3D\Lesson06 folder.

Figure 6.054

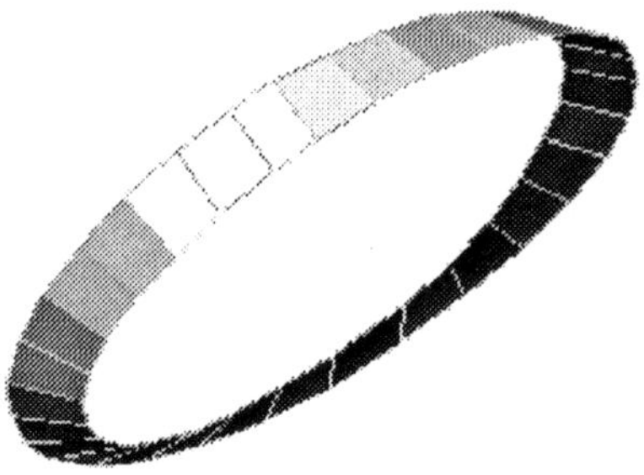

Figure 6.055

6. Here's another challenge! Create the curl drawing shown in Figure 6.055. Follow these guidelines:
   6.1. I started with a 1" line.
   6.2. There are 30 faces in all.
   6.3. The ring is ~4¼" diameter (I started with a line at 1,1 and arrayed it about point 4,4).
   6.4. I used three layers and two colors.
   6.5. The faces rotate 180°.
   6.6. Save the drawing as *MyCurl.dwg* in the C:\Steps3D\Lesson06 folder.

7. Create the sailboat drawing shown in Figure 6.056. Follow these guidelines:
   7.1. This is a toy sailboat. The boat itself is 6 "x 2" x ¾".
   7.2. The keel is 3" below the bottom of the boat.
   7.3. The mast is 7" long x 1/8" diameter.
   7.4. The boom is 5" long x 1/8" diameter.
   7.5. Save the drawing as *MySBoat.dwg* in the C:\Steps3D\Lesson06 folder.

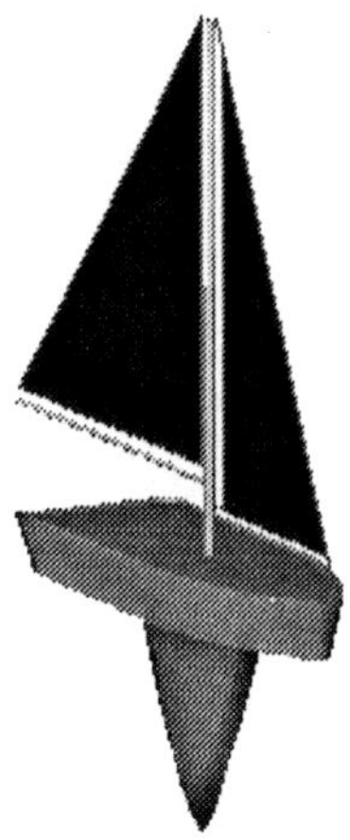

Figure 6.056

6.6	**For Web-Based Review Questions, visit: http://www.uneedcad.com/2006/Files/06R6-3D.pdf**

Lesson 7

Following this lesson, you will:

- ✓ *Know how to create AutoCAD's Solid Modeling Building Blocks*
  - ***Box***
  - ***Wedge***
  - ***Cone***
  - ***Sphere***
  - ***Cylinder***
  - ***Torus***
  - ***Extrude***
  - ***Revolve***
- ✓ *Know how and why to use AutoCAD's Isolines system variable*

Solid Modeling Building Blocks

Understanding some of the history of three-dimensional AutoCAD might help you prepare for this lesson.

AutoCAD began its trek into Z-Space by creating the Z-axis. The Z-axis gave us the ability to create three-dimensional lines and circles for the first time. AutoCAD called this development Wireframe Modeling.

But although the creation of a Z-axis was no small feat for programmers, Wireframe Modeling came up short in its usefulness to draftsmen. After all, a skeleton without skin is a fairly transparent accomplishment.

AutoCAD "covered" the need by developing Surface Modeling. Here, we gained the 3DFace (and related) commands that could be used to "stretch a blanket" over the wireframe. This appeared to solidify AutoCAD's three-dimensional experiment. But the success, like its models, was hollow.

AutoCAD programmers still dreamed of a model that would be "just like the real thing." That is, they wanted the computer to be able to reflect mass properties – solids where the object was solid, and spaces where the solid was empty. They wanted a solid object to be *a solid object – not a loose conglomeration of circles and lines. So AutoCAD developed Solid Modeling.*

Obviously, I couldn't give you the full history in these few paragraphs. The reason for these paragraphs, then, is to let you know that developers of Solid Modeling had Wireframe and Surface Modeling on which to build.

What does that mean to you now? Simply that having studied the intricacies of the more primitive modeling techniques, you're well prepared (better, perhaps, than you might think) for tackling this newest – and most remarkable – of AutoCAD's modeling tools.

7.1 What Are Solid Modeling Building Blocks?

Most people refer to Solid Modeling building blocks as primitive solids. But frankly, that term isn't as descriptive as it might be. Building blocks are toys with which we all played as children. You're already familiar with their basic shapes – box, wedge, cone, cylinder, sphere, and torus. You studied all these, except cylinder, as part of your predefined surface models. (A cylinder is simply a cone with equal radii at both ends.)

Additionally, we'll include homemade shapes as part of our building blocks (didn't you wish you could do that when you were a child?). To create these, we'll use the solids equivalent of the ***Revsurf*** command – ***Revolve*** – and a command that turns 2D objects into 3D solids – ***Extrude***.

Therefore, to answer the question, "What are Solid Modeling building blocks?" (for the test), let me give you a quick definition. Solid Modeling building blocks are predefined and user-defined solid shapes with which you build your model.

Let's look at each of them.

7.2 Extruding 2D Regions and Solids

One of the easiest ways to create a three-dimensional solid is simply to *extrude* a two-dimensional object. This means that AutoCAD will take the two-dimensional object and "stretch" it or "pull" it into Z-Space. The objects on which AutoCAD can perform this engineering marvel are 3D faces, closed polylines, circles, ellipses, closed splines, donuts, regions, and 2D solids. The results may surprise you!

The command sequence looks like this:

Command: ***extrude*** **(or *ext*)**

Current wire frame density: ISOLINES=4

Select objects: ***[select the object(s) you want to extrude]***

Select objects: ***[confirm the selection set]***

Specify height of extrusion or [Path]: ***[tell AutoCAD how far into Z-Space you want to "pull" the object]***

Specify angle of taper for extrusion <0>: ***[specify a taper angle, if desired]***

There aren't many options to confuse you, but what they can do will astound you. Let's look at each line.

- AutoCAD first lets you know how many *isolines* it'll use to display the object.

  Let me explain isolines.

  Remember when you drew surface models? You had to identify the number of faces to use by answering some prompts or adjusting the values of the **Surftab1** and **Surftab2** system variables.

  When drawing a solid object, the shape is unaffected by the surftab settings. A round solid object is round regardless of the number of lines AutoCAD uses to show that it's round. But using a large number of lines to show something is round takes a bit more memory and regeneration time, so AutoCAD allows you to control the number.

  You'll control the number of lines used to show a rounded solid object with the **Isolines** system variable.

  This will become clearer in our next exercise.
- The first option occurs right after the **Select objects** prompts. Here you can tell AutoCAD how "tall" to make the object (how far to stretch it into Z-Space), or you can select an object that'll define the extrusion **Path**. The path object can be a line, arc, or 3DPoly. The results can be quite elaborate.
- Another option that can produce elaborate results is the **taper for extrusion** option. The default (**0**) produces a nice straight extrusion. An angle entry, however, can turn a box into a pyramid!

Let's extrude some objects.

Do This: 7.2.1	**Extruding into Z-Space**

I. Open the *regions & solids.dwg* file in the C:\Steps3D\Lesson07 folder. The drawing looks like Figure 7.2.1a. (The top two I-Beam are regions, the third I-Beam is a polyline, the square is a solid, and the circle is a circle.)

II. Set the **obj1** layer current.

III. Set the **Isolines** system variable is set to **4**.

IV. Follow these steps.

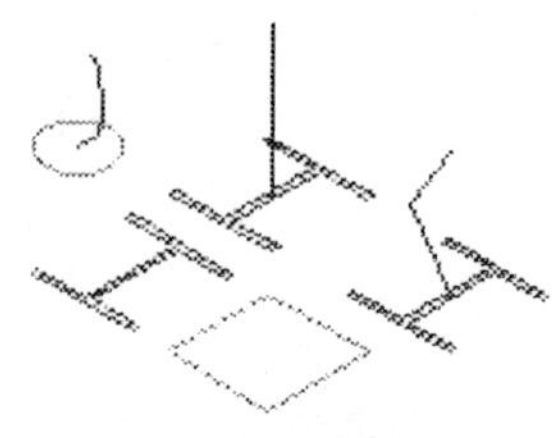

Figure 7.001

TOOLS	COMMAND SEQUENCE	STEPS
Extrude Button	**Command:** ***ext***	1. Enter the ***Extrude*** command. Alternately, you can pick the **Extrude** button on the Solids toolbar.

Tools	Command Sequence	Steps
	Current wire frame density: ISOLINES=4 **Select objects:** **Select objects:** ***[enter]***	2. Select the lower-left I-Beam.
	Specify height of extrusion or [Path]: ***5*** **Specify angle of taper for extrusion <0>:** ***[enter]***	3. Enter a **height of extrusion** of **5** and accept the default **angle of taper for extrusion**. The I-Beam looks like Figure 7.002 (I've removed hidden lines for clarity). Notice that the original object disappears and that the new 3D solid object is created on the current layer.

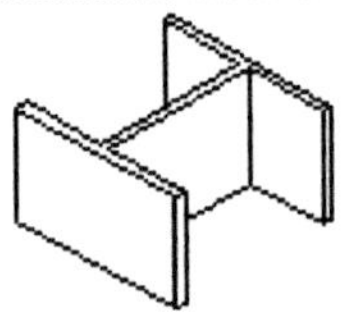

Figure 7.002

Tools	Command Sequence	Steps
	Command: ***[enter]***	4. Repeat the ***Extrude*** command.
	Select objects: **Select objects:** ***[enter]***	5. Select the I-Beam directly behind the first (the one with the straight line rising from it).
Path	**Specify height of extrusion or [Path]:** ***p***	6. Tell AutoCAD to use a **Path** to guide the extrusion …
	Select extrusion path or [Taper angle]:	7. … and select the line in the center of the I-Beam. The beams look like Figure 7.003.

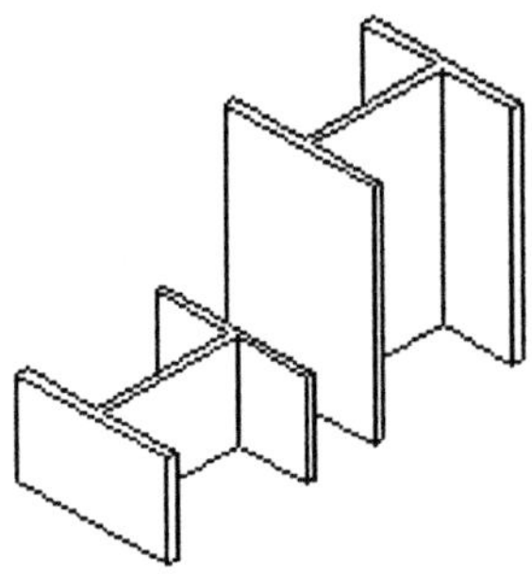

Figure 7.003

Tools	Command Sequence	Steps
	Command: ***[enter]***	8. Repeat Steps 4 through 7 for the other I-Beam using the 3D polyline as the path. The beams look like Figure 7.004. Notice the difference when the path isn't straight.

Tools	Command Sequence	Steps

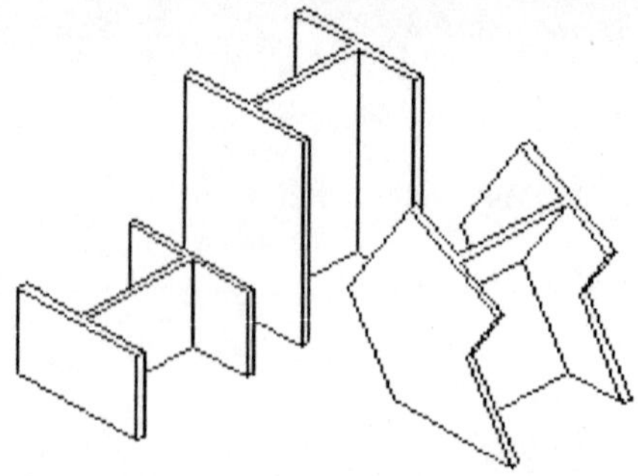

Figure 7.004

Tools	Command Sequence	Steps
		9. Repeat Step 8 on the circle. Notice how difficult it is to tell what you've drawn (Figure 7.005).

Figure 7.005

Tools	Command Sequence	Steps
	Command: ***isolines*** **Enter new value for ISOLINES <4>:** ***24*** **Command:** ***re*** **Command:** ***hide***	10. Set the **Isolines** system variable to **24** and regenerate the drawing. Remove the hidden lines. The object now looks like Figure 7.006.

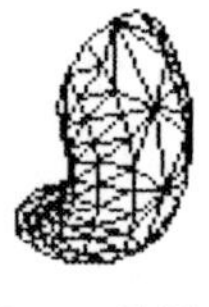

Figure 7.006

Tools	Command Sequence	Steps
	Command: ***ext*** **Select objects:** **Select objects:** ***[enter]***	11. Now let's look at the last prompt. Repeat the ***Extrude*** command and select the solid (the square).
	Specify height of extrusion or [Path]: ***5***	12. Use **5** as the **height of extrusion** …
	Specify angle of taper for extrusion <0>: ***30***	13. …but give it an **angle of taper** of **30°**. The object looks like Figure 7.007.

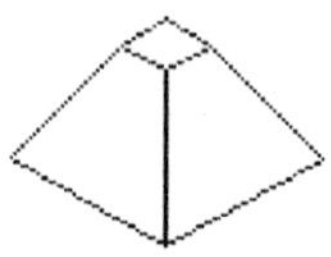

Figure 7.007

TOOLS	COMMAND SEQUENCE	STEPS
	**Command:** ***qsave***	14. Save the drawing.

Are you beginning to see why solids are the tool of choice for most three-dimensional work? But wait! We have much, much more to cover!

7.3 Drawing the Solid Modeling Building Blocks

Extruding two-dimensional objects into Z-Space is handy. But solids offer many of the same predefined shapes that you used in Surface Modeling – plus one additional shape. We'll look at these now, and then we'll look at the ***Revolve*** command. But the real marvels of Solid Modeling – the nifty tricks that make it so very valuable – will have to wait for Lessons 10 and 11.

Let's look at the predefined Solid Modeling shapes and their similarities to (and differences from) their Surface Modeling counterparts.

7.3.1 Box

Use the ***Box*** command to draw any size box whose sides are parallel or perpendicular to the current UCS. This is the command sequence:

Command: ***box***

Specify corner of box or [CEnter] <0,0,0>: ***[identify the first corner of the box]***

Specify corner or [Cube/Length]: ***[identify the opposite corner of the box]***

Specify height: ***tell AutoCAD how tall to make the box]***

The first thing you probably noticed is that the command sequence is shorter than the ***3D*** command's **Box** option. Indeed, the ***Box*** command has only three prompts to the ***AI_Box***'s five. But the ***Box*** command's prompts offer more options.

Let's look at each line.

- The first prompt asks for a **corner** of the box. Satisfy this prompt by picking a point on the screen or entering a coordinate.

  The alternative to specifying the first corner is to specify the **CEnter** of the box. Access this option by typing ***C*** or ***CE***. AutoCAD will ask you to specify the center of the box:

  Specify center of box <0,0,0>:

  Once you locate the center of the box, AutoCAD will continue with the remaining prompts.

- Next, AutoCAD asks you to specify the opposite corner of the box. You may notice a programming flaw when it does – normally, a corner prompt uses a rubber band box on the screen to help you select corners (as it does with the ***Rectangle*** command). Here you'll use the rubber band line normally used when drawing lines or polylines. AutoCAD will use the point you select to determine the length and width of the box. It'll then prompt you for the height.

  The **Cube** option on this line will prompt you to **Specify length** and use the value you enter as length, width, and height for the cube.

  The **Length** option will also ask you to **Specify length** but will follow that request with prompts for **width** and **height** as well.

We'll draw some boxes in our exercise to see these options.

You can also access the ***Box*** command (as well as the other commands in this lesson) using the Draw pull-down menu. Follow this path:

Draw – Solids – [command]

Do This: 7.3.1.1	Drawing Solid Boxes

I. Start a new drawing from scratch.

II. Adjust the viewpoint to see the drawing from a SE isometric view (1,-1,1).

III. Follow these steps.

TOOLS	COMMAND SEQUENCE	STEPS
 Box Button	**Command: *box***	1. Enter the ***Box*** command. Alternately, you can pick the **Box** button on the Solids toolbar.
	Specify corner of box or [CEnter] <0,0,0>: *4,4* **Specify corner or [Cube/Length]: *@4,2***	2. Specify the corners of the box as shown.
	Specify height: *1.5*	3. Give it a **height** of **1.5** units. That was simple, wasn't it? Your drawing looks like Figure 7.008.

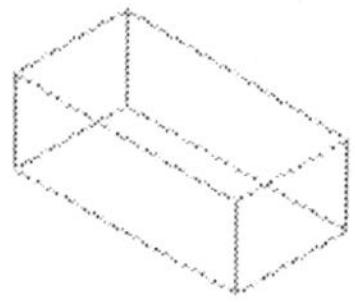

Figure 7.008

TOOLS	COMMAND SEQUENCE	STEPS
	**Command: *[enter]***	4. This time, let's draw a cube. Repeat the command.
• 0,0,0 CEnter	**Specify corner of box or [CEnter] <0,0,0>: *CE***	5. We'll use the **CEnter** option.
	Specify center of box <0,0,0>: *5,5,2.5*	6. And place the center as indicated.
Cube Length	**Specify corner or [Cube/Length]: *c***	7. Tell AutoCAD to draw a **Cube**.
	Specify length: *2*	8. Make the sides of the cube **2** units. Your drawing looks like Figure 7.009. (Again, I've removed hidden lines for clarity). Notice that the center we indicated is the center of the box along all three axes – X, Y, and Z.

TOOLS	COMMAND SEQUENCE	STEPS
	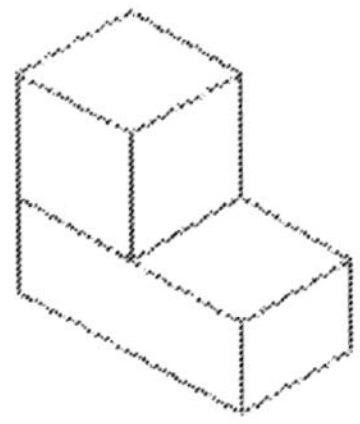 Figure 7.009	
	Command: ***[enter]***	9. We'll draw one more to see the **Length** option. Repeat the command.
	Specify corner of box or [CEnter] <0,0,0>:	10. Pick the bottom corner of the upper box (at coordinates 6,4,1.5) as the first **corner**.
Length	**Specify corner or [Cube/Length]:** ***l***	11. Use the **Length** option.
	Specify length: 2 **Specify width: 2** **Specify height: 2**	12. Finally, specify the length, width, and height as shown. Your drawing looks like Figure 7.010.
	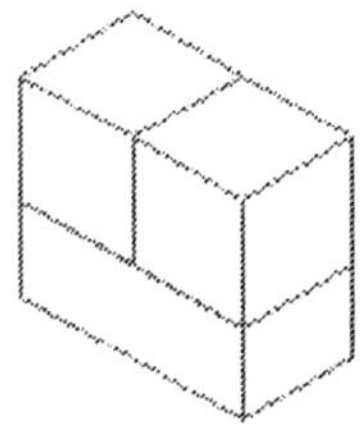 Figure 7.010	
	Command: ***save***	13. Save the drawing as *MyBlocks-Boxes.dwg* in the C:\Steps3D\Lesson07 folder.

Does it remind you of playing with blocks when you were a child? Well, now you can make a living playing with those blocks!

7.3.2 Wedge

The similarities between the ***3D*** command's **Box** and **Wedge** options hold true for the ***Box*** and ***Wedge*** commands as well. The command sequence for the ***Wedge*** command looks like this:

> **Command:** ***wedge*** **(or** ***we*****)**
>
> **Specify first corner of wedge or [CEnter] <0,0,0>:** ***[identify the first corner of the wedge (this will be the right-angled corner)]***
>
> **Specify corner or [Cube/Length]:** ***[identify the opposite corner of the wedge]***
>
> **Specify height:** ***[tell AutoCAD how tall to make the wedge]***

Look familiar? The prompts and the options are identical to those of the ***Box*** command. The only additional information you need to know is that the first corner of the wedge identifies the right angle.

We'll draw a couple of wedges for practice.

Do This: 7.3.2.1	Drawing Solid Wedges

I. Start a new drawing from scratch.

II. Adjust the viewpoint to see the drawing from a SE isometric view (1,-1,1).

III. Follow these steps.

TOOLS	COMMAND SEQUENCE	STEPS
Wedge Button	**Command:** ***we***	1. Enter the ***Wedge*** command. Alternately, you can pick the **Wedge** button on the Solids toolbar.
	Specify first corner of wedge or [CEnter] <0,0,0>: ***1,1***	2. Start the wedge as shown.
	Specify corner or [Cube/Length]: ***@-4,2*** **Specify height:** ***1.5***	3. Point the wedge away from the screen by using a negative X value, as shown, and give it a height of 1.5 units. Your wedge looks like Figure 7.011 (hidden lines removed for clarity – I'll do this for the remainder of the lesson).

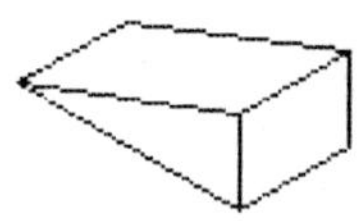

Figure 7.011

TOOLS	COMMAND SEQUENCE	STEPS
	Command: ***[enter]***	4. Let's try the **Cube** option. Repeat the command.
	Specify first corner of wedge or [CEnter] <0,0,0>:	5. Pick the bottom corner (at coordinate 1,1) as the **first corner of wedge**.
Cube	**Specify corner or [Cube/Length]:** ***C***	6. Use the **Cube** option…
	Specify length: ***2***	7. … and give it a length of **2**. Your drawing looks like Figure 7.012. Obviously, you haven't drawn a cube but a wedge (the diagonal half of a cube) based on the cube you specified.

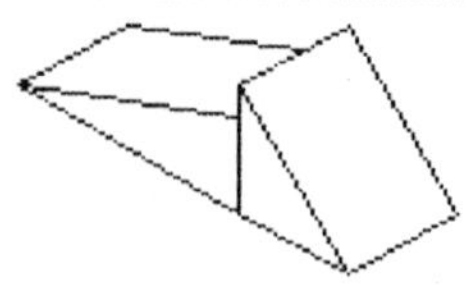

Figure 7.012

TOOLS	COMMAND SEQUENCE	STEPS
	Command: ***save***	8. Save the drawing as *MyBlocks-Wedges.dwg* in the C:\Steps3D\Lesson07 folder.

You probably noticed that the ***Wedge*** command doesn't have a rotation angle prompt like the surface method did. You can orient the solid wedge as you draw it (using positive or negative numbers), or you can use the ***Rotate*** or ***Rotate3d*** command to point it in the desired direction.

7.3.3 Cones and Cylinders

Although the command sequence for solid cones is shorter (and easier) than its surface model counterpart, it was not shortened without sacrifice. But the sacrifices are offset by some new opportunities.

The difference between the two mean that

- you can't draw a solid cylinder with the cone command (you'll use the ***Cylinder*** command instead);
- all solid cones come to a point at one end, so creating a megaphone shape becomes more difficult;
- it's possible to draw an *elliptical* solid cone but not an elliptical surface model cone;
- drawing solid cones and cylinders is often faster and produces more rounded shapes.

The ***Cone*** and ***Cylinder*** command sequences are almost identical. The ***Cone*** command looks like this:

> **Command:** ***cone***
>
> **Current wire frame density: ISOLINES=4**
>
> **Specify center point for base of cone or [Elliptical] <0,0,0>:** ***[identify the center point for the base of the cone]***
>
> **Specify radius for base of cone or [Diameter]:** ***[identify the radius of the cone's base]***
>
> **Specify height of cone or [Apex]:** ***[tell AutoCAD how tall to make the cone]***

And the ***Cylinder*** command sequence looks like this:

> **Command:** ***cylinder***
>
> **Current wire frame density: ISOLINES=4**
>
> **Specify center point for base of cylinder or [Elliptical] <0,0,0>:** ***[identify the center point for the base of the cylinder]***
>
> **Specify radius for base of cylinder or [Diameter]:** ***[identify the radius of the cylinder's base]***
>
> **Specify height of cylinder or [Center of other end]:** ***[tell AutoCAD how tall the make the cylinder]***

The options are fairly straightforward. Let's take a look.

- Before starting, AutoCAD reminds you of the **Isolines** settings.
- The first option of both commands is an opportunity to draw elliptical cones or cylinders. When you select this option, AutoCAD's prompts will change slightly to resemble the standard ***Ellipse*** command prompts:

  > **Specify axis endpoint of ellipse for base of cylinder or [Center]:**
  >
  > **Specify second axis endpoint of ellipse for base of cylinder:**
  >
  > **Specify length of other axis for base of cylinder:**

- The next option allows you to specify a base radius or diameter.
- The last lines of each command's prompts ask for the same thing, although they're worded a bit differently. An **Apex** or **Center of other end** allows you to draw cones or cylinders that

aren't straight up and down. Use one of these options and pick an off-center point to change the direction to which the cone or cylinder points.

Try your hand at the solid approach to cones and cylinders in an exercise.

Do This: 7.3.3.1	Drawing Solid Cones and Cylinders

I. Start a new drawing from scratch.

II. Adjust the viewpoint to see the drawing from a SE isometric view (1,-1,1).

III. Set the **Isolines** system variable to **24** for clarity.

IV. Follow these steps.

TOOLS	COMMAND SEQUENCE	STEPS
 Cone Button	**Command:** ***cone***	1. Enter the ***Cone*** command. Alternately, you can pick the **Cone** button on the Solids toolbar. (We'll complete the sequence with the ***Cone*** command. Then we'll repeat it using the ***Cylinder*** command.)
	Current wire frame density: ISOLINES=24 **Specify center point for base of cone or [Elliptical] <0,0,0>:** ***2,8***	2. We'll use default options on the first cone/cylinder. Place the first cone at coordinate **2,8** (the first cylinder at coordinate **8,8**).
	Specify radius for base of cone or [Diameter]: ***2*** **Specify height of cone or [Apex]:** ***4***	3. Give the cone/cylinder a base radius of **2** and a height of **4**.
 Cylinder Button	**Command:** ***cylinder***	4. Repeat Steps 1 through 3 using the ***Cylinder*** command. Your drawing looks like Figure 7.013.

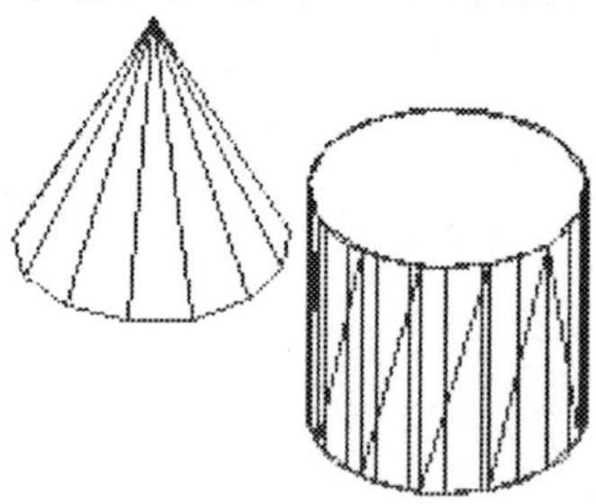

Figure 7.013

	**Command:** ***cone***	5. Let's use our new commands to draw an elliptical cone and then an elliptical cylinder. Repeat the ***Cone*** command.

TOOLS	COMMAND SEQUENCE	STEPS
Elliptical	**Current wire frame density: ISOLINES=24** **Specify center point for base of cone or [Elliptical] <0,0,0>:** ***e***	6. Choose the **Elliptical** option.
	Specify axis endpoint of ellipse for base of cone or [Center]: ***0,4***	7. Place the **axis endpoint of ellipse for base of cone** at coordinate **0,4**. (Place the **axis endpoint of ellipse for base of cylinder** at **6,4**.)
	Specify second axis endpoint of ellipse for base of cone: ***4,4***	8. Place the **second axis endpoint of ellipse for base of cone** at coordinate **4,4**. (Place the **second axis endpoint fof ellipse for base of cylinder** at **10,4**.)
	Specify length of other axis for base of cone: ***1***	9. The length of other axis should be **1** …
	Specify height of cone or [Apex]: ***4***	10. … and the height should be **4**.
	Command: ***cylinder***	11. Repeat Steps 5 through 10 for the ***Cylinder*** command as detailed. Your drawing looks like Figure 7.014.

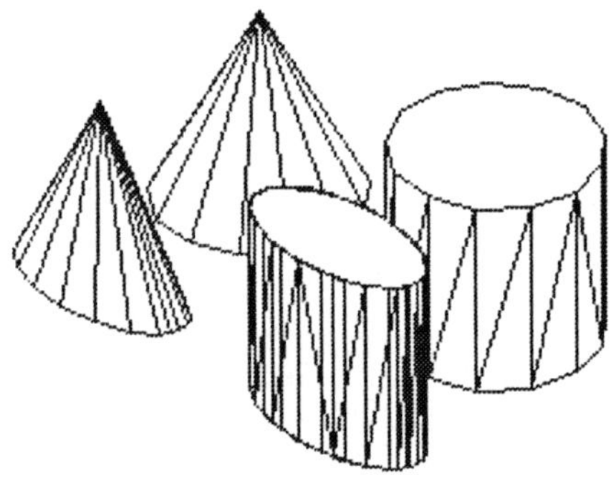

Figure 7.014

	Command: ***cone***	12. Now use the **Apex/Center of other end** option to change the direction of the cone/cylinder. Repeat the ***Cone*** command.
	Current wire frame density: ISOLINES=24 **Specify center point for base of cone or [Elliptical] <0,0,0>:** ***2,0,2***	13. Place the base point of the cone at **2,0,2**. (Place the base point of the cylinder at **8,0,2**.)
Diameter	**Specify radius for base of cone or [Diameter]:** ***d*** **Specify diameter for base of cone:** ***4***	14. Set the **Diameter** of the base at **4**.

Tools	Command Sequence	Steps
	**Specify height of cone or [Apex]:** *a* **Specify apex point:** *2,-4,2*	15. This time, use the **Apex** (or **Center of other end**) option. Place the **Apex** of the cone at **2,-4,2**. (Place the **Center of other end** of the cylinder at **8,-4,2**.)
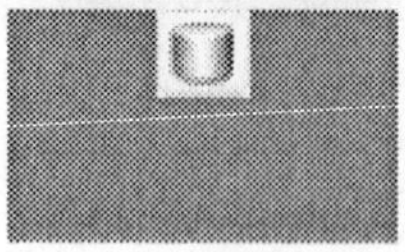		16. Repeat Steps 12 through 15 using the ***Cylinder*** command. Your drawing looks like Figure 7.015.

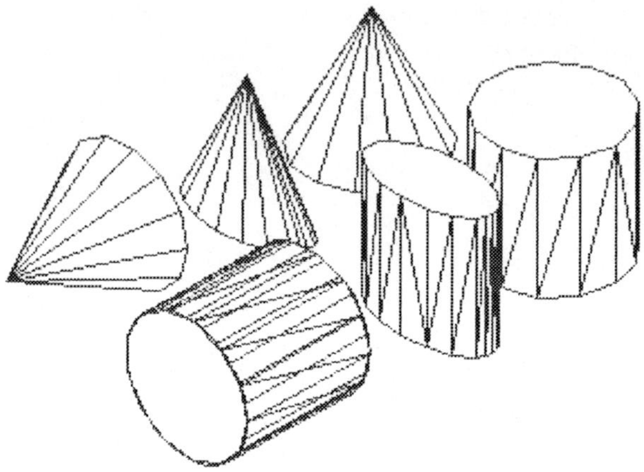

Figure 7.015

Tools	Command Sequence	Steps
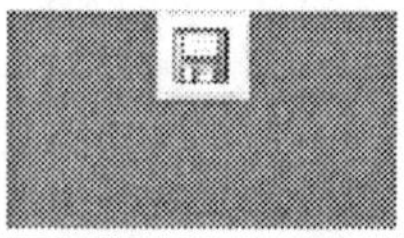	**Command:** ***save***	17. Save the drawing as *MyBlocks-Cones.dwg* in the C:\Steps3D\Lesson07 folder.

It's interesting that these commands use the same procedures to produce such similar objects. Perhaps, in the future, AutoCAD will reduce them to one command with a **Cone/Cylinder** option.

7.3.4 Sphere

The sphere is another object whose production was greatly simplified between creation of the ***3D*** command's **Sphere** option and the solid ***Sphere*** command. Indeed, we've gone from five prompts and three options to two prompts and one option! And the option is the common radius/diameter choice available in so many commands.

Here's the solid ***Sphere*** command sequence:

> **Command:** ***sphere***
>
> **Current wire frame density: ISOLINES=4**
>
> **Specify center of sphere <0,0,0>:** ***[locate the center of the sphere]***
>
> **Specify radius of sphere or [Diameter]:** ***[how big do you want it to be?]***

Although AutoCAD doesn't prompt for the number of longitudinal or latitudinal segments, it's a good idea to set the **Isolines** system variable to a large enough number for proper viewing. But that's something you should do early in the drawing session. It doesn't have to be repeated for each command.

Draw a sphere.

Do This: 7.3.4.1	Drawing a Solid Sphere

I. Start a new drawing from scratch.

II. Adjust the viewpoint to see the drawing from a SE isometric view (1,-1,1).

III. Set the **Isolines** system variable to **64** for clarity.

IV. Follow these steps.

TOOLS	COMMAND SEQUENCE	STEPS
 Sphere Button	**Command:** ***sphere***	1. Enter the ***Sphere*** command. Alternately, you can pick the **Sphere** button on the Solids toolbar.
	Current wire frame density: ISOLINES=64 **Specify center of sphere <0,0,0>:** ***4,4,4***	2. Locate the **center of sphere** as indicated …
	Specify radius of sphere or [Diameter]: ***2***	3. … and give it a **radius** of **2**. Your drawing looks like Figure 7.016.

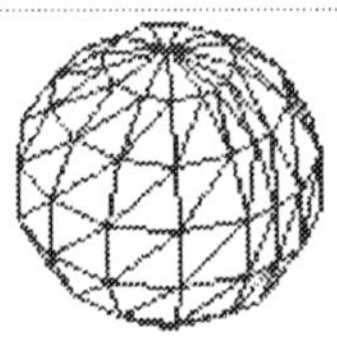

Figure 7.016

	**Command:** ***save***	4. Save the drawing as *MyBlocks-Sphere.dwg* in the C:\Steps3D\Lesson07 folder.

There's nothing else to show you about spheres. AutoCAD simplicity – what a marvel!

7.3.5	Torus

There's a subtle difference in the way you drew the surface model torus and how you'll draw a solid torus. But the difference will drive you crazy if you're not aware of it.

The difference lies in the way you size the torus itself (as opposed to sizing the tube of the torus). When you gave a radius or diameter for the surface model torus, you were indicating how large it would be from the center to the outer edge of the torus. When you give a radius or diameter for a solid torus, you're indicating the distance from the center of the torus to the center of the tube that forms it.

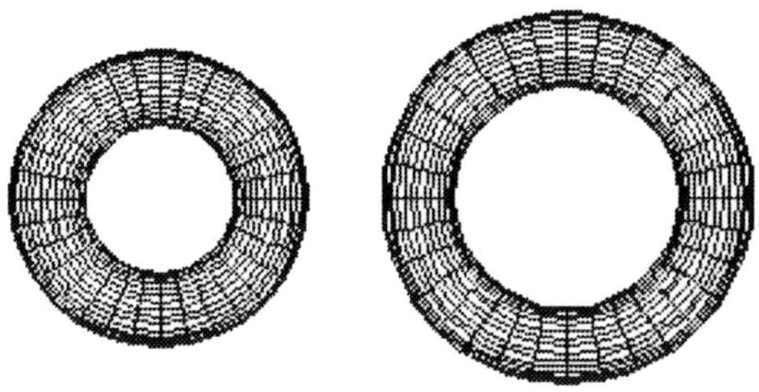

Figure 7.017

Consider the tori in Figure 7.017. The torus on the left is a surface model; the torus on the right is a solid model. Both have a torus diameter of 4 and a tube diameter of 1. But the torus diameter of the surface model measures the distance to the outer edge of the tube, whereas the diameter of the solid model measures the distance to the center of the tube. Bear this in mind when drawing surface or solid tori.

The command sequence for a solid torus (like the other solid sequences) is shorter than its surface model counterpart. It looks like this:

> **Command:** ***torus*** **(or** ***tor*****)**
>
> **Current wire frame density: ISOLINES=4**
>
> **Specify center of torus <0,0,0>:** ***[locate the center of the torus]***
>
> **Specify radius of torus or [Diameter]:** ***[indicate the size of the torus]***
>
> **Specify radius of tube or [Diameter]:** ***[indicate the size of the tube that will make up the torus]***

As with the ***Sphere*** command, it's a good idea to set the **Isolines** system variable to a large enough number for proper viewing.

Draw a torus.

Do This: 7.3.5.1	**Drawing a Solid Torus**

I. Start a new drawing from scratch.
II. Adjust the viewpoint to see the drawing from a SE isometric view (1,-1,1).
III. Set the **Isolines** system variable to **64** for clarity.
IV. Follow these steps.

TOOLS	COMMAND SEQUENCE	STEPS
 Torus Button	**Command:** ***tor***	1. Enter the ***Torus*** command. Alternately, you can pick the **Torus** button on the Solids toolbar.
	Current wire frame density: ISOLINES=64 **Specify center of torus <0,0,0>:** ***4,4,1***	2. Locate the torus as indicated.
	Specify radius of torus or [Diameter]: ***3*** **Specify radius of tube or [Diameter]:** ***.5***	3. Size the torus and the tube as indicated. Your drawing looks like Figure 7.018.

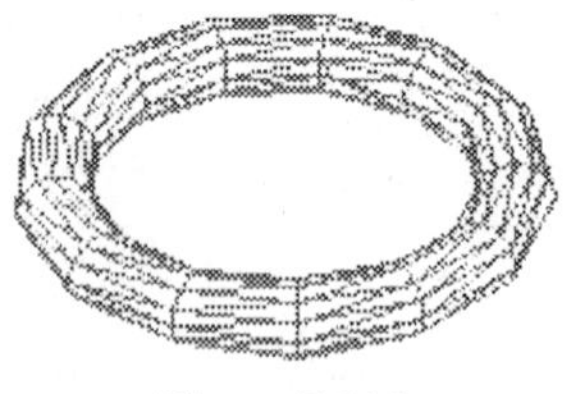

Figure 7.018

	**Command:** ***[enter]***	4. The ***Torus*** command cries for experimentation. Let's play a little. What happens when the tube diameter is larger than the torus diameter? Repeat the ***Torus*** command (let's find out).

Tools	Command Sequence	Steps
	Current wire frame density: ISOLINES=64 **Specify center of torus <0,0,0>:** ***4,4,1***	5. Let's put this torus in the center of the first one.
	Specify radius of torus or [Diameter]: ***-3***	6. Just for fun, let's give the radius of the torus a negative number …
	Specify radius of tube or [Diameter]: ***6***	7. … and the tube a larger (absolute) number. Your drawing looks like Figure 7.019. (Okay, I used the Shademode to enhance the image, but you can, too). (Whoa, cool! See what you can discover with a bit of experimentation!)

Figure 7.019

Tools	Command Sequence	Steps
	Command: ***save***	8. Save the drawing as *MyBlocks-Torus.dwg* in the C:\Steps3D\Lesson07 folder.

Oh, the fun you can have with a computer, AutoCAD, time, and a little imagination!

7.4 Creating More Complex Solids Using the *Revolve* Command

After we studied the predefined surface model objects in Lesson 4, we spent another lesson studying more complex surface models. We covered six commands: ***Rulesurf***, ***Revsurf***, ***Tabsurf***, ***Edgesurf***, ***3DMesh***, and ***PFace***.

We replaced ***Tabsurf*** easily with the ***Extrude*** command that we saw at the beginning of this lesson. But it may please you know that, with one exception, the rest of the surface-specific commands have no solid equivalents. After all, surfaces are only two-dimensional objects.

The exception involves your favorite command (and mine) – ***Revsurf***. Remember the nifty shapes we created on our train back in Lesson 5 (the top of the smokestack, the wheels, and the bell)? It'd be a shame not to be able to create such objects as solids.

For that reason, AutoCAD has provided the ***Revolve*** command. But unlike the other solid commands and their Surface Modeling counterparts, ***Revolve*** is just a bit more difficult to use than ***Revsurf***. But this is mostly because of the additional options involved. Here's the command sequence:

Command: ***revolve*** **(or *rev*)**

Current wire frame density: ISOLINES=4

Select objects: ***[select the object that defines the basic shape of the object you wish to create – you may select multiple objects, but each object must be closed]***

Select objects: ***[confirm the selection set]***

Specify start point for axis of revolution or define axis by [Object/X (axis)/Y (axis)]: ***[select a point on the axis of revolution]***

Specify endpoint of axis: ***[select another point to define the axis]***

Specify angle of revolution <360>: ***[tell AutoCAD how much of a revolution you want]***

> One of the main differences between ***Revsurf*** and ***Revolve*** is that ***Revsurf*** will create an object through revolution *without regard to the UCS*. ***Revolve*** requires that the axis of rotation exist in the current UCS.
>
> Another important difference is that the shape being revolved with the ***Revolve*** command must be a *closed* shape. Ideal objects to revolve include polylines, polygons, rectangles, circles, ellipses, and regions.

The first options don't occur until AutoCAD prompts you for an **axis of revolution**. Then you have four!

- The default option requires that you specify a point on the axis. AutoCAD will then ask you to specify another point to define the axis.
- You can also define the axis by **Object**. When you choose this option, AutoCAD asks you to **Select an object**. Select an object that exists in the current XY plane and AutoCAD will do the rest.
- The **X (axis)** or **Y (axis)** option will revolve the object about the selected axis using coordinate 0,0 as the center of the revolution.
- AutoCAD presents the last option after you've made the **axis of revolution** decision. This option allows you to control the **angle of revolution** (how much of a revolution do you want?). Simply enter an angle in degrees.

Let's see the ***Revolve*** command in action.

Do This: 7.4.1	Drawing a 3D Solid with the *Revolve* Command

I. Open the *finial.dwg* file in the C:\Steps3D\Lesson07 folder. The drawing looks like Figure 7.020.

II. Notice the orientation of the UCS and where it's centered; then turn off the UCS icon.

III. Be sure the **obj1** layer and the **Gouraud** Shademode are current.

IV. Follow these steps.

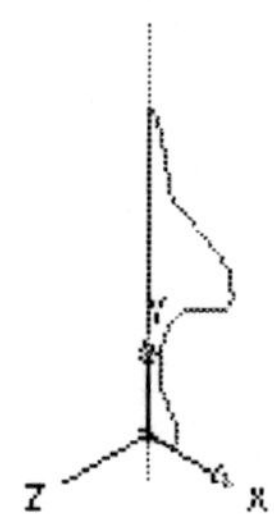

Figure 7.020

TOOLS	COMMAND SEQUENCE	STEPS
 Revolve Button	**Command:** ***rev***	1. Enter the ***Revolve*** command. Alternately, you can pick the **Revolve** button on the Solids toolbar.

Tools	Command Sequence	Steps
	Select objects: **Select objects:** ***[enter]***	2. Select the shape.
	Specify start point for axis of revolution or define axis by [Object/X (axis)/Y (axis)]: **Specify endpoint of axis:**	3. Using OSNAPs, pick the endpoints of the vertical line to define your **axis of revolution**. (Pick the bottom endpoint first.)
	Specify angle of revolution <360>: ***270***	4. Revolve the object **270°**. Your drawing looks like Figure 7.021. Notice that AutoCAD creates the solid on the current layer.

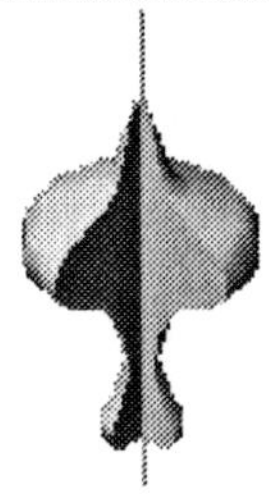

Figure 7.021

Tools	Command Sequence	Steps
	Command: ***u***	5. Undo the change.
	Command: ***rev***	6. Let's use an object to define our axis. Repeat Steps 1 and 2.
Object X (axis) Y (axis)	**Specify start point for axis of revolution or** **define axis by [Object/X (axis)/Y (axis)]:** ***O***	7. Tell AutoCAD you'll use an **Object** to define the **axis of revolution**.
	Select an object:	8. Then select the vertical line.
	Specify angle of revolution <360>: ***[enter]***	9. Accept the default **360°** this time. Your drawing looks like Figure 7.022.

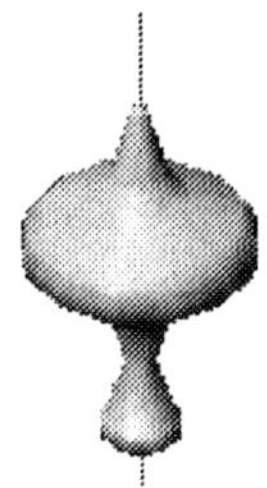

Figure 7.022

Tools	Command Sequence	Steps
	Command: ***u***	10. Undo the change.
	Command: ***rev***	11. Let's use the Y-axis to define our axis of revolution. Repeat Steps 1 and 2.

Tools	Command Sequence	Steps
Y (axis)	**Specify start point for axis of revolution or** **define axis by [Object/X (axis)/Y (axis)]:** *y*	12. Tell AutoCAD to revolve the objects about the Y-axis.
	Specify angle of revolution <360>:	13. Accept the **360°** default rotation. Your drawing again looks like Figure 7.022 (Step 9).
	Command: *u*	14. Undo the change.
	Command: *rev*	15. Let's see what happens when we use the X-axis to define our axis or revolution. Repeat Steps 1 and 2.
X (axis)	**Specify start point for axis of revolution or** **define axis by [Object/X (axis)/Y (axis)]:** *x*	16. This time, tell AutoCAD to revolve the objects about the X-axis ...
	Specify angle of revolution <360>:	17. ... and accept the **360°** default rotation. Your drawing now looks like Figure 7.023. (Okay. See if you can rotate it to look like that. Hint: Use the ***3DOrbit*** command.) Your finial has become an ashtray! Doesn't that deserve another "Whoa, cool"?).

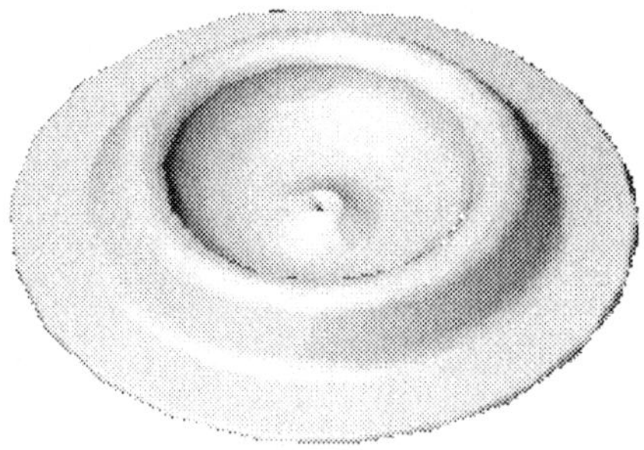

Figure 7.023

	Command: *saveas*	18. Save the drawing as *MyAshtray.dwg* in the C:\Steps3D\Lesson07 folder.

7.5 Extra Steps

Read through the *3D Solids – Creating* section of the *User's Guide*. (Follow this path: Help – Help; then on the **Index** tab, enter **3D Solids** and double-click on **Creating**.) You can pick up some tips here on each of the tools we've covered in this lesson ... and some that are yet to come. Don't forget to try some of the procedures outlined on the **Procedures** tab.

7.6 What Have We Learned?

Items covered in this lesson include:

AutoCAD's Solid Modeling building blocks

- ***Extrude***
- ***Box***
- ***Wedge***
- ***Cone***
- ***Sphere***
- ***Cylinder***
- ***Torus***
- ***Revolve***

- *Support for the building blocks*
  - ***Isolines***

This has been another fun lesson! (We need these occasionally.) The commands have been simple and straightforward.

In Lesson 7, you saw how to draw familiar shapes as solids rather than simple surface models. You also had the opportunity to use AutoCAD's ***Extrude*** command – this one is the basis for most 2D-to-3D conversion packages. Did you feel like a kid again – opening a new box of blocks for the first time and exploring each wooden shape? Did your mind slip ever so slightly into that thin mist that inevitably precedes any great discovery? Did you start to create vague mental objects using the shapes as building blocks? How many times did you begin a thought with the words, "I can use this for …" or "This is a lot easier than …"?

The Solid Modeling bug has bitten you!

Actually, you may not be quite bit … yet. But wait until you finish Lessons 8 and 9! There you'll get to play with your new building blocks in ways you never dreamed possible back in your nursery. You'll see ways to combine blocks … ways that were simply not possible until the advent of the computer. Wait until you see …

But I'm getting ahead of myself. First, we must finish this lesson. Do the exercises. Get some practice. Answer the review questions. And then we can proceed!

7.7 Exercises

To get the opportunity to compare Solid Modeling with Surface Modeling, repeat the exercises you did in Lesson 3, 4, 5, and 6. Use solids whenever possible.

1. through 6. Create the drawings in Exercises 4.6.9 through 4.6.14 using solids instead of surface models. Save the drawings to the C:\Steps3D\Lesson07 folder.

7. through 12. Create the drawings in Exercises 5.6.1 through 5.6.3, and Exercises 5.6.6 through 5.6.8 using solids instead of surface models. Save the drawings to the C:\Steps3D\Lesson07 folder.

13. Create the paper clip drawing shown in Figure 7.024.
    13.1. The drawing is set up and dimensioned using metrics.
    13.2. The paper clip is a single solid object.
    13.3. Save the drawing as *MyClip.dwg* in the C:\Steps3D\Lesson07 folder.

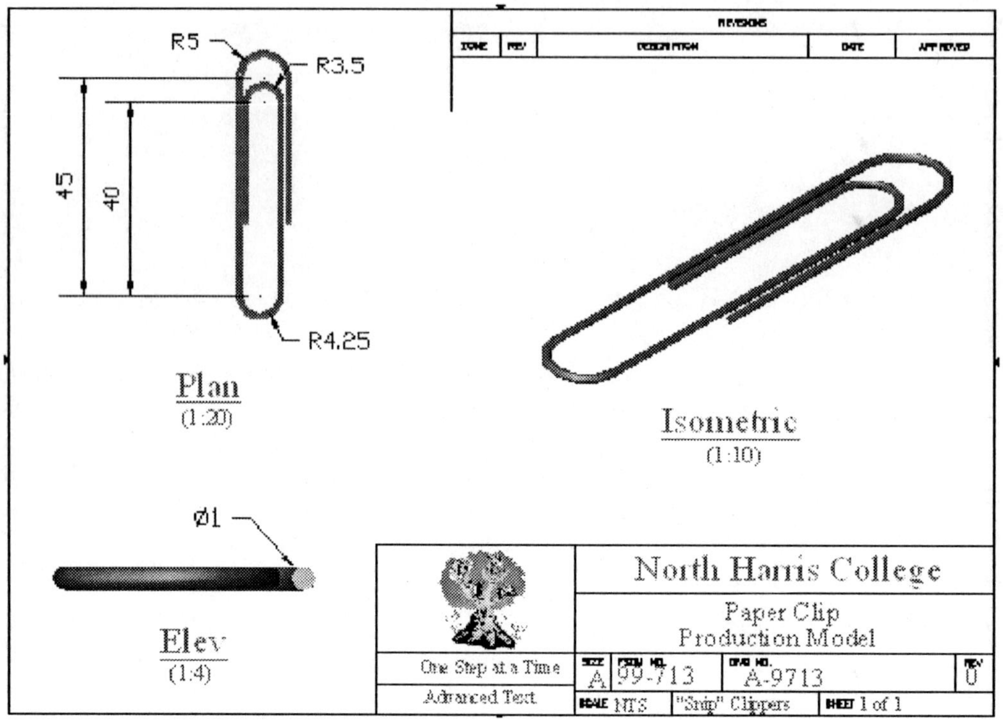

Figure 7.024

14. Create the fan cover drawing in Figure 7.025.
    14.1. Each of the wires is 1/16" in diameter (including the torus around the frame).
    14.2. The center plate is 1/8" thick.
    14.3. Save the drawing as *MyFanCover.dwg* in the C:\Steps3D\Lesson07 folder.

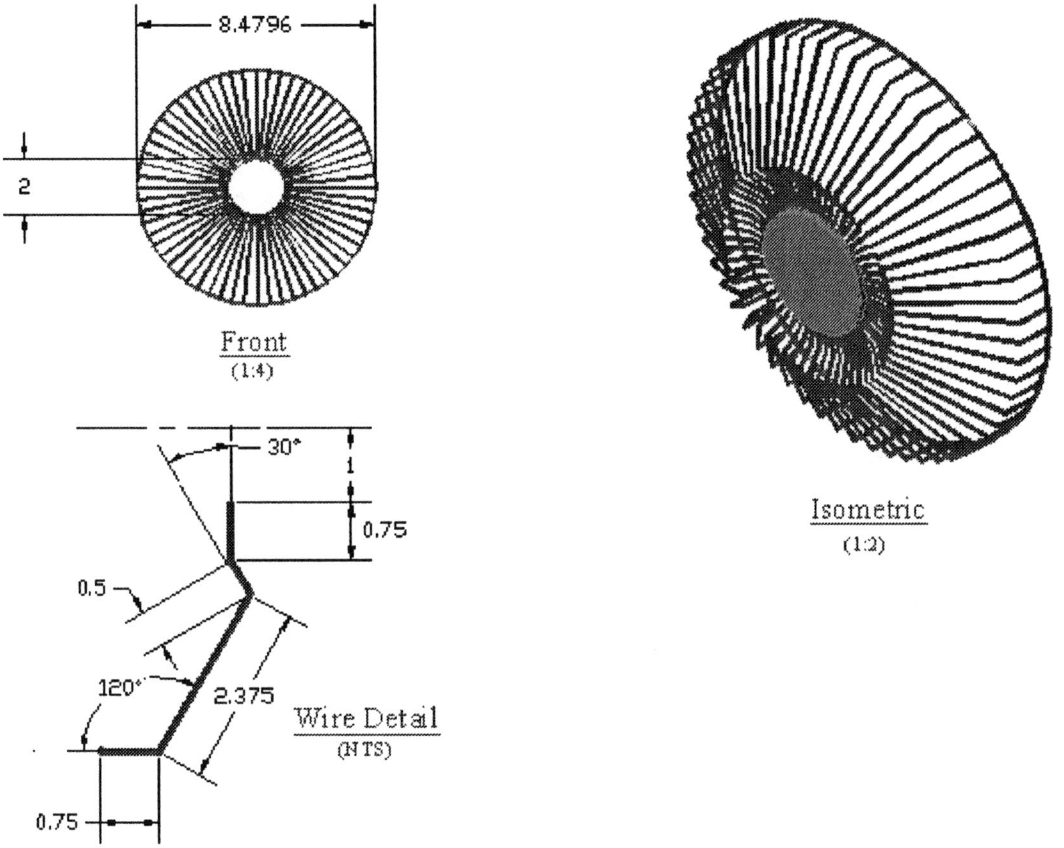

Figure 7.025

15. Create the round planter drawing shown in Figure 7.026. Follow these guidelines.
    15.1. Use the *ANSI A Title Block* in AutoCAD's template folder.
    15.2. Text size is 3/16" and 1/8".
    15.3. Save the drawing as *MyPlanter.dwg* in the C:\Steps3D\Lesson07 folder.

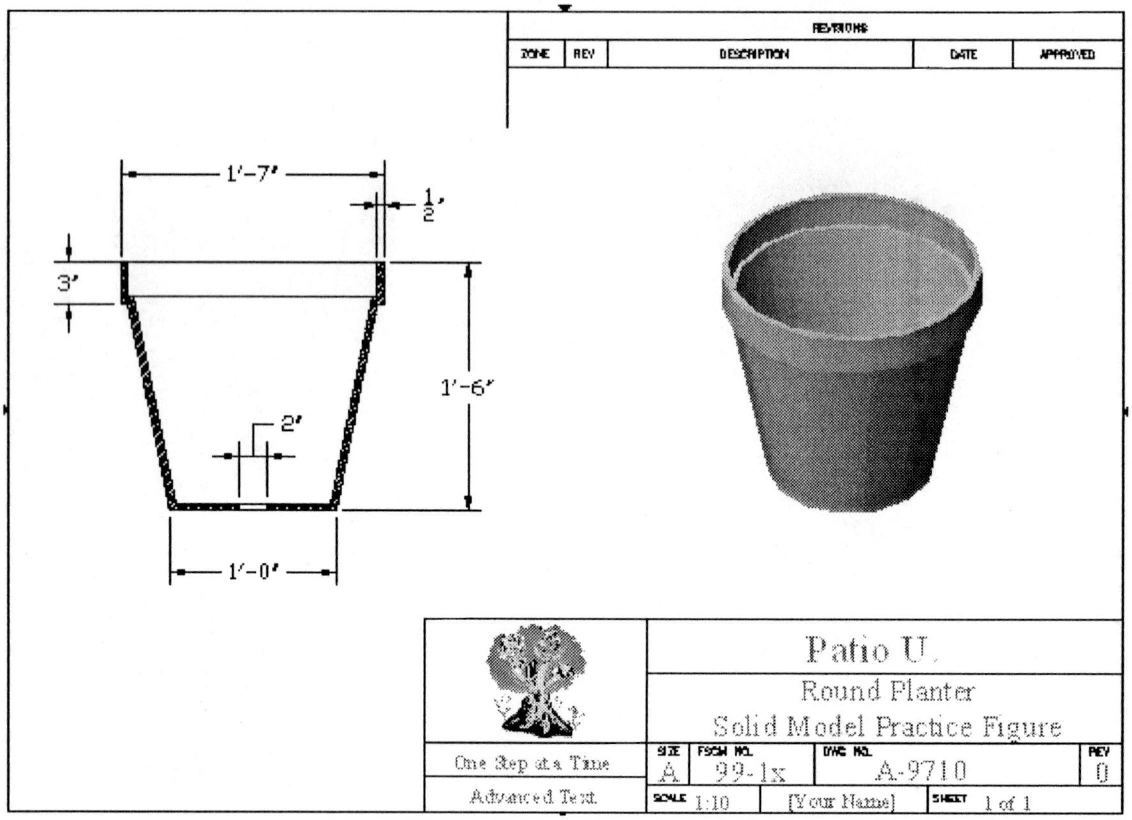

Figure 7.026

16. Create the double helix drawing shown in Figure 7.027. Follow these guidelines.
    16.1. The balls are 1" diameter.
    16.2. The rods are 1/8" diameter x 1" long.
    16.3. Each pairing rotates 15°.
    16.4. Save the drawing as *MyGenes.dwg* in the C:\Steps3D\Lesson07 folder.

Figure 7.027

17. Create the fence drawing shown in Figure 7.028. Follow these guidelines.
    17.1. The posts are 4 x 4s (3½" x 3½").
    17.2. The slats are 1 x 4s (¾" x 3½"). (I started with splines and turned them into regions.)
    17.3. The fence is 6" above the ground.
    17.4. The center rail is a 1 x 2 (¾" x 1½").
    17.5. Save the drawing as *MyFence.dwg* in the C:\Steps3D\Lesson07 folder.

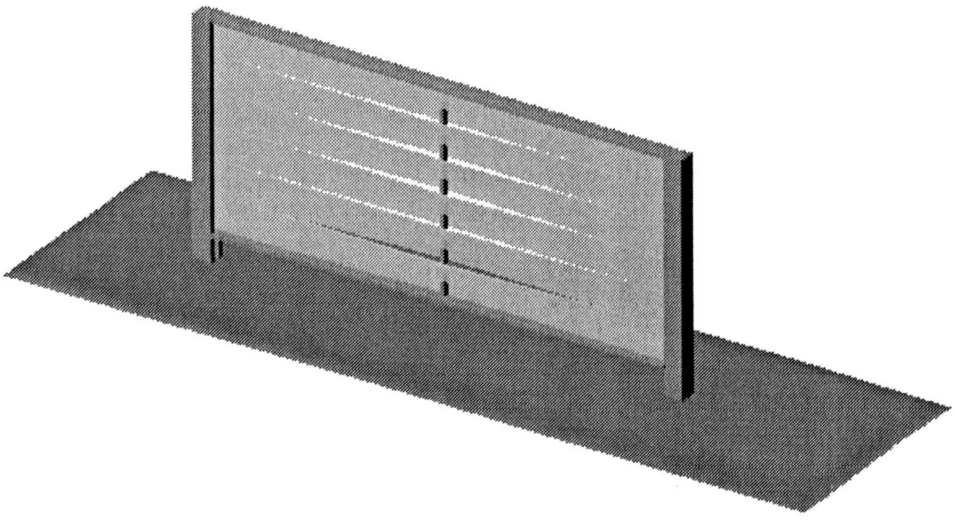

Figure 7.028

18. Create the patio planter box drawing shown in Figure 7.029. Follow these guidelines.
    18.1. Use the *ANSI A Title Block* found in AutoCAD's template folder.
    18.2. Text size is 3/16" and 1/8".
    18.3. Save the drawing as *MyPlanterBox.dwg* in the C:\Steps3D\Lesson07 folder.

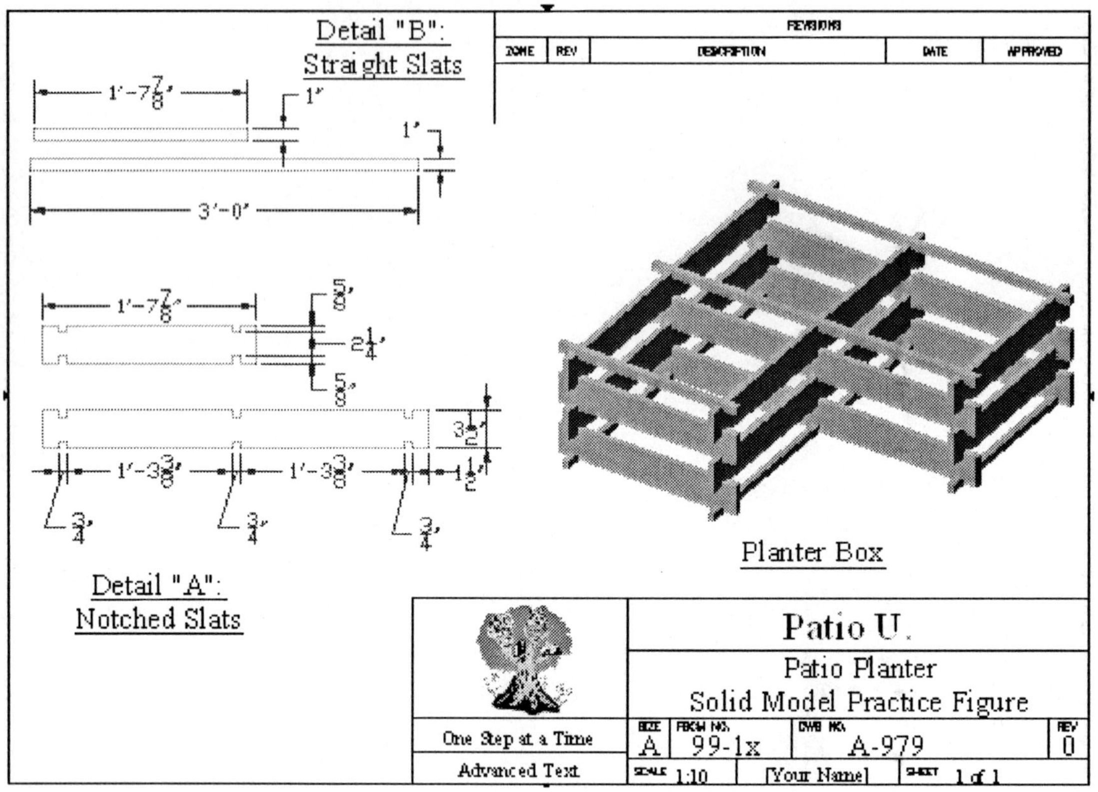

Figure 7.029

7.8	For Web-Based Review Questions, visit: http://www.uneedcad.com/2006/Files/06R7-3D.pdf

Lesson 8

Following this lesson, you will:

- ✓ *Know how to create composite solids from AutoCAD's solid building blocks using these commands:*
  - ***Union***
  - ***Subtract***
  - ***Intersect***
  - ***Slice***
  - ***Interfere***
- ✓ *Know how to calculate mass properties of a solid*
- ✓ *Know how to create a cross section using the **Section** command*
- ✓ *Know how to shape solids using these commands:*
  - ***Fillet***
  - ***Chamfer***

Composite Solids

I think I can … I think I can … I think I can …

Watty Piper's *The Little Engine That Could*

You've made it through the beginnings of Solid Modeling. You've experienced successes and near misses throughout your study, but the semester is half over. You may be tired and thinking more about Christmas or Easter or Labor Day than AutoCAD. You may be wondering, "Why AutoCAD … why school … why spend all this time and money to educate (or reeducate) myself?"

Let's pause for a paragraph or two for some words of encouragement. Let me tell you where you are in the overall scheme of (AutoCAD) things.

In all professions, there's a turning point – the point where the draftsman becomes the engineer, where the painter becomes the artist, the idea becomes the design. In every life there is (hopefully) a time where adolescence gives way to adulthood. In the world of computer drafting, you're at that turning point. You're about to leave the CAD Draftsman designation behind and become a true CAD Operator.

In the next few lessons, you'll discover how to take the building blocks – 3D solids and all of the basic and advanced modifying tools you've learned (and some you'll learn now) – and create *objects*. (Notice I didn't say, "create *drawings*.") You'll show the objects you create *in* drawings, but be assured that you'll be creating objects.

Once you've accomplished that very doable goal, I'll show you how to apply materials to those objects – how to show the wood grain on a table or make glass transparent. We'll cover this in the lesson on rendering.

But for now, let me offer these words of encouragement: Approach these lessons with the confidence of a graduate moving into graduate school. You've many successes under your belt, but the best is yet to come!

---

*You may have noticed that AutoCAD has two Solids toolbars – the first is simply Solids and the second is Solids Editing. Between them, there are quite a few opportunities for creating or modifying three-dimensional solid objects. We've already discussed eight of these – the Object Creation commands (**Box**, **Wedge**, **Cone**, **Cylinder**, **Sphere**, **Torus**, **Extrude**, and **Revolve**). I'll group the remaining tools into loose categories to help your understanding. These categories include Construction/Shaping Tools (**Slice**, **Section**, **Interfere**, **Union**, **Subtract**, **Intersect**, **Fillet**, and **Chamfer**), Solid Editing Tools (the multifaceted **Solidedit** command), and some Print Setup commands exclusively for use with solid objects (**Solprof**, **Soldraw**, and **Solview**).*

*We'll consider the Construction/Shaping tools in Lesson 8, and the many tools hidden within the **Solidedit** command in Lesson 9. (An entire lesson for one command? Doesn't that fill you with dread!?) We'll save the Print Setup commands for Lesson 10.*

8.1 Solid Construction Tools

When you were a child, did you ever wish you could fuse your playing blocks together in order to preserve a particularly clever building effort? You wanted to keep that castle or tower forever to demonstrate your prowess with the tools of your trade. You wanted your family and friends to be able to see – years from now – how you built the perfect model!

And then your sister rode through on her tricycle and your dreams were shattered.

Well, the programmers at AutoCAD had sisters, too. So they created a way to permanently fuse their computer blocks so that no one could ever disassemble them. Then they followed their childhood fantasies and created ways to remove parts of their blocks by using different shapes to define the carving. And they invented ways to find interferences between their blocks and to create cross sections …

… and their childhood dreams became reality in Solid Modeling Construction Tools.

Let's see how they work.

8.1.1 Union

The ***Union*** command behaves very much like a computer-controlled welding rod. It combines two solid objects into one. But unlike the welding rod, it leaves no seams that can break!

It's one of the simplest tools you'll ever hope to find. The command sequence is

> **Command: *union* (or *uni*)**
> **Select objects: *[select the solids you wish to weld]***
> **Select objects: *[confirm the selection]***

They just don't come any easier. But the value of the ***Union*** command can't be overstated. It turns simple objects like cylinders and boxes into production models like flanges, tools, doorstops, doorknobs, and much, much more.

We have to try this one. We'll create a flange over the next few exercises by using two construction tools and several cylinders.

> You can also access the ***Union*** command (as well as the ***Subtract*** and ***Intersect*** commands) using the Modify pull-down menu. Follow this path:
> *Modify – Solids Editing – Union*

Do This: 8.1.1.1 Welding Solid Objects with the *Union* Command

I. Open the *flange.dwg* file in the C:\Steps3D\Lesson08 folder. The drawing looks like Figure 8.001.

II. Follow these steps.

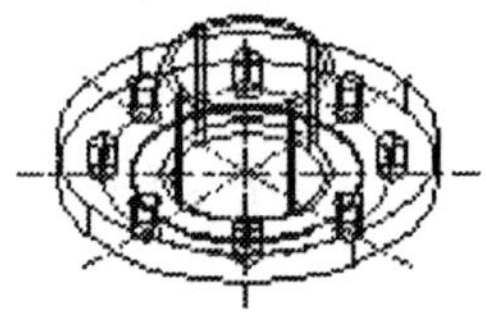

Figure 8.001

TOOLS	COMMAND SEQUENCE	STEPS
	Command: *hide*	1. Remove hidden lines. Your drawing looks like Figure 8.002. Notice that the center cylinder disappears into the bottom cylinder with no visible connection.

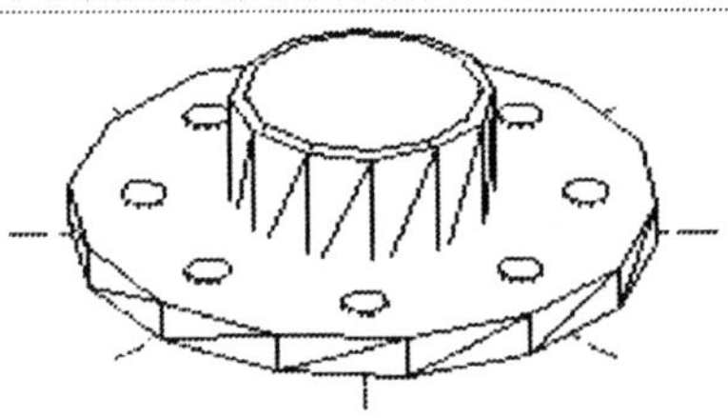

Figure 8.002

Tools	Command Sequence	Steps
	Command: ***re***	2. Regenerate the drawing.
Union Button	**Command:** ***uni***	3. Enter the ***Union*** command. Alternately, you can pick the **Union** button on the Solids Editing toolbar.
	Select objects:	4. Select the three cylinders indicated in Figure 8.003.

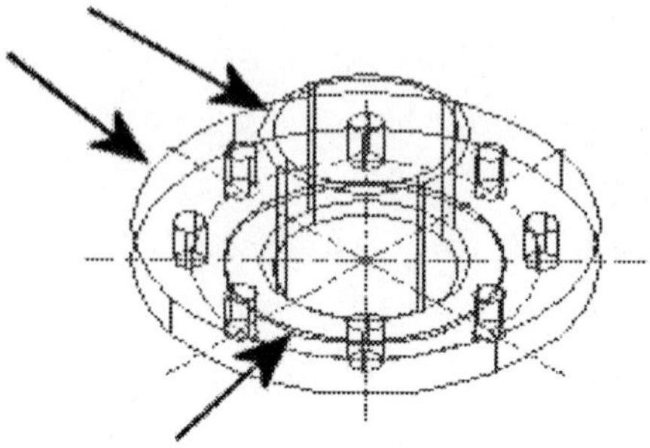

Figure 8.003

Tools	Command Sequence	Steps
	Select objects: ***[enter]*** **Command:** ***hide***	5. Confirm the selection and remove the hidden lines. Your drawing looks like Figure 8.004. The selected cylinders have become a single unit. (You can try to erase one of them to verify this.)

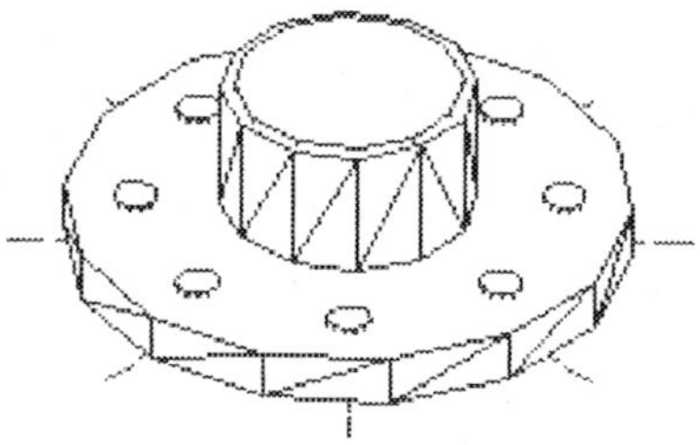

Figure 8.004

Tools	Command Sequence	Steps
	Command: ***qsave***	6. Save the drawing, but don't exit.

Of course, the ***Union*** command is only one side of the coin. If you can add objects to each other, you should be able to remove one object from another as well.

Let's look at the ***Subtract*** command.

8.1.2 Subtract

We created a solid flange in our last exercise, but a flange has little use if nothing can flow through it. Enter the ***Subtract*** command.

You're already familiar with the ***Subtract*** command from Lesson 3. We used it in Section 3.3.3 to put windows in our walls. Subtract works the same on 3D solids as it did on regions.

We'll use it to remove the boltholes and the core of our flange.

Do This: 8.1.2.1	**Removing One Solid from Another**

I. Be sure you're still in the *flange.dwg* file in the C:\Steps3D\Lesson08 folder. If not, please open it now.

II. Follow these steps.

TOOLS	COMMAND SEQUENCE	STEPS
Subtract Button	**Command:** ***su***	1. Enter the ***Subtract*** command. Alternately, you can pick the **Subtract** button on the Solids Editing toolbar.
	Select solids and regions to subtract from .. **Select objects:** **Select objects:** ***[enter]***	2. AutoCAD asks you to select the object from which you wish to remove something. Select the solid you created in our last exercise (select the outermost cylinder). Confirm the selection.
	Select solids and regions to subtract .. **Select objects:** **Select objects:** ***[enter]***	3. AutoCAD now wants to know what to remove. Select the innermost cylinder and each of the smaller cylinders arrayed about the flange (the bolt holes). Confirm this selection as well, and remove the hidden lines. Your drawing looks like Figure 8.005.
	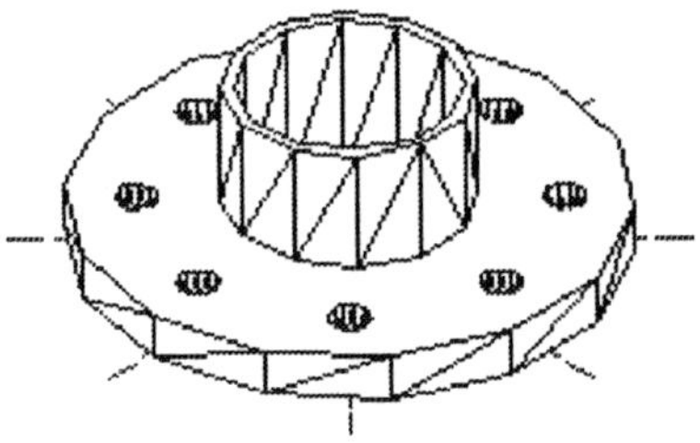 Figure 8.005	
	Command: ***qsave***	4. Save and close the drawing.

These first two exercises have been fairly easy and straightforward. But take a moment and fill in the blanks in the following sentences with as many answers as you can. (Time yourself and see how many answers you can find in 60 seconds.)

I can use the ***Union*** command to create __________ out of __________.

I can use the ***Subtract*** command to create __________ out of __________.

Here are some hints:

- Look about the room and consider objects on the desk, floor, walls, and shelves.
- Imagine that you have the blocks with which you played as a child, but now you have a bottle of glue, a drill, and a chisel, as well.

Are you beginning to see the possibilities?

8.1.3 Intersect

The ***Intersect*** command comes in handy when creating intricate multisided figures. It works by removing everything that doesn't intersect something else. This will become clearer with an exercise, but first look at the command sequence.

Command: *intersect* (or *in*)

Select objects: *[select objects that intersect each other]*

Select objects: *[confirm the selection set]*

If only all commands accomplished as much – as easily!

Let's see what we can do with the ***Intersect*** command.

Do This: 8.1.3.1 Creating Objects at Intersections

I. Open the *emerald.dwg* file in the C:\Steps3D\Lesson08 folder. The drawing looks like Figure 8.006. (It's two octogons extruded at 30° to become solid objects. We mirrored the objects to have the four solids you see now.)

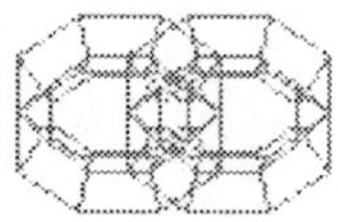

Figure 8.006

II. Follow these steps.

TOOLS	COMMAND SEQUENCE	STEPS
 Intersect Button	**Command: *in***	1. Enter the ***Intersect*** command. Alternately, you can pick the **Intersect** button on the Solids Editing toolbar.
	Select objects: **Select objects: *[enter]***	2. Select the four octagon solids.
	Command: *vp* **Command: *shademode***	3. Adjust the viewpoint for clarity. (I used a viewpoint of 1,-1,1.) Set the Shademode to **Gouraud**. Your drawing looks like Figure 8.007.

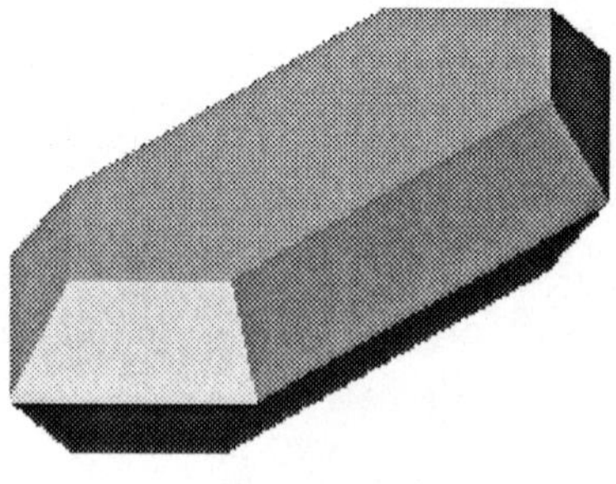

Figure 8.007

	**Command: *qsave***	4. Save the drawing, but don't exit.

Wow! What else can we do?!

8.1.4	Slice

Did you know that the value of precious stones often increases when they're cut just right? Let's cut our emerald.

The command we'll use is called ***Slice***. Use the ***Slice*** command to make a straight cut or remove a piece of an object as if cutting it away with a knife. The sequence offers more options than the others we've seen in this lesson, but that means more opportunities to cut it the way you want it cut.

It looks like this:

> **Command:** ***slice*** **(or** ***sl*****)**
>
> **Select objects:** ***[select the solid object to cut]***
>
> **Select objects:** ***[confirm the selection set]***
>
> **Specify first point on slicing plane by [Object/Zaxis/View/XY/YZ/ZX/3points]**
>
> **<3points>:** ***[pick three points on the object to define the slicing plane]***
>
> **Specify second point on plane:**
>
> **Specify third point on plane:**
>
> **Specify a point on desired side of the plane or [keep Both sides]:** ***[pick a point on the side of the object you wish to keep]***

The options should be familiar from your study of the ***Rotate3d*** and ***Mirror3d*** commands, but let's go over them again to be sure.

- The default option requires that you select **3points** to define a slicing plane. This is like drawing the knife blade that'll be slicing through the object.
- The **Object** option is still the easiest. If you have an object (circle, ellipse, arc, spline, or polyline) drawn through the object you want to slice, you can select it as your slicing plane.
- The **Zaxis** option is difficult to follow. It prompts like this:

  > **Specify a point on the section plane:**
  >
  > **Specify a point on the Z-axis (normal) of the plane:**

  - The first prompt is asking for a point on the slicing plane.
  - The next prompt is asking for a point on the Z-axis of the slicing plane. Use this point to orient the slicing plane – essentially by picking a point to define which way is "up" if you're standing on the slicing plane.
- The **XY/YZ/ZX** options allow you to define the slicing plane by identifying a single point on the chosen plane. AutoCAD defines these planes according to the current UCS.
- The last option occurs at the final prompt. It allows you to keep either a selected piece (the default) or both pieces of the object after you've sliced it.

Let's cut our gemstone.

> You can also access the ***Slice*** command (as well as the ***Interfere*** and ***Section*** commands) using the Draw pull-down menu. Follow this path:
>
> *Draw – Solids – Slice*

Do This: 8.1.4.1	Slicing 3D Solid Objects

I. Be sure you're still in the *emerald.dwg* file in the C:\Steps3D\Lesson08 folder. If not, please open it now.

II. Set the Shademode to **Hidden**.

III. Follow these steps.

TOOLS	COMMAND SEQUENCE	STEPS
 Slice Button	**Command:** ***sl***	1. Enter the ***Slice*** command. Alternately, you can pick the **Slice** button on the Solids toolbar.
	Select objects: **Select objects:** ***[enter]***	2. Select the emerald.
	Specify first point on slicing plane by [Object/Zaxis/View/XY/YZ/ZX/3points] **<3points>:** ***[select point 1]*** **Specify second point on plane:** ***[select point 2]*** **Specify third point on plane:** ***[select point 3]***	3. (Refer to Figure 8.008.) Select the points indicated.

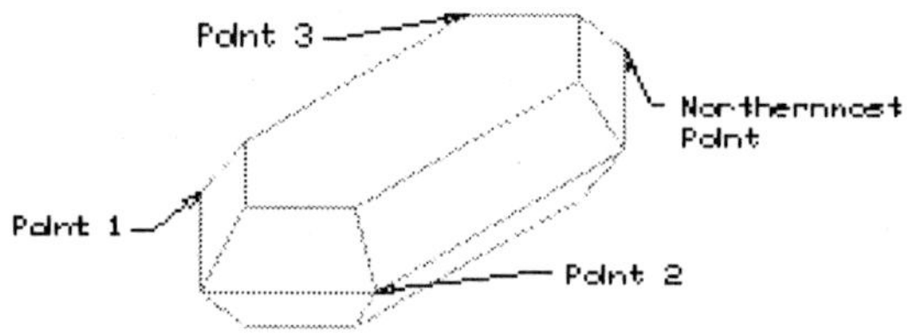

Figure 8.008

TOOLS	COMMAND SEQUENCE	STEPS
	Specify a point on desired side of the plane or [keep Both sides]:	4. Pick the northernmost endpoint on the emerald (indicating that you want to keep that section of the gem). Your drawing looks like Figure 8.009.

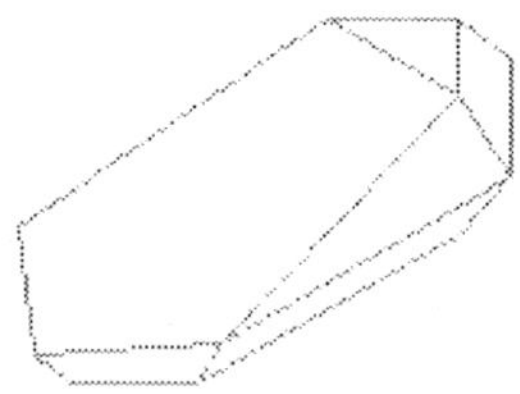

Figure 8.009

TOOLS	COMMAND SEQUENCE	STEPS
	**Command:** ***u***	5. Undo the changes.
		6. Let's try the **Object** option. First, thaw the **obj1** layer. Notice the circle that intersects the emerald (Figure 8.010).

Tools	Command Sequence	Steps
	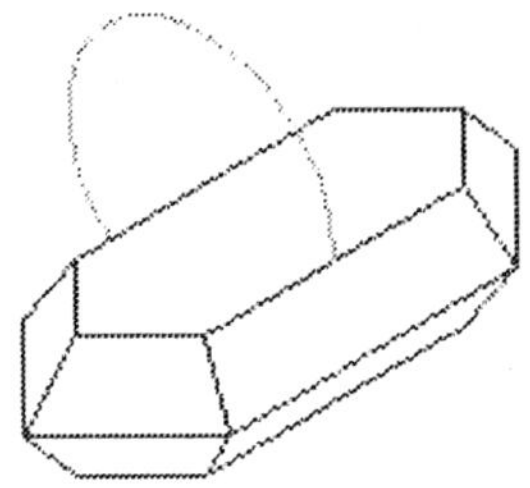 Figure 8.010	
	**Command:** ***sl***	7. Repeat Steps 1 and 2.
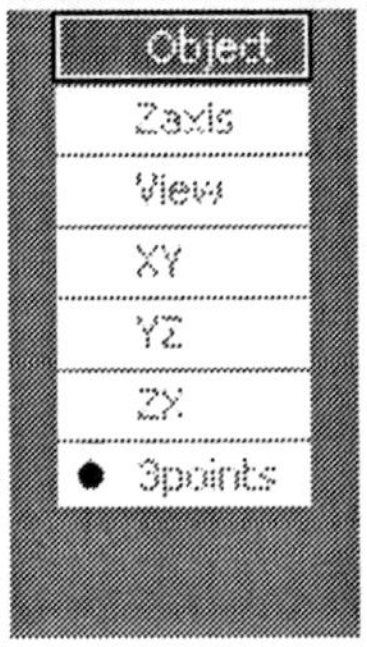	**Specify first point on slicing plane by [Object/Zaxis/View/XY/YZ/ZX/ 3points] <3points>:** ***O***	8. Choose the **Object** option.
	Select a circle, ellipse, arc, 2D-spline, or 2D-polyline:	9. Select the circle …
	Specify a point on desired side of the plane or [keep Both sides]:	10. … and pick the northernmost point on the emerald. Your drawing looks like Figure 8.011.
	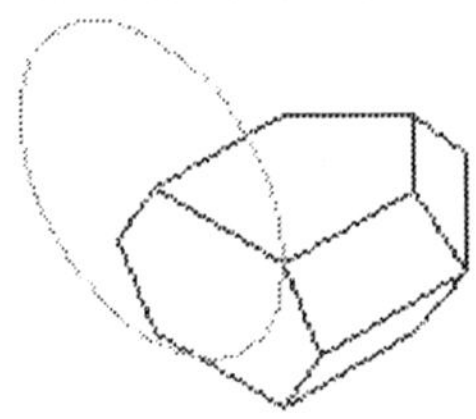 Figure 8.011	
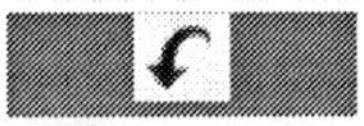	**Command:** ***u***	11. Undo the changes.
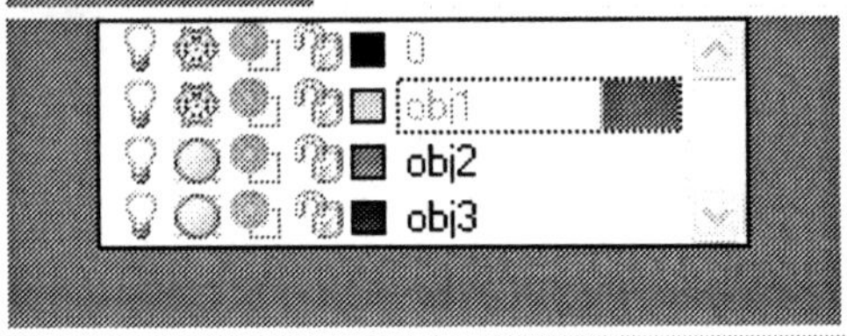		12. Now we'll try the **Zaxis** option. First, freeze the **obj1** layer and thaw the **obj3** layer. Notice the line that retreats from the center of the object (Figure 8.012).
	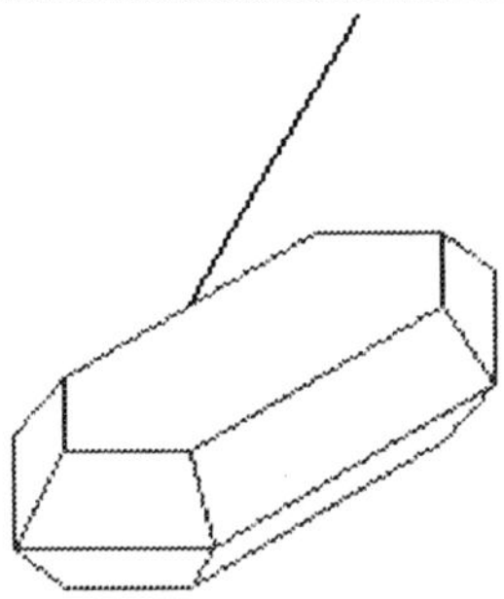 Figure 8.012	
	**Command:** ***sl***	13. Repeat Steps 1 and 2.

TOOLS	COMMAND SEQUENCE	STEPS
Zaxis	**Specify first point on slicing plane by [Object/Zaxis/View/XY/ YZ/ZX/3points] <3points>:** *z*	14. Choose the **Zaxis** option.
	Specify a point on the section plane:	15. Pick the endpoint of the line where it meets the emerald. The slicing plane will pass through this point.
	Specify a point on the Z-axis (normal) of the plane:	16. Pick the other endpoint of the line. The slicing plane will be *perpendicular* to the line you identified with the last pick.
	Specify a point on desired side of the plane or [keep Both sides]: *b*	17. This time, let's **keep Both sides.** A line appears through the emerald.
	Command: *m*	18. Use the ***Move*** command to separate the two pieces of the emerald. Your drawing looks like Figure 8.013.

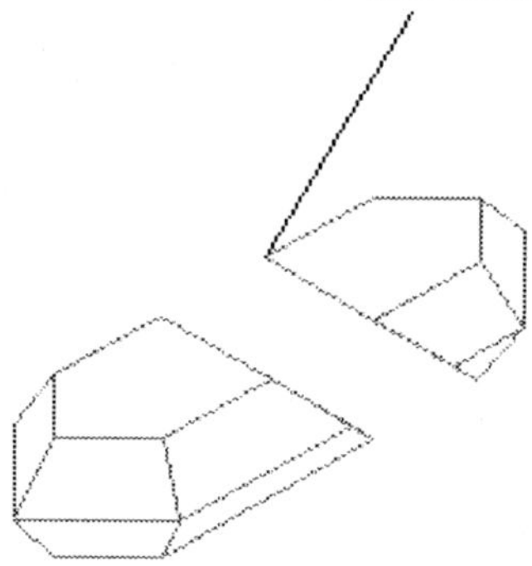

Figure 8.013

TOOLS	COMMAND SEQUENCE	STEPS
	Command: *u*	19. Undo the changes until the emerald is one piece again.
	Command: *sl*	20. Repeat Steps 1 and 2.
View	**Specify first point on slicing plane by [Object/Zaxis/View/XY/YZ/ZX/ 3points] <3points>:** *v*	21. Use the **View** option.
	Specify a point on the current view plane <0,0,0>:	22. Pick the endpoint of the line where it meets the emerald …
	Specify a point on desired side of the plane or [keep Both sides]:	23. …and keep the back part of the gemstone. Your drawing looks like Figure 8.014. AutoCAD has created a slicing plane parallel to your view and through the point you specified in Step 22.

TOOLS	COMMAND SEQUENCE	STEPS
	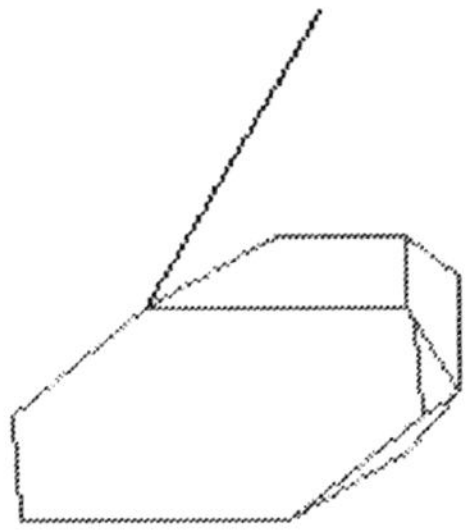 Figure 8.014	
	**Command:** ***u***	24. Undo the changes and freeze the **obj3** layer.
	**Command:** ***sl***	25. Repeat Steps 1 and 2.
	**Specify first point on slicing plane by [Object/Zaxis/View/XY/YZ/ZX/ 3points] <3points>:** ***xy***	26. Now we'll slice along a UCS plane. Use the **XY** plane.
	Specify a point on the XY-plane <0,0,0>:	27. Select the front point (southernmost) on the emerald. AutoCAD will cut through this point.
	Specify a point on desired side of the plane or [keep Both sides]:	28. Keep the bottom of the emerald. Your drawing looks like Figure 8.015.
	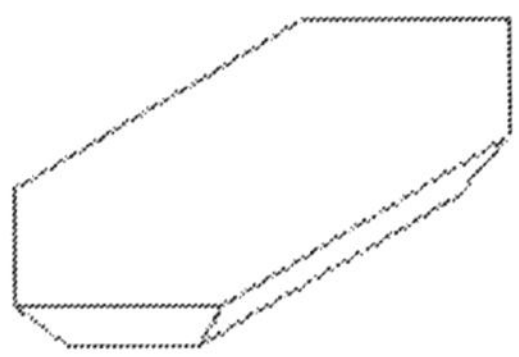 Figure 8.015	
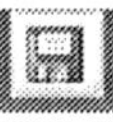	**Command:** ***qsave***	29. Save the drawing and exit.

As I said, the ***Slice*** command is slightly more involved than the commands we learned earlier, but you can see why.

Most people find one method of doing things easier than other methods. With the ***Slice*** command, AutoCAD gives you plenty of methods from which to choose. Knowing all the ways to accomplish an intended goal, however, may save you some time and hassle later when the preferred method refuses to work.

> Suppose you had to determine the volume of the object we created in our last exercise (the cut emerald). Can you think of an easy way?
>
> AutoCAD provides a tool to make the calculations downright easy – ***Massprop***. The ***Massprop*** command will compute not only the volume of the selected object but also the mass, bounding box, centroid, moments and products of inertia, radii of gyration, and principle moments and directions about the centroid. All you have to do is enter the command and select the object! (And your boss spent all those years in engineering school learning how to do this on a slide rule!)

8.1.5 Interfere

Using the ***Interfere*** command, you can identify problems cheaply and easily *before* construction finds them.

Interfere identifies places in a drawing where one solid interferes with another. You'll use it more as a checking tool once you've completed the drawing than as a drawing tool itself – although you can use it to draw an interference.

The command sequence looks like this:

> **Command: *interfere* (or *inf*)**
>
> **Select first set of solids: *[identify the solids you want to check]***
>
> **Select objects: *[confirm the selection]***
>
> **Select second set of solids: *[identify a second set of solids if you wish to check one against the other (otherwise, AutoCAD will check all the objects in the first set against each other)]***
>
> **Select objects: *[confirm the selection]***
>
> **Interfering solids: *[AutoCAD tells you how many objects interfere with other objects]***
>
> **Interfering pairs : *[AutoCAD tells you how many occurrences of interference it has discovered]***
>
> **Create interference solids? [Yes/No] <N>: *[if you wish, AutoCAD will create a solid from the interference(s) – much as it does with the* Intersect *command – except that nothing is removed]***
>
> **Highlight pairs of interfering solids? [Yes/No] <N>: *[if you wish, AutoCAD will show you each interference one at a time]***

AutoCAD presents no options that might confuse you, but there is a quirk that can be useful. If you select all the objects to check at the first selection prompt (**Select first set of solids**) and simply hit ***enter*** at the second prompt (**Select second set of solids**) AutoCAD will check all of the solids in the selection set against each other. Otherwise, AutoCAD checks objects in the first selection set against objects in the second selection set only.

Let's see this one in action.

Do This: 8.1.5.1 Interference Detection

I. Open the *pipe08.dwg* file in the C:\Steps3D\Lesson08 folder. The drawing looks like Figure 8.016. (It's a simple piping plan with a two-level piperack. Can you see any interferences?)

II. Follow these steps.

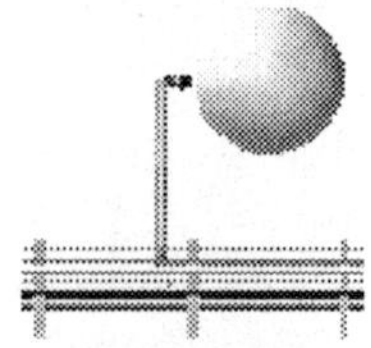

Figure 8.016

TOOLS	COMMAND SEQUENCE	STEPS
Interfere Button	**Command: *inf***	1. Begin the ***Interfere*** command. (The button is on the Solids toolbar.)
	Select first set of solids: **Select objects: *all*** **Select objects: *[enter]***	2. We'll check the entire drawing for interferences. At the **Select first set of solids** prompt, type ***all***, and then confirm the selection set.

TOOLS	COMMAND SEQUENCE	STEPS
	Select second set of solids: **Select objects:** ***[enter]*** **No solids selected.** **Comparing 38 solids with each other.**	3. We'll check all of the solids against each other, so hit ***enter*** at the **Select second set of solids** prompt. AutoCAD tells you that it's comparing the solids with each other (this may take a few moments).
	Interfering solids: 3 **Interfering pairs : 2**	4. AutoCAD found three solids hitting each other in two instances of interference. Notice that it highlights the interfering solids (Figure 8.017).

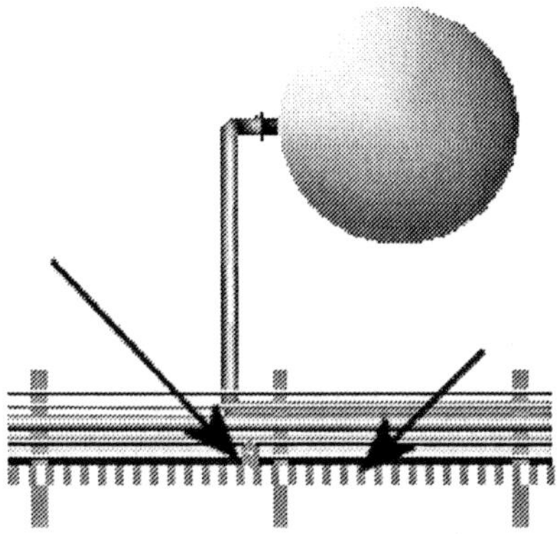

Figure 8.017

	Create interference solids? [Yes/No] <N>: ***[enter]***	5. We don't need to create a solid at the interference …
	**Highlight pairs of interfering solids? [Yes/No] <N>:** ***y***	6. … but we're not sure which objects interfere (flange against flange or two instances of flange against pipe). So we'll ask AutoCAD to highlight the interfering pairs. AutoCAD highlights a flange and the pipe.
	**Enter an option [Next pair/ eXit] <Next>:** **Enter an option [Next pair/ eXit] <Next>:** ***x***	7. Hit ***enter*** to see the other interference. AutoCAD highlights the other flange and the pipe. ***Exit*** the command.
	Command: ***quit***	8. Exit the drawing without saving it.

In a few simple steps, you've located a problem that might have cost tons of money in redesign and construction costs.

We have another timesaver to see, but first, let's visit some old friends.

8.2 Using Some Old Friends on Solids – *Fillet* and *Chamfer*

Remember how much fun you had drawing the outhouse door back in Lesson 8 of the basic text? You used the ***Fillet*** and ***Chamfer*** commands to make the corners (twice). Where would you be if you couldn't use those convenient tools on 3D solids?

Luckily, AutoCAD saw the need to round and mitre corners on solid objects and made the tools available – with a few necessary adjustments. Look at the command sequences when these two are used on solids (we'll begin with the ***Fillet*** command):

> **Command:** ***fillet*** **(or** ***f*****)**
>
> **Current settings: Mode = TRIM, Radius = 0.5000**
>
> **Select first object or [Undo/Polyline/Radius/Trim/Multiple]:** ***[select a solid]***
>
> **Enter fillet radius:** ***[enter the desired radius]***
>
> **Select an edge or [Chain/Radius]:** ***[select the edge to fillet]***

The command begins just as it did when you studied it in the basic text. But when you select a solid object at the **Select first object** prompt, AutoCAD recognizes the solid and asks for some different input.

- It immediately prompts for a **radius** – something it didn't do for a two-dimensional object.
- You can hit ***enter*** at the **Select an edge** prompt and AutoCAD will assume that you intend to fillet the edge you selected at the **Select first object** prompt. It'll then proceed to fillet that edge. Alternately, you can choose one of the other options:
  - When you pick a single edge on the surface of a solid while using the **Chain** option, AutoCAD should pick the other lines on that surface that are sequential and tangential to the one you selected. [Frankly, I've never been impressed by the way this works (or doesn't work).]
  - The **edge** option (the default) simply allows you to select the edges to fillet one at a time.
  - The **Radius** option allows you to change the radius of each edge you select.

The command sequence for the ***Chamfer*** command is

> **Command:** ***chamfer*** **(or** ***cha*****)**
>
> **(TRIM mode) Current chamfer Dist1 = 0.5000, Dist2 = 0.5000**
>
> **Select first line or [Undo/Polyline/Distance/Angle/Trim/mEthod/Multiple]:** ***[select a solid]***
>
> **Base surface selection...**
>
> **Enter surface selection option [Next/OK (current)] <OK>:** ***[each edge naturally has two surfaces that are next to it; hit*** **enter** ***if the correct surface is highlighted or type*** **N** ***to toggle between the surfaces until the appropriate one highlights]***
>
> **Specify base surface chamfer distance <0.5000>:** ***[enter the chamfer distances]***
>
> **Specify other surface chamfer distance <0.5000>:**
>
> **Select an edge or [Loop]:** ***[select the edge to chamfer]***

- The first option, as explained, allows you to select the correct surface to chamfer.
- The next two options – **Specify base surface chamfer distance** and **specify other surface chamfer distance** – allow you to accept or change the chamfer distances (notice that there's no **Angle** option here as there is for two-dimensional objects).
- The last option – **Select an edge or [Loop]** – allows you to pick the edges to chamfer individually or, when you use the **Loop** option, collectively around the entire surface.

Let's try the ***Fillet*** and ***Chamfer*** commands on a solid.

Do This: 8.2.1	Solid Fillets and Chamfers

I. Open the *flange.dwg* file in the C:\Steps3D\Lesson08 folder. If you haven't completed it yet, open the *Flange-done.dwg* file instead. We'll countersink the boltholes, mitre the weld neck (the upper cylinder), and fillet the edge.

II. Set the Shademode to **Hidden**.

III. Follow these steps.

TOOLS	COMMAND SEQUENCE	STEPS
	**Command:** ***f***	1. Let's begin with a simple fillet. Enter the ***Fillet*** command.
	Current settings: Mode = TRIM, Radius = 0.5000 **Select first object or Undo/ Polyline/ Radius/rim/Multiple]:**	2. Select the top surface of the base of the flange.
	Enter fillet radius <0.5000>: .25	3. Set the radius to ¼ unit.
	Select an edge or [Chain/Radius]: ***[enter]***	4. Complete the command. Your drawing looks like Figure 8.018.

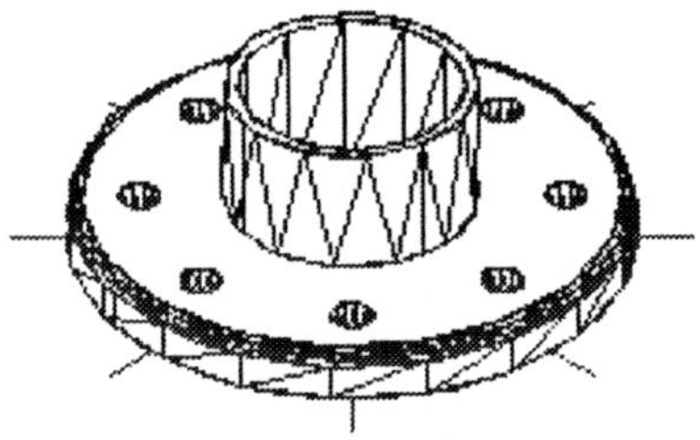

Figure 8.018

TOOLS	COMMAND SEQUENCE	STEPS
	**Command:** ***cha***	5. Now we'll mitre the weld neck. Enter the ***Chamfer*** command.
	(TRIM mode) Current chamfer Dist1 = 0.5000, Dist2 = 0.5000 **Select first line or [Undo/Polyline/ Distance/Angle/Trim/mEthod/ Multiple]:**	6. Select the upper cylinder …
	Base surface selection... **Enter surface selection option [Next/OK (current)] <OK>:** ***n*** **Enter surface selection option [Next/OK (current)] <OK>:** ***[enter]***	7. … and adjust the surface selected until just the outer circle of the cylinder is highlighted.
	Specify base surface chamfer distance <0.5000>: ***.125*** **Specify other surface chamfer distance <0.5000>:** ***.125***	8. Set the chamfer distances to 1/8 unit.

Tools	Command Sequence	Steps
	Select an edge or [Loop]: **Select an edge or [Loop]:** ***[enter]***	9. Select the outer edge of the upper cylinder and then confirm the selection. Your drawing looks like Figure 8.019.

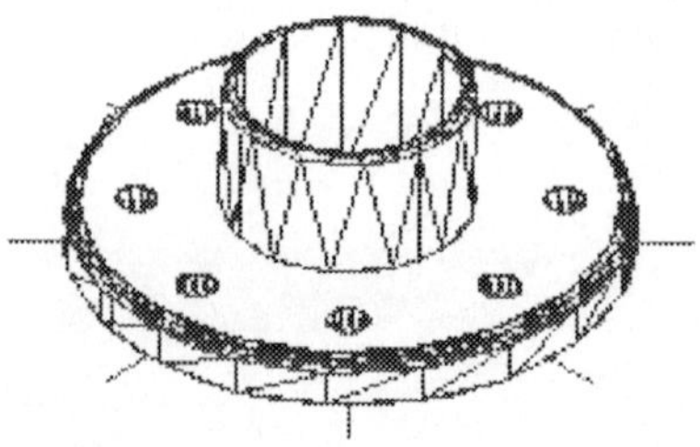

Figure 8.019

Tools	Command Sequence	Steps
	Command: ***[enter]***	10. Now countersink the bolt holes. Repeat the ***Chamfer*** command.
	(TRIM mode) Current chamfer Dist1 = 0.1250, Dist2 = 0.1250 **Select first line or [Undo/Polyline/ Distance/Angle/Trim/mEthod/ Multiple]:** **Base surface selection...** **Enter surface selection option [Next/OK (current)] <OK>:** ***[enter]***	11. Select the upper surface of the base of the flange.
	Specify base surface chamfer distance <0.1250>: ***.25*** **Specify other surface chamfer distance <0.1250>:** ***.25***	12. Set the chamfer distances to ¼ unit.
	Select an edge or [Loop]:	13. Select the upper circle around each of the bolt holes.
	Select an edge or [Loop]: ***[enter]***	14. Complete the command and set the Shademode to **Gouraud**. Your drawing looks like Figure 8.020.

Figure 8.020

Tools	Command Sequence	Steps
	Command: ***qsave*** **Command:** ***saveas***	15. Save the drawing, but don't exit. Save it a second time as *MyFlange10* in the C:\Steps3D\Lesson10 folder. (We'll do more with it later). Close the drawing.

Now let's look at that other timesaver – the ***Section*** command.

8.3 Creating Cross Sections the Easy Way – The *Section* Command

Have you ever completed the tedious cross section of an object only to discover that you missed something (perhaps a line or an arc that was difficult to see)? Well, AutoCAD has just the tool for you!

The ***Section*** command creates cross sections of solid objects. And it's one of AutoCAD's easiest commands to master!

The command sequence is

Command: ***section*** **(or** ***sec*****)**

Select objects: ***[select one of more solids to section]***

Select objects: ***[confirm the selection]***

Specify first point on Section plane by [Object/Zaxis/View/XY/YZ/ZX/3points]

<3points>: ***[identify three points on the section plane (to define it)]***

Specify second point on plane:

Specify third point on plane:

It's really just that simple. AutoCAD does the rest and places the section inside the object(s) being section. Move it to a suitable place on the drawing, add section (hatch) lines, and you're finished!

Some things to note about the ***Section*** command:

- The options for defining the section plane are identical to those used to define a slice plane.
- The section that AutoCAD creates is a region. You can hatch a region, or you can explode it into lines and arcs.
- The section created is aligned with the section plane – rotate it as necessary to align it to the UCS.
- AutoCAD creates the section on the current layer.

We'll create a cross section of our flange.

Do This: 8.3.1 Creating Cross Sections

I. Reopen the *flange.dwg* file (or the *flange-done.dwg* file) in the C:\Steps3D\Lesson08 folder.

II. Set the Shademode to **2D,** but remove hidden lines with the ***Hide*** command.

III. Set the **obj1** layer current.

IV. Follow these steps.

TOOLS	COMMAND SEQUENCE	STEPS
 Section Button	**Command:** ***sec***	1. Enter the ***Section*** command. The button is on the Solids toolbar.
	Select objects: **Select objects:** ***[enter]***	2. Select the flange.

Tools	Command Sequence	Steps
	Specify first point on Section plane by [Object/Zaxis/View/ XY/YZ/ZX/3points] <3points>:	3. Pick the westernmost endpoint of the east-west centerline.
	Specify second point on plane:	4. Pick the other endpoint of the same line.
	Specify third point on plane:	5. Pick the center of the top of the flange (use OSNAPs). Notice that AutoCAD creates a section inside the flange (Figure 8.021).

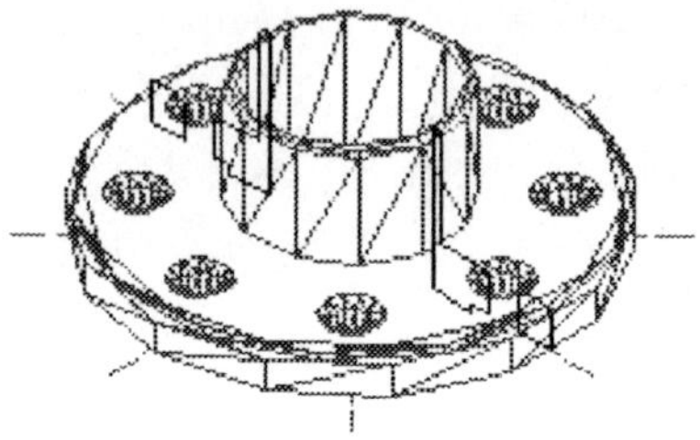

Figure 8.021

Tools	Command Sequence	Steps
	Command: ***m*** **Select objects:** ***l*** **Select objects:** ***[enter]*** **Specify base point or displacement:** ***12,0*** **Specify second point of displacement or <use first point as displacement>:** ***[enter]***	6. Move the section, as indicated, to see it better. (I selected the last item created and moved it using the displacement method.)
	Command: ***-vp*** **Current view direction: VIEWDIR=1.0000,-1.0000,1.0000** **Specify a view point or [Rotate] <display compass and tripod>:** ***0,-1,0***	7. Adjust the view to see the objects from the front. Your drawing looks like Figure 8.022.

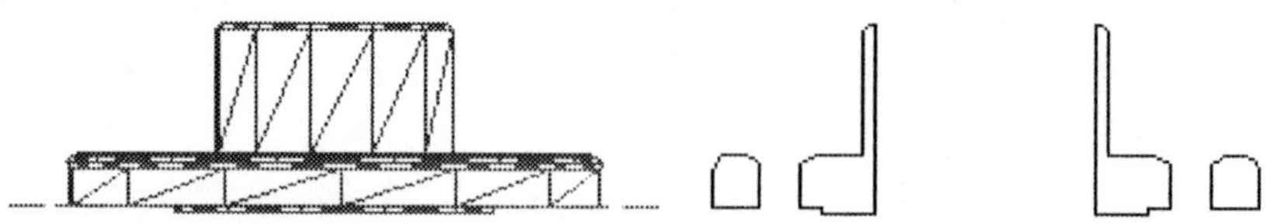

Figure 8.022

Tools	Command Sequence	Steps
	Command: ***qsave***	8. Now you can add hatching and centerlines to your cross section if you wish. Save and close the drawing.

How does that compare to drawing a cross section from scratch?

8.4 Extra Steps

Go back to the list of items you created at the end of Section 8.1.2. Take a few hours (or an afternoon) and see how many of them you can create in AutoCAD using the tools you learned in this lesson. I can't think of a better way to gain experience (or identify questions).

Some hints for this exercise:

- Don't attempt to draw anything that won't fit in a shoebox.
- One of those flexible, 6" rulers with inches on one side and millimeters on the other will serve you well (now and in the future).
- Try (at first) to limit yourself to objects that'll require no more than three of the basic shapes you've learned.
- You'll find other terrific objects to draw in garages and kitchens.

8.5 What Have We Learned?

Items covered in this lesson include:

- *Tools used to create composite solids*
  - ***Union***
  - ***Subtract***
  - ***Intersect***
  - ***Slice***
  - ***Interfere***
- *Tools used to shape solids*
  - ***Fillet***
  - ***Chamfer***
- *Other Solid Modeling tools*
  - ***Section***
  - ***Massprop***

Well, what do you think? Wouldn't it have been fun to have these tools when you were playing with blocks as a child?

As I promised, you've stopped drawing pictures of things and have actually begun creating objects using the tools in AutoCAD's "shop." I hope you can sense the potential of these tools from what you've seen here.

> When I was in junior high school, I read a book by Jack London called *Call of the Wild.* It was about a dog that was taken from an easy life in the Northwest and forced to pull a sled in the Klondike during the Gold Rush. Buck (the dog) had many adventures (learning experiences) as he adapted to the wild frontier life of the arctic. But all the while – with increasing intensity – he felt a call from the wild to move out on his own. He experimented with the urging – often leaving camp for days at a time to explore the wilderness. In the end, after learning all that he could in the safety and comfort of the camps, he answered the call and moved out to live with the other wild creatures.
>
> At this point in your AutoCAD training, you should be experimenting on your own, just as Buck did. You'll find thrills – and chills – as you discover things about the software that even the masters don't know. You'll make some mistakes, but the adventure lies in overcoming the mistakes (that's what makes learning fun!).
>
> In a few short chapters, you'll be on your own (with the other wild creatures in the design world). Learn all that you can now!

8.6 Exercises

1. through 8. Create the "su" drawings in Appendix B using solids. Use solid primitives and the composite solid creation tools you learned in this lesson to make each drawing a single object. Save the drawings in the C:\Steps3D\Lesson08 folder.

9. Create the hinge shown in Figure 8.023. Refer to the following guidelines.
    9.1. The hinge is a single solid object.
    9.2. Fully dimension the hinge as shown.
    9.3. Place it with the title block of your choice on an 11" x 8½" sheet of paper.
    9.4. Save the drawing as *MyHinge.dwg* in the C:\Steps3D\Lesson08 folder.

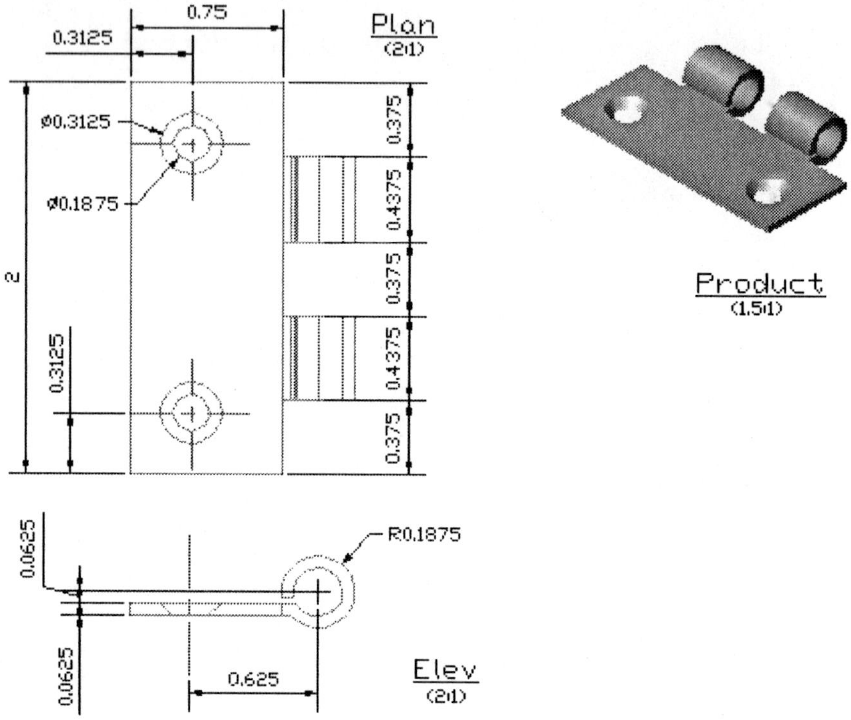

Figure 8.023

10. Create the flange shown in Figure 8.024. Refer to the following guidelines.
    10.1. The flange is a single solid object.
    10.2. Place it with the title block of your choice on an 11" x 8½" sheet of paper.
    10.3. Bold holes are 3/8" diameter.
    10.4. The center hole is 2¼" diameter.
    10.5. Save the drawing as *MyFlange.dwg* in the C:\Steps3D\Lesson08 folder.

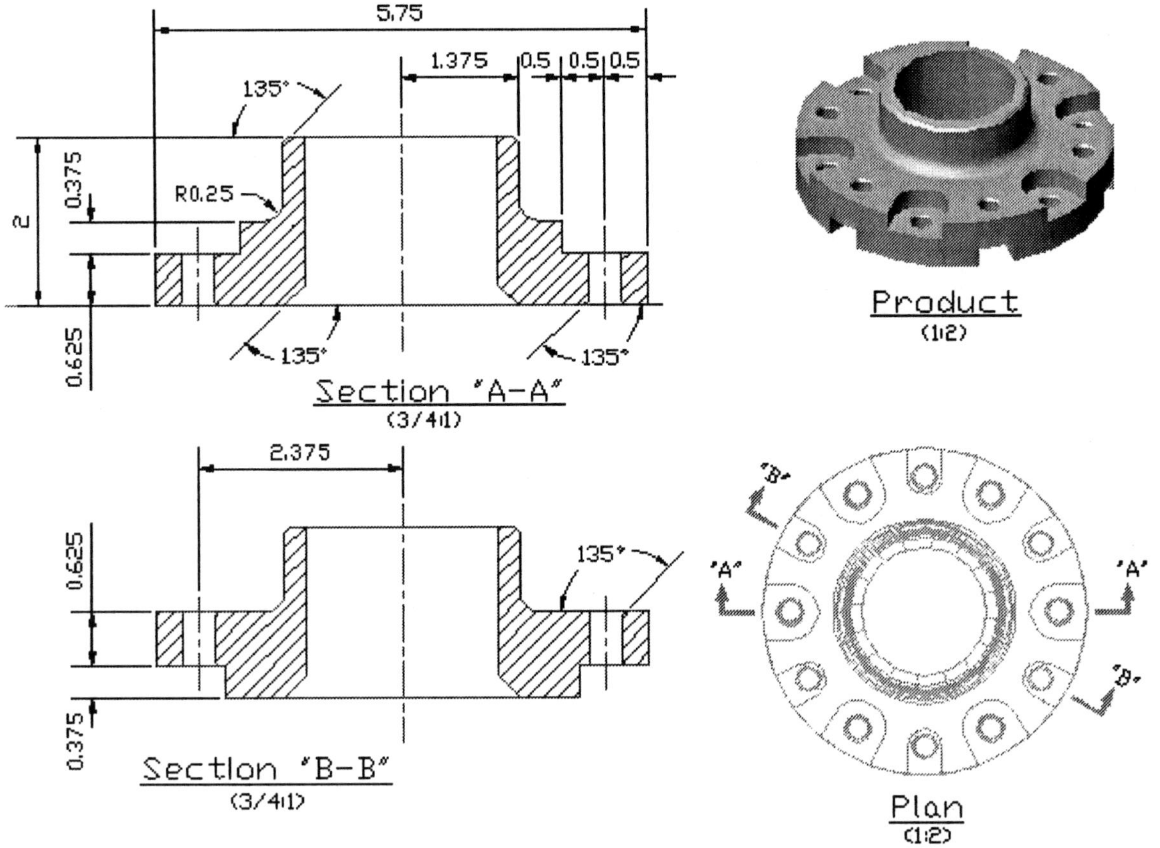

Figure 8.024

11. Create the flange gear shown in Figure 8.025. Refer to the following guidelines.
    11.1. The flange gear is a single solid object. (Hint: I created the top of the brace with wedges whose length and height were 4".)
    11.2. Fully dimension the object as shown.
    11.3. Place it with the title block of your choice on an 11" x 8½" sheet of paper.
    11.4. Save the drawing as *MyAnchor.dwg* in the C:\Steps3D\Lesson08 folder.

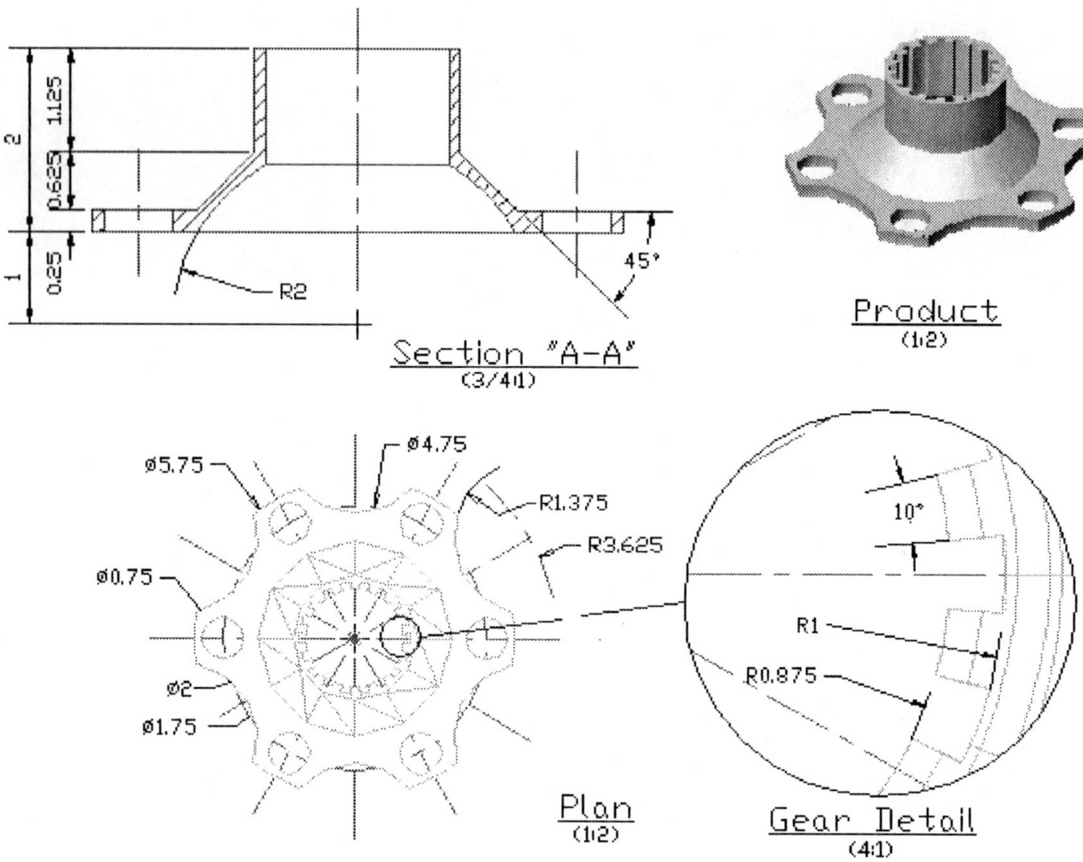

Figure 8.025

12. Create the floating support and anchor shown in Figure 8.026. Refer to the following guidelines.
    12.1. Each piece is a single solid object.
    12.2. Fully dimension the object as shown.
    12.3. Place it with the title block of your choice on an 11" x 8½" sheet of paper.
    12.4. Save the drawing as *MyFlangeGear.dwg* in the C:\Steps3D\Lesson08 folder.

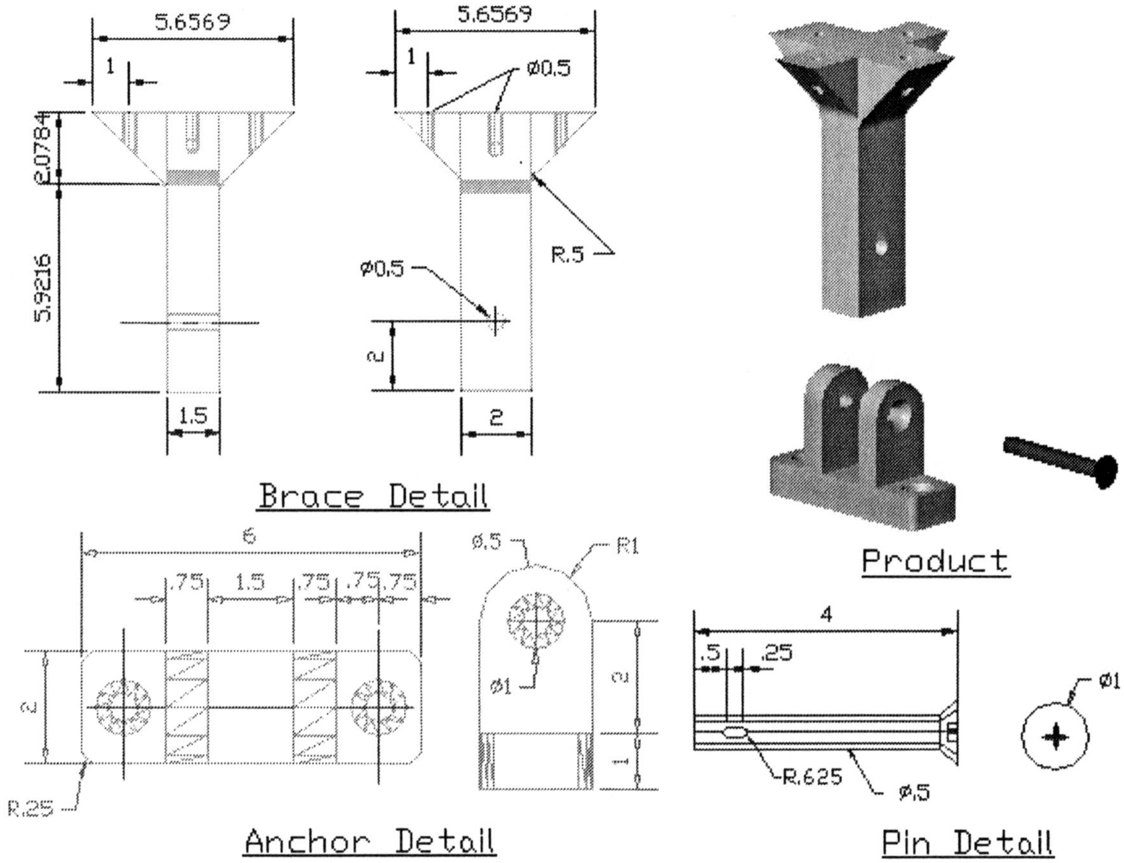

Figure 8.026

13. Create the dining chair shown in Figure 8.027. Refer to the following guidelines.
    13.1. Each leg begins at 1" diameter, but balloons to 1½" in the middle.
    13.2. The leg bracing is ¾" diameter; the back dowels are ½" diameter.
    13.3. The back support is 1" squared.
    13.4. Fully dimension the object as shown.
    13.5. Place it with the title block of your choice on an 11" x 8½" sheet of paper.
    13.6. Save the drawing as *MyDiningChair.dwg* in the C:\Steps3D\Lesson08 folder.

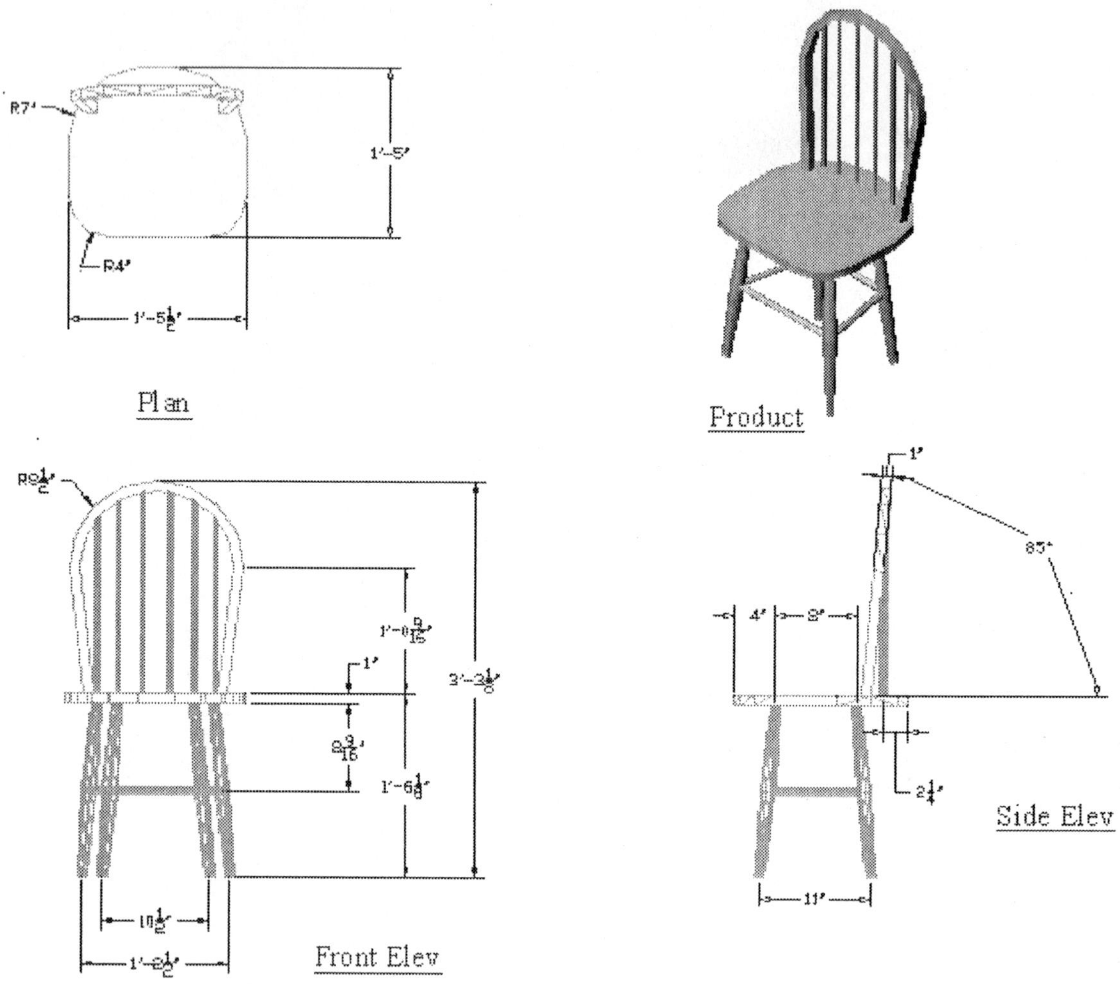

Figure 8.027

14. Create the table lamp shown in Figure 8.028. Refer to the following guidelines.
    14.1. The base is a solid object.
    14.2. The top is 1/8" thick hollow glass ball.
    14.3. Fully dimension the object as shown.
    14.4. Place it with the title block of your choice on an 11" x 8½" sheet of paper.
    14.5. Save the drawing as *MyTableLamp.dwg* in the C:\Steps3D\Lesson08 folder.

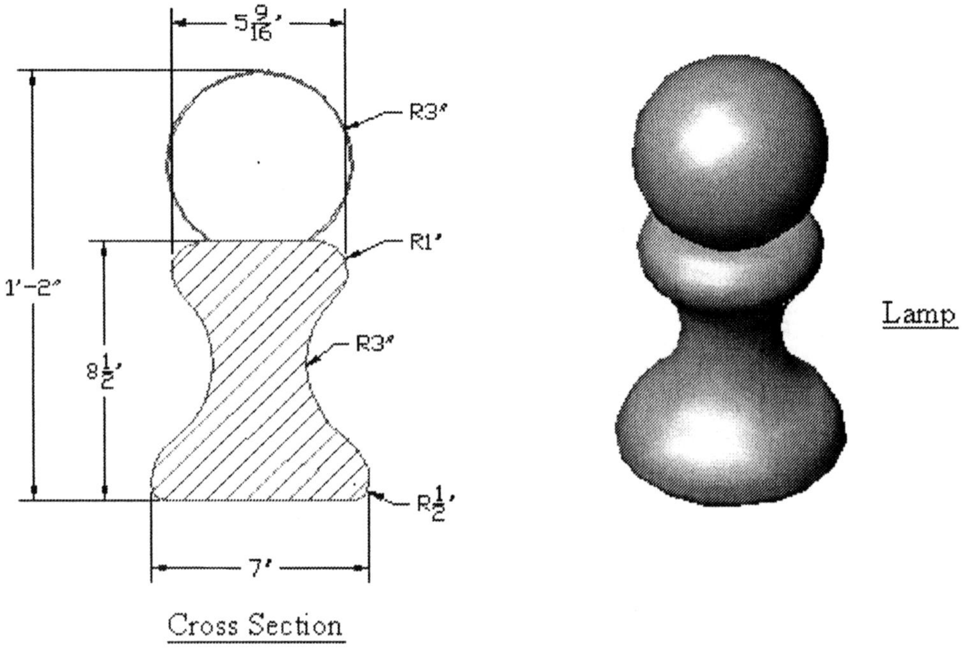

Figure 8.028

15. Create the fountain shown in Figure 8.029. Refer to the following guidelines.
    15.1. Each piece (including the water) is a separate solid object.
    15.2. Fully dimension the objects as shown.
    15.3. Save the drawing as *MyFountain.dwg* in the C:\Steps3D\Lesson08 folder.

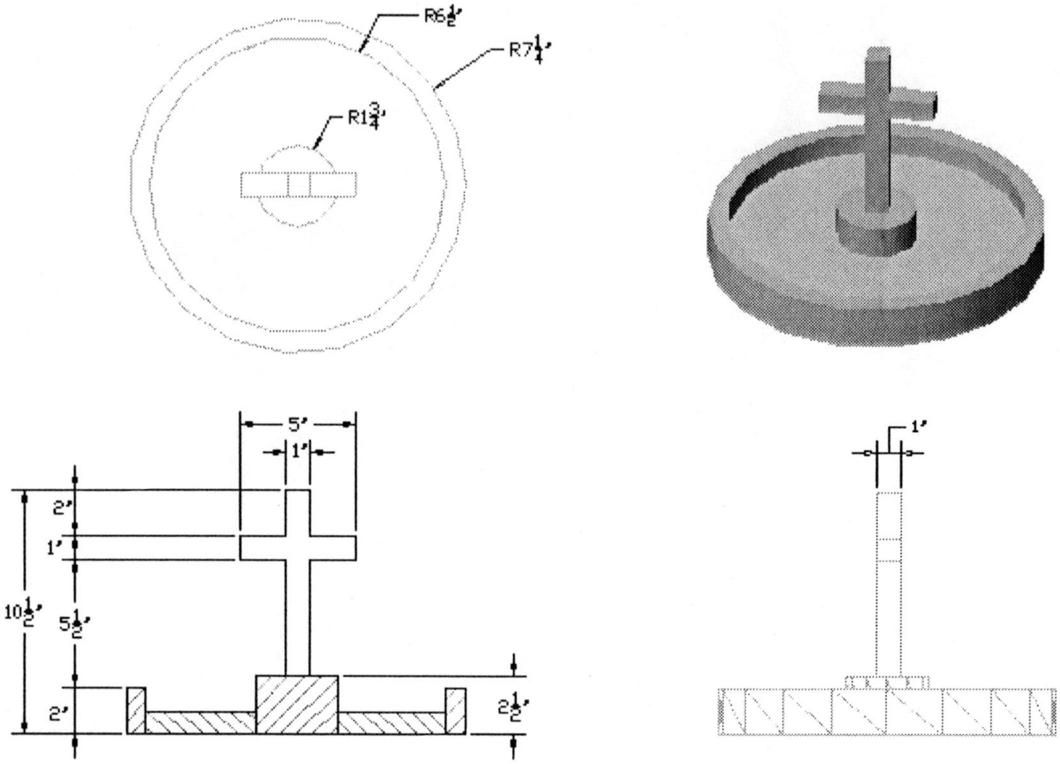

Figure 8.029
Thanks to Casey Peel for permission to use this drawing.

8.7	**For Web-Based Review Questions, visit: http://www.uneedcad.com/2006/Files/06R8-3D.pdf**

Lesson 9

Following this lesson, you will:

✓ *Know how to use the subcommands of AutoCAD's* ***SolidEdit*** *command:*

- ***Face***
  - ***Extrude***
  - ***Move***
  - ***Rotate***
  - ***Offset***
  - ***Taper***
  - ***Delete***
  - ***Copy***
  - ***coLor***
- ***Edge***
  - ***Copy***
  - ***coLor***
- ***Body***
  - ***Imprint***
  - ***seParate solids***
  - ***Clean***
  - ***Shell***

✓ *Know how to use the* ***SolidCheck*** *system variable*

Editing 3D Solids

*Over the course of your studies, I've pointed out many of AutoCAD's redundant features. (Indeed, by now you know me to be a great advocate of AutoCAD redundancy.) In this lesson, we'll present a feature that duplicates some other features of 3D solid editing. But this command – **SolidEdit** – has some routines and twists that make it a favorite to solid modelers everywhere.*

*Unfortunately, **SolidEdit** is not a simple command. In fact, it's a command in the tradition of **PEdit** or **Splinedit**. In other words, expect a multitiered command with multiple options per tier. But the wonders of those options will make Solid Modeling easier (and more fun) than you ever thought possible.*

Let's take a look.

9.1 A Single Command, But It Does So Much - *SolidEdit*

Actually, as a command by itself, ***SolidEdit*** doesn't accomplish a thing. The ***SolidEdit*** command should be considered a ticket into a realm where 15 new commands dwell (17 if you consider **Undo** and **eXit**) – each capable of something beneficial to the solid modeler.

If we tried to study ***SolidEdit*** as a single command, we might find it somewhat overwhelming. But luckily, AutoCAD divided the command options into three categories – **Face**, **Edge**, and **Body**. In fact, AutoCAD's ***SolidEdit*** command prompt looks like this:

Enter a solids editing option [Face/Edge/Body/Undo/eXit] <eXit>:

Each of the options (categories) presents a separate tier of choices designed to help modify a solid object. (The other two options – **Undo** and **eXit** – are the standard options for most commands.) We'll use these natural divisions to study each option as a category, or grouping of several routines.

AutoCAD also provides a Solids Editing toolbar with buttons that quickly access each of the commands in the categories. We'll use these buttons throughout our lesson.

> Most of you won't remember that AutoCAD's original dimensioning tool called a Dim prompt. From there, you issued one of the many dimension commands (linear, angular, etc.). Think of what dimensioning looks like today, and then imagine what the future might hold for the ***SolidEdit*** command.

9.2 Changing Faces – The Face Category

The Face Category contains the bulk of ***SolidEdit***'s commands. This category includes routines (or command options) designed to alter the faces of a solid.

> Don't confuse the face on a 3D solid with a 3D face. Remember, you'll find a 3D face on a *surface model*. A solid model has a 3D *solid* face.

To get to the Face Category's options, follow this sequence:

Command: ***Solidedit***

Solids editing automatic checking: SOLIDCHECK=1

Enter a solids editing option [Face/Edge/Body/Undo/eXit] <eXit>: ***f***

Enter a face editing option

[Extrude/Move/Rotate/Offset/Taper/Delete/Copy/coLor/Undo/eXit] <eXit>:

You can then select the routine you wish to use.

Of course, an easier way to access a specific ***SolidEdit*** routine would be simply to pick the desired choice on the Solids Editing toolbar.

Let's look at each of the routines as though they were individual commands.

9.2.1 Changing the Thickness of a 3D Solid Face – the Extrude Option

Have you tried to stretch a 3D solid? If so, you may have noticed that, once created, you can't change the shape or dimensions of individual sides (faces) of the 3D solid. And, if you've tried to extrude a 3D solid, you've seen that you can't.

So how do you change the individual faces of a 3D solid?

Well, one way is by using the **Extrude** option of the Face Category. The option works identically to the ***Extrude*** command – but only on selected faces.

Here's the sequence:

> **Command:** ***solidedit***
>
> **Solids editing automatic checking: SOLIDCHECK=1**
>
> **Enter a solids editing option [Face/Edge/Body/Undo/eXit] <eXit>:** ***f***
>
> **Enter a face editing option**
>
> **[Extrude/Move/Rotate/Offset/Taper/Delete/Copy/coLor/Undo/eXit] <eXit>:** ***e [select the Extrude option]***
>
> **Select faces or [Undo/Remove]:** ***[pick a face or an edge(s) of the face you wish to extrude – AutoCAD will highlight the two faces that form that edge]***
>
> **Select faces or [Undo/Remove/ALL]:** ***r [unless you wish to extrude both faces, tell AutoCAD you wish to Remove a face]***
>
> **Remove faces or [Undo/Add/ALL]:** ***[select the face you don't wish to extrude]***
>
> **Remove faces or [Undo/Add/ALL]:** ***[confirm the selection set]***
>
> **Specify height of extrusion or [Path]:** ***[these last two options are identical to the*** **Extrude** ***command's sequence]***
>
> **Specify angle of taper for extrusion <0>:**

AutoCAD will extrude the selected face and return to the Face Category of the ***SolidEdit*** command.

Let's give it a try.

> The Extrude option of the ***SolidEdit*** command (as well as the other options discussed in this lesson) can also be reached via the Modify pull-down menu. Follow this path:
>
> *Modify – Solids Editing – Extrude Faces (or the desired option)*

Do This: 9.2.1.1 Extruding a 3D Solid Face

I. Open the *SE-Box.dwg* file in the C:\Steps3D\Lesson09 folder. The drawing is a simple box. (The current viewpoint is 1, -2,1).

II. Follow these steps.

TOOLS	COMMAND SEQUENCE	STEPS
 Extrude Faces Button	**Command:** ***solidedit*** **Solids editing automatic checking: SOLIDCHECK=1** **Enter a solids editing option**	1. Enter the command sequence shown to access the **Extrude** option of the Face Category. Alternately, you can pick the **Extrude Faces** button on the Solids Editing toolbar. (Note:

TOOLS	COMMAND SEQUENCE	STEPS
	[Face/Edge/Body/Undo/eXit] <eXit>: *f* **Enter a face editing option** **[Extrude/Move/Rotate/Offset/ Taper/Delete/Copy/coLor/Undo/ eXit] <eXit>:** *e*	Picking the **Extrude Faces** button replaces the entire sequence shown in this step.)
	Select faces or [Undo/Remove]:	2. Select the upper-east edge (Figure 9.001).

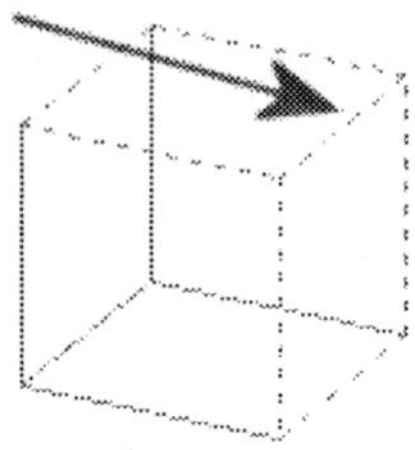

Figure 9.001

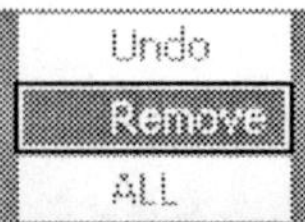	**Select faces or [Undo/Remove/ ALL]:** *r*	3. Tell AutoCAD to remove a face from the selection set.
	Remove faces or [Undo/Add/ALL]: **Remove faces or [Undo/Add/ALL]:** *[enter]*	4. Select an edge on the eastern face.
	Specify height of extrusion or [Path]: *1* **Specify angle of taper for extrusion <0>:** *30*	5. Tell AutoCAD to use an extrusion height of **1** and a taper angle of **30°**.
	Solid validation started. **Solid validation completed.** **Enter a face editing option** **[Extrude/Move/Rotate/Offset/ Taper/Delete/Copy/coLor/Undo/ eXit] <eXit>:** *[enter]* **Solids editing automatic checking: SOLIDCHECK=1** **Enter a solids editing option [Face/Edge/Body/Undo/eXit] <eXit>:** *[enter]*	6. By default, AutoCAD validates that the task you've outlined is possible and then extrudes the object. Hit ***enter*** twice to exit the command. Remove hidden lines. Your drawing looks like Figure 9.002.

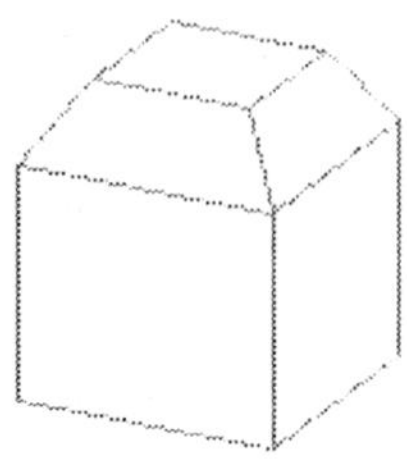

Figure 9.002

TOOLS	COMMAND SEQUENCE	STEPS
	Command: ***qsave***	7. Save the drawing, but don't exit.

You can select a face without using the edge approach we used here (selecting an edge and then deselecting one of the faces). But that ability requires that the face be reachable. (You couldn't select a face in the back.) We used the select-edge approach to show you how to reach any face.

Let's look at a similar option.

9.2.2 Moving a Face on a 3D Solid

The **Move** routine of the Face Category proves to be quite handy when it becomes necessary to relocate part of a 3D solid – such as a bolt hole – that was improperly placed.

The command sequence looks like this:

Command: ***solidedit***

Solids editing automatic checking: SOLIDCHECK=1

Enter a solids editing option [Face/Edge/Body/Undo/eXit] <eXit>: ***f***

Enter a face editing option

[Extrude/Move/Rotate/Offset/Taper/Delete/ Copy/coLor/Undo/eXit] <eXit>: ***m [select the Move option]***

Select faces or [Undo/Remove]: ***[the face selection options are the same as those in the Extrude option]***

Select faces or [Undo/Remove/ALL]:

Specify a base point or displacement: ***[the next options are identical to the basic two-dimensional* Move *command's options]***

Specify a second point of displacement:

AutoCAD will then move the selected face and return to the Face Category of the ***SolidEdit*** command.

Let's give it a try.

Do This: 9.2.2.1 Extruding a 3D Solid Face

I. Be sure you're still in the *SE-Box.dwg* file in the C:\Steps3D\Lesson09 folder. If not, please open it now.

II. Thaw the **obj2** layer. Notice the cylinder that appears inside the box.

III. Follow these steps.

TOOLS	COMMAND SEQUENCE	STEPS
Subtract Button	**Command:** ***su***	1. Use the ***Subtract*** command to subtract the cylinder from the box. Your drawing looks like Figure 9.003 (hidden lines removed for clarity).

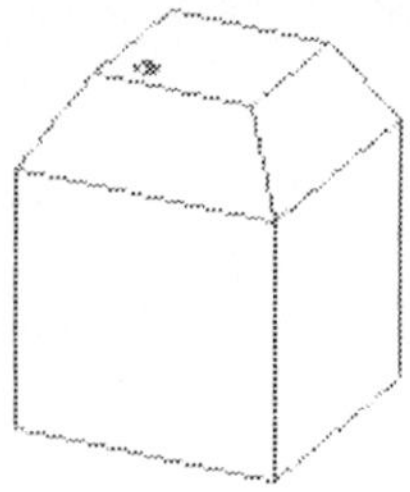

Figure 9.003

Tools	Command Sequence	Steps
Move Faces Button	**Command:** ***solidedit*** **Solids editing automatic checking: SOLIDCHECK=1** **Enter a solids editing option [Face/ Edge/Body /Undo/eXit] <eXit>:** ***f*** **Enter a face editing option** **[Extrude/Move/Rotate/Offset/ Taper/Delete/Copy/coLor/Undo/ eXit] <eXit>:** ***m***	2. Enter the command sequence shown to access the **Move** option of the Face Category. Alternately, you can pick the **Move Faces** button on the Solids Editing toolbar.
	Select faces or [Undo/Remove]: **Select faces or [Undo/Remove/ ALL]:** ***[enter]***	3. Select the cylinder inside the box (restore hidden lines if necessary).
	Specify a base point or displacement: ***1,1*** **Specify a second point of displacement:** ***[enter]***	4. I'll use the displacement method to move the hole **1** unit east and **1** unit north on the 3D solid.
	Solid validation started. **Solid validation completed.** **Enter a face editing option** **[Extrude/Move/Rotate/ Offset/ Taper/Delete/Copy/coLor/Undo/ eXit] <eXit>:** ***[enter]*** **Solids editing automatic checking: SOLIDCHECK=1** **Enter a solids editing option [Face/ Edge/Body/Undo/eXit] <eXit>:** ***[enter]***	5. As with the **Extrude** routine, AutoCAD validates the procedure before actually moving the hole. Hit ***enter*** twice to exit the command. Your drawing looks like Figure 9.004 (hidden lines remove).

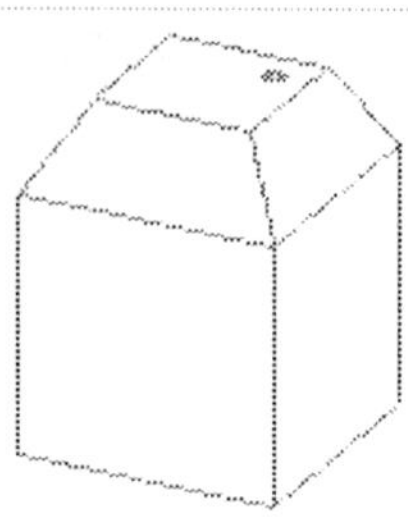

Figure 9.004

TOOLS	COMMAND SEQUENCE	STEPS
	Command: ***qsave***	6. Save the drawing, but don't exit.

9.2.3 Rotating Faces on a 3D Solid

Like the other Face Category options, **Rotate** emulates another command. But in the case of the **Rotate** routine, it doesn't emulate the ***Rotate*** command but rather the ***Rotate3d*** command. In other words, you'll have the opportunity to rotate a 3D solid face about an axis (as opposed to a point).

The command sequence looks like this:

Command: ***solidedit***

Solids editing automatic checking: SOLIDCHECK=1

Enter a solids editing option [Face/Edge/Body/Undo/eXit] <eXit>: ***f***

Enter a face editing option

[Extrude/Move/Rotate/Offset/Taper/Delete/Copy/coLor/Undo/eXit] <eXit>: ***r***

Select faces or [Undo/Remove]: ***[the*** **Select faces** ***routine is identical to that of the*** **Extrude** ***and*** **Move** ***routines]***

Select faces or [Undo/Remove/ALL]:

Specify an axis point or [Axis by object/View/Xaxis/Yaxis/Zaxis] <2points>: ***[the rest of the options are identical to the*** **Rotate3d** ***command's options]***

Specify the second point on the rotation axis:

Specify a rotation angle or [Reference]:

(Don't you just love commands that are based on routines you already know?!) Let's give it a try.

Do This: 9.2.3.1 Extruding a 3D Solid Face

I. Be sure you're still in the *SE-Box.dwg* file in the C:\Steps3D\Lesson09 folder. If not, please open it now.

II. Thaw the **obj3** layer. Notice the object that appears inside the box.

III. Use the ***Subtract*** command to remove the new object from the box. (It'll leave a slot.)

IV. Follow these steps.

TOOLS	COMMAND SEQUENCE	STEPS
Rotate Faces Button	**Command:** ***solidedit*** **Solids editing automatic checking: SOLIDCHECK=1** **Enter a solids editing option [Face/ Edge/Body/Undo/eXit] <eXit>:** ***f*** **Enter a face editing option** **[Extrude/Move/Rotate/Offset/ Taper/Delete/Copy/coLor/Undo/ eXit] <eXit>:** ***r***	1. Enter the command sequence shown to access the **Rotate** option of the Face Category. Alternately, you can pick the **Rotate Faces** button on the Solids Editing toolbar.

TOOLS	COMMAND SEQUENCE	STEPS
	Select faces or [Undo/Remove]: **Select faces or [Undo/Remove/ALL]:** ***[enter]***	2. Select the four faces forming the slot.
	Specify an axis point or [Axis by object/View/Xaxis/Yaxis/Zaxis] <2points>: **Specify the second point on the rotation axis:**	3. Select the upper- and lower-center points indicated in Figure 9.005. (Pick the lower-center point first to properly set the Z-axis.)

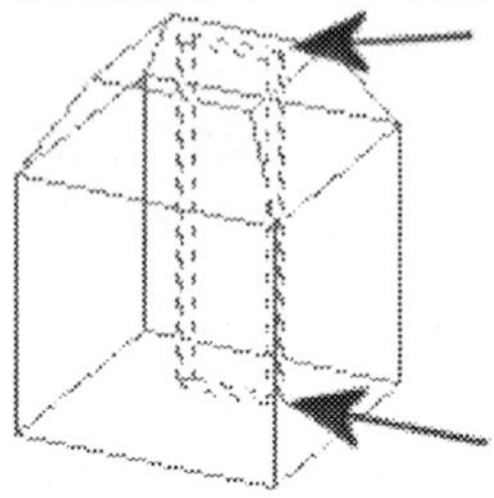

Figure 9005

	Specify a rotation angle or [Reference]: ***45*** **Solid validation started.** **Solid validation completed.**	4. Rotate the slot 45°.
	Enter a face editing option **[Extrude/Move/Rotate/Offset/Taper/Delete/Copy/coLor/Undo/eXit] <eXit>:** ***[enter]*** **Solids editing automatic checking: SOLIDCHECK=1** **Enter a solids editing option [Face/Edge/Body/Undo/eXit] <eXit>:** ***[enter]***	5. Complete the command. Your drawing looks like Figure 9.006.

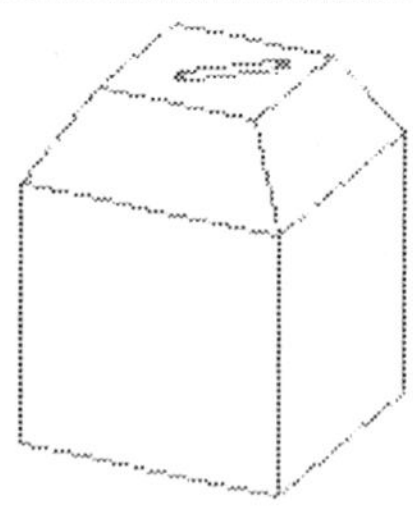

Figure 9.006

	Command: ***qsave***	6. Save the drawing, but don't exit.

Imagine the trouble you'd have had accomplishing this task without the ***SolidEdit*** command!

9.2.4	Offsetting Faces on a 3D Solid

The **Offset** routine of the Face Category works very much like the two-dimensional command. There are, however, some quirks about it that you must know to avoid frustration.

The first of these quirks lies in the direction of the offset. When using the two-dimensional command, you pick the direction of the offset on the screen or by coordinate input. In Z-Space, this might present problems (since picking a point on the screen doesn't work well in Z-Space). Instead, you'll control the direction of the offset by using a positive or negative number to identify the distance of the offset. But here's the quirk: the positive number doesn't increase the size of the face being offset (it won't increase the size of the slot, as you'll see). Rather, it increases the *volume of the solid* (thus *decreasing* the size of the slot). A negative number, of course, has the opposite effect.

Another quirk is actually an omission on AutoCAD's part. Whereas you can offset an object through a point in two-dimensional space (using the ***Offset*** command), no such option is presented when using the ***SolidEdit*** command. You'll miss this convenience. (Hopefully, AutoCAD will remedy this oversight in the future.)

The command sequence for the **Offset** option is

Command: ***solidedit***

Solids editing automatic checking: SOLIDCHECK=1

Enter a solids editing option [Face/Edge/Body/Undo/eXit] <eXit>: ***f***

Enter a face editing option

[Extrude/Move/Rotate/Offset/Taper/Delete/Copy/coLor/Undo/eXit] <eXit>: ***o***

Select faces or [Undo/Remove]: ***[the*** **Select faces** ***routine is identical to that of the SolidEdit routines you've already seen]***

Select faces or [Undo/Remove/ALL]:

Specify the offset distance: ***[enter the distance you wish to offset the selected faces]***

Try enlarging the slot a bit.

Do This: 9.2.4.1	Offsetting a 3D Solid Face

I. Be sure you're still in the *SE-Box.dwg* file in the C:\Steps3D\Lesson09 folder. If not, please open it now.

II. Follow these steps.

TOOLS	COMMAND SEQUENCE	STEPS
 Offset Faces Button	**Command:** ***solidedit*** **Solids editing automatic checking: SOLIDCHECK=1** **Enter a solids editing option [Face/Edge/Body/Undo/eXit] <eXit>:** ***f*** **Enter a face editing option** **[Extrude/Move/Rotate/Offset/Taper/Delete/Copy/coLor/Undo/eXit] <eXit>:** ***o***	1. Enter the command sequence shown to access the **Offset** option of the Face Category. Alternately, you can pick the **Offset Faces** button on the Solids Editing toolbar.

Tools	Command Sequence	Steps
	Select faces or [Undo/Remove]: **Select faces or [Undo/Remove/ ALL]:** ***[enter]***	2. Select the four faces forming the slot.
	Specify the offset distance: -.125	3. We want to increase the size of the slot (decreasing the volume of the solid). We'll enter a negative number. Offset the slot by 1/8" as indicated.
	Solid validation started. **Solid validation completed.** **Enter a face editing option [Extrude/Move/Rotate/Offset/ Taper/Delete/ Copy/coLor/Undo/ eXit] <eXit>:** ***[enter]*** **Solids editing automatic checking: SOLIDCHECK=1** **Enter a solids editing option [Face/ Edge/Body/Undo/eXit] <eXit>:** ***[enter]***	4. Complete the command. Your drawing looks like Figure 9.007 (hidden lines removed).

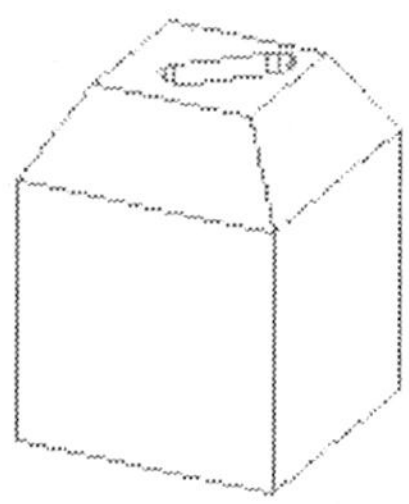

Figure 9.007

Tools	Command Sequence	Steps
	Command: ***qsave***	5. Save the drawing, but don't exit.

9.2.5 Tapering Faces on a 3D Solid

Tapering a face on a 3D solid isn't difficult, but you'll need to watch the positive and negative numbers just as you did when you offset a face. The command sequence is

Command: ***solidedit***

Solids editing automatic checking: SOLIDCHECK=1

Enter a solids editing option [Face/Edge/Body/Undo/eXit] <eXit>: ***f***

Enter a face editing option

[Extrude/Move/Rotate/Offset/Taper/Delete/Copy/coLor/Undo/eXit] <eXit>: ***t***

Select faces or [Undo/Remove]: ***[the*** **Select faces** ***routine is identical to that of the SolidEdit routines you've already seen]***

Select faces or [Undo/Remove/ALL]:

Specify the base point: ***[the base point doesn't have to be on the face itself; essentially, you're using the base point and second point to identify a direction, or axis, for the taper]***

Specify another point along the axis of tapering: ***[identify a second point along the axis]***

Specify the taper angle: ***[tell AutoCAD how much of an angle you wish to create – remember, a positive number will enlarge a hole in the direction of the axis you indicated; a negative number, of course, will reduce the hole]***

Let's taper a hole in our solid.

Do This: 9.2.5.1	Tapering a 3D Solid Face

I. Be sure you're still in the *SE-Box.dwg* file in the C:\Steps3D\Lesson09 folder. If not, please open it now.

II. Thaw the **obj4** layer. Notice the cylinder that appears inside the box.

III. Subtract the new cylinder from the box.

IV. Follow these steps.

TOOLS	COMMAND SEQUENCE	STEPS
 Taper Faces Button	**Command:** *solidedit* **Solids editing automatic checking: SOLIDCHECK=1** **Enter a solids editing option [Face/Edge/ Body/Undo/eXit] <eXit>:** ***f*** **Enter a face editing option** **[Extrude/Move/Rotate/Offset/ Taper/Delete/Copy/coLor/Undo/ eXit] <eXit>:** ***t***	1. Enter the command sequence shown to access the **Taper** routine of the Face Category. Alternately, you can pick the **Taper Faces** button on the Solids Editing toolbar.
	Select faces or [Undo/Remove]: **Select faces or [Undo/Remove/ ALL]:** ***[enter]***	2. Select one of the isolines defining the large hole.
	Specify the base point: **Specify another point along the axis of tapering:**	3. Use the eastern (right) center of the large hole as the base point and the center of the other end as the other **point along the axis of tapering**.
	Specify the taper angle: ***-5***	4. We'll reduce the size of the hole as it moves westward. Enter a negative **taper angle** of ***5°*** as indicated.
	Solid validation started. **Solid validation completed.** **Enter a face editing option** **[Extrude/Move/Rotate/Offset/ Taper/Delete/Copy/coLor/Undo/ eXit] <eXit>:** ***[enter]*** **Solids editing automatic checking: SOLIDCHECK=1** **Enter a solids editing option**	5. Complete the command. Your drawing looks like Figure 9.008.

TOOLS	COMMAND SEQUENCE	STEPS
	[Face/Edge/Body/Undo/eXit] <eXit>: *[enter]*	

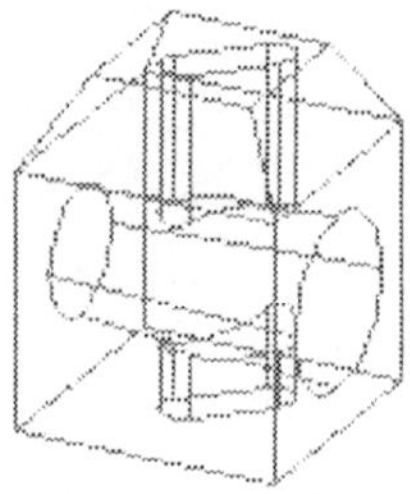

Figure 9.0087a

	Command: *qsave*	6. Save the drawing, but don't exit.

9.2.6 Deleting 3D Solid Faces

Our next routing provides a method for removing some faces (like fillets, chamfers, holes, etc.) from a 3D solid. The command sequence is one of the easiest (no points or axes to identify).

Command: *solidedit*

Solids editing automatic checking: SOLIDCHECK=1

Enter a solids editing option [Face/Edge/Body/Undo/eXit] <eXit>: *f*

Enter a face editing option

[Extrude/Move/Rotate/Offset/Taper/Delete/Copy/coLor/Undo/eXit] <eXit>: *d*

Select faces or [Undo/Remove]: *[the* **Select faces** *routine is identical to that of the SolidEdit routines you've already seen]*

Select faces or [Undo/Remove/ALL]: *[enter]*

Suppose we want to get rid of the large hole altogether. We'll use the **Delete** routine.

Do This: 9.2.6.1 Deleting a 3D Solid Face

I. Be sure you're still in the *SE-Box.dwg* file in the C:\Steps3D\Lesson09 folder. If not, please open it now.

II. Follow these steps.

TOOLS	COMMAND SEQUENCE	STEPS
Delete Faces Button	**Command:** *solidedit* **Solids editing automatic checking: SOLIDCHECK=1** **Enter a solids editing option [Face/ Edge/ Body/Undo/eXit] <eXit>:** *f* **Enter a face editing option** **[Extrude/Move/Rotate/Offset/ Taper/Delete/Copy/coLor/Undo/ eXit] <eXit>:** *d*	1. Enter the command sequence shown to access the **Delete** routine of the Face Category. Alternately, you can pick the **Delete Faces** button on the Solids Editing toolbar.

Tools	Command Sequence	Steps
	Select faces or [Undo/Remove]:	2. Select the large, tapered hole (you'll fine it easier to either select inside the hole or the select one of the internal isolines defining the inside face of the hole).
	Select faces or [Undo/Remove/ALL]: ***[enter]*** **Solid validation started.** **Solid validation completed.** **Enter a face editing option** **[Extrude/Move/Rotate/Offset/Taper/Delete/Copy/coLor/Undo/eXit] <eXit>:** ***[enter]*** **Solids editing automatic checking: SOLIDCHECK=1** **Enter a solids editing option [Face/Edge/Body/Undo/eXit] <eXit>:** ***[enter]***	3. Complete the command. Your drawing looks like Figure 9.009.

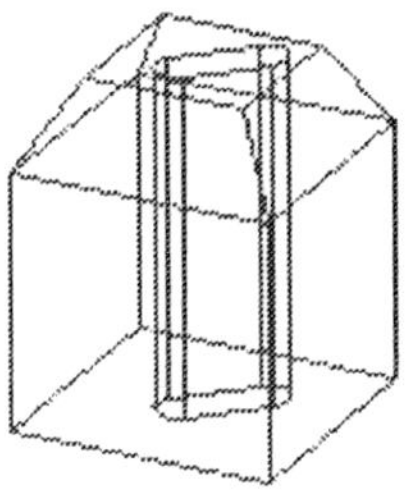

Figure 9.009

Tools	Command Sequence	Steps
	Command: ***qsave***	4. Save the drawing, but don't exit.

9.2.7 Copying 3D Solid Faces as Regions or Bodies

Use the **Copy** routine of the Face Category to create copies of one or more faces of a 3D solid. AutoCAD creates the copies as regions or bodies, which you can explode into individual lines, arcs, circles, and so forth, or extrude into new 3D solid objects. Unlike the results of the ***Copy*** command, however, the new objects exist on the layer that was current when the copies were created.

> A *body* is any structure that represents a solid or a Non-Uniform Rational B-Spline (NURBS) surface.

It's as simple to use as the ***Copy*** command. The command sequence is

Command: ***solidedit***

Solids editing automatic checking: SOLIDCHECK=1

Enter a solids editing option [Face/Edge/Body/Undo/eXit] <eXit>: ***f***

Enter a face editing option

[Extrude/Move/Rotate/Offset/Taper/Delete/Copy/coLor/Undo/eXit] <eXit>: ***c***

Select faces or [Undo/Remove]: ***[the*** **Select faces** ***routine is identical to that of the SolidEdit routines you've already seen]***

Select faces or [Undo/Remove/ALL]: ***[enter]***

Specify a base point or displacement: ***[the last prompts are the same as the*** **Copy** ***command's prompts]***

Specify a second point of displacement:

We'll copy the top faces of our solid.

Do This: 9.2.7.1	Copying a 3D Solid Face

I. Be sure you're still in the *SE-Box.dwg* file in the C:\Steps3D\Lesson09 folder. If not, please open it now.

II. Follow these steps.

TOOLS	COMMAND SEQUENCE	STEPS
 Copy Faces Button	**Command:** ***solidedit*** **Solids editing automatic checking: SOLIDCHECK=1** **Enter a solids editing option [Face/ Edge/ Body/Undo/eXit] <eXit>:** ***f*** **Enter a face editing option** **[Extrude/Move/Rotate/Offset/ Taper/Delete/Copy/coLor/Undo/ eXit] <eXit>:** ***c***	1. Enter the command sequence shown to access the **Copy** routine of the Face Category. Alternately, you can pick the **Copy Faces** button on the Solids Editing toolbar.
	Select faces or [Undo/Remove]: **Select faces or [Undo/Remove/ ALL]:** ***[enter]***	2. Select the top face and the four angled faces around it.
	Specify a base point or displacement: ***5,0*** **Specify a second point of displacement:** ***[enter]***	3. Use the displacement method to copy the faces five units to the right as indicated.
	Enter a face editing option **[Extrude/Move/Rotate/Offset/ Taper/Delete/Copy/coLor/Undo/ eXit] <eXit>:** ***[enter]*** **Solids editing automatic checking: SOLIDCHECK=1** **Enter a solids editing option [Face/ Edge/Body/Undo/eXit] <eXit>:** ***[enter]***	4. Complete the command. Your drawing looks like Figure 9.010.

Tools	Command Sequence	Steps

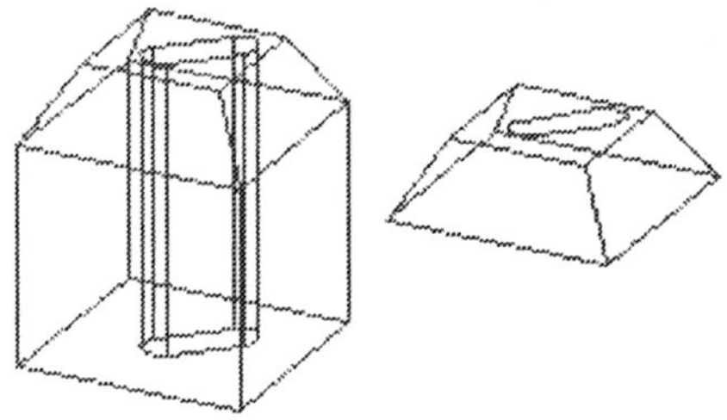

Figure 9.010

Tools	Command Sequence	Steps
	Command: ***qsave***	5. Save the drawing, but don't exit.

9.2.8 Changing the Color of a Single Face

The **coLor** routine of the Face Category is useful if you intend to shade or remove hidden lines from your drawing. It helps to distinguish between the different faces.

The command sequence is

> **Command:** ***solidedit***
> **Solids editing automatic checking: SOLIDCHECK=1**
> **Enter a solids editing option [Face/Edge/Body/Undo/eXit] <eXit>:** ***f***
> **Enter a face editing option**
> **[Extrude/Move/Rotate/Offset/Taper/Delete/Copy/coLor/Undo/eXit] <eXit>:** ***l***
> **Select faces or [Undo/Remove]:** ***[the*** **Select faces** ***routine is identical to that of the SolidEdit routines you've already seen]***
> **Select faces or [Undo/Remove/ALL]:** ***[enter]***
> ***[AutoCAD presents the Color Selection dialog box; select the color you wish the face(s) to be]***

Let's make the slot a different color and view our 3D solid using the **Gouraud** Shademode.

Do This: 9.2.8.1 Changing the Color of a 3D Solid Face

I. Be sure you're still in the *SE-Box.dwg* file in the C:\Steps3D\Lesson09 folder. If not, please open it now.

II. Follow these steps.

Tools	Command Sequence	Steps
Color Faces Button	**Command:** ***solidedit*** **Solids editing automatic checking: SOLIDCHECK=1** **Enter a solids editing option [Face/ Edge/ Body/Undo/eXit] <eXit>:** ***f*** **Enter a face editing option** **[Extrude/Move/Rotate/Offset/ Taper/Delete/Copy/coLor/Undo/ eXit] <eXit>:** ***c***	1. Enter the command sequence shown to access the **coLor** routine of the Face Category. Alternately, you can pick the **Color Faces** button on the Solids Editing toolbar.

Tools	Command Sequence	Steps
	Select faces or [Undo/Remove]: **Select faces or [Undo/Remove/ ALL]:** *[enter]*	2. Select the faces that form the slot.
		3. AutoCAD presents the Select Color dialog box. Select **Green**.
	Enter a face editing option [Extrude/Move/Rotate/Offset/Taper/ Delete/Copy/coLor/Undo/eXit] <eXit>: *[enter]* **Solids editing automatic checking: SOLIDCHECK=1** **Enter a solids editing option [Face/Edge/Body/Undo/eXit] <eXit>:** *[enter]*	4. Complete the command.
	Command: *shademode* **Current mode: 2D wireframe** **Enter option [2D wireframe/3D wireframe/Hidden/Flat/Gouraud/ fLat+edges/gOuraud+edges] <2D wireframe>:** *g*	5. Set the Shademode system variable to **Gouraud**. Your drawing looks like Figure 9.011.

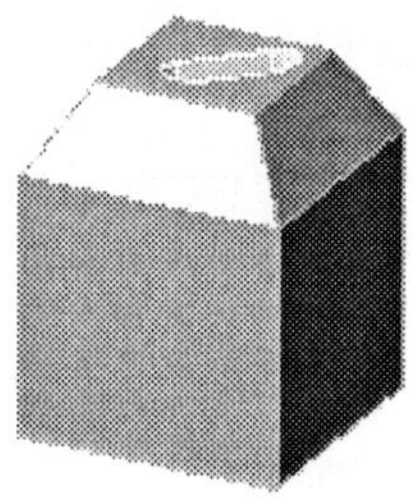

Figure 9.011

	Command: *qsave*	6. Save the drawing, but don't exit.

9.3 Modifying Edges – The Edge Category

The Edge Category contains only two real options (besides the **Undo/eXit** options). Both of these – **Copy** and **coLor** – repeat options found in the Face Category. But here they're for use on single edges rather than entire faces.

The command sequence to access the **Edge** options of the ***SolidEdit*** command is

Command: *solidedit*

Solids editing automatic checking: SOLIDCHECK=1

Enter a solids editing option [Face/Edge/Body/Undo/eXit] <eXit>: *e*

Enter an edge editing option [Copy/coLor/Undo/eXit] <eXit>:

Prompts for both the **Copy** option and the **coLor** option are identical to their counterparts in the Face Category.

Let's look at each in an exercise.

Do This: 9.3.1	Changing Edges on a 3D Solid

I. Be sure you're still in the *SE-Box.dwg* file in the C:\Steps3D\Lesson09 folder. If not, please open it now.

II. Change the Shademode setting back to **2D**.

III. Set **const** as the current layer.

IV. Follow these steps.

TOOLS	COMMAND SEQUENCE	STEPS
Color Edges Button	**Command:** ***solidedit*** **Solids editing automatic checking: SOLIDCHECK=1** **Enter a solids editing option [Face/ Edge/Body/Undo/eXit] <eXit>:** ***e*** **Enter an edge editing option [Copy/ coLor/Undo/eXit] <eXit>:** ***l***	1. Enter the command sequence shown to access the **coLor** routine of the Edge Category. Alternately, you can pick the **Color Edges** button on the Solids Editing toolbar.
	Select edges or [Undo/Remove]: **Select edges or [Undo/Remove]:** ***[enter]***	2. Select the lowermost and easternmost edge of the 3D solid (Figure 9.012).

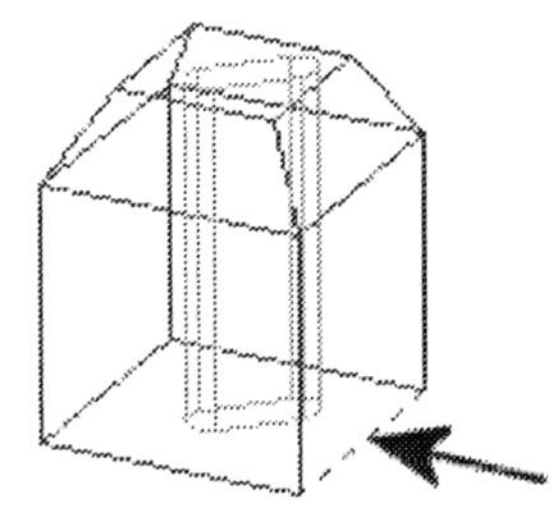

Figure 9.012

TOOLS	COMMAND SEQUENCE	STEPS
		3. AutoCAD presents the Select Color dialog box. Select **Red**.
Copy Edges Button	**Enter an edge editing option [Copy/coLor/Undo/eXit] <eXit>:** ***c***	4. AutoCAD changes the color of the selected line and then returns to the **edge editing option** prompt. Tell it you want to **Copy** an edge as shown. (Alternately, you can pick the **Copy Edges** button on the Solids Editing toolbar).
	Select edges or [Undo/Remove]: **Select edges or [Undo/Remove]:**	5. Select the same edge as in Step 2 and one of the adjoining edges (Figure 9.013).

Tools	Command Sequence	Steps

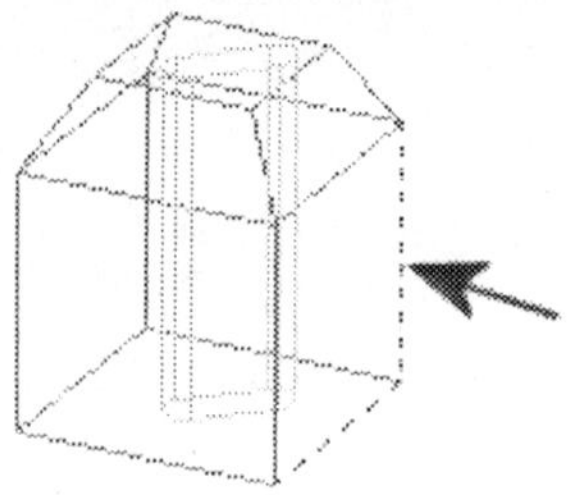

Figure 9.013

Tools	Command Sequence	Steps
	Specify a base point or displacement: ***2,0*** **Specify a second point of displacement:** ***[enter]***	6. Use the displacement method to copy the edges two units to the east, as indicated.
	Enter an edge editing option [Copy/ coLor/Undo/eXit] <eXit>: ***[enter]*** **Solids editing automatic checking: SOLIDCHECK=1** **Enter a solids editing option [Face/ Edge/Body/Undo/eXit] <eXit>:** ***[enter]***	7. Complete the command. Your drawing looks like Figure 9.014. Notice that AutoCAD has placed both the copies on the current layer. Notice also that both objects assumed the characteristic color of that layer regardless of their layer and color on the 3D solid.

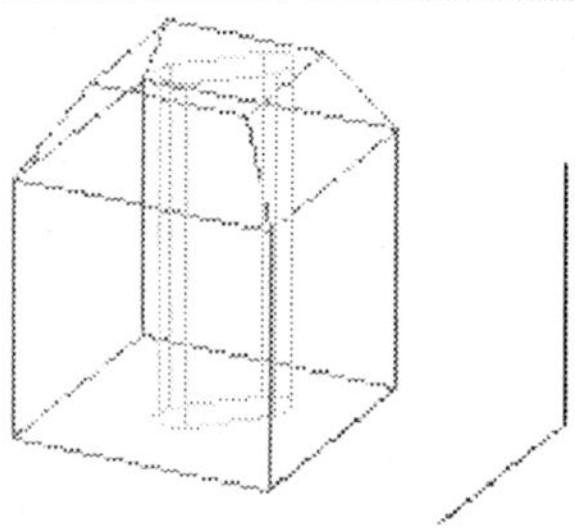

Figure 9.014

Tools	Command Sequence	Steps
	Command: ***u***	8. Undo these changes.

So you see that not much difference occurs between these procedures and their counterparts in the Face Category – except for the obvious effect on edges rather than faces.

Our next category, however, will be quite different!

9.4 Changing the Whole 3D Solid – The Body Category

The Body Category includes routines to modify a 3D solid as a whole. None of its five options (except the **Undo/eXit** options) has a counterpart in the 2D or 3D modification worlds. We'll look at each.

To access the Body Category's options, follow this sequence:

Command: ***solidedit***

Solids editing automatic checking: SOLIDCHECK=1

Enter a solids editing option [Face/Edge/Body/Undo/eXit] <eXit>: *b*
Enter a body editing option
[Imprint/seParate solids/Shell/cLean/Check/Undo/eXit] <eXit>:

9.4.1 Imprinting an Image onto a 3D Solid

The **Imprint** routine of the Body Category "imprints" (or draws) an image of a selected object – arc, circle, line, 2D or 3D polyline, ellipse, spline, region, or body – onto a 3D solid. Essentially, the imprint is a two-dimensional representation of the object on one of the faces of the 3D solid.

Once the impression has been made, *the line (or arc, spline, etc.) actually becomes the defining edge of a new face on the 3D solid.* So you can use this to help create new faces where they're needed.

The command sequence to use the Imprint option is

Command: *solidedit*
Solids editing automatic checking: SOLIDCHECK=1
Enter a solids editing option [Face/Edge/Body/Undo/eXit] <eXit>: *b*
Enter a body editing option
[Imprint/seParate solids/Shell/cLean/Check/Undo/eXit] <eXit>: *i*
Select a 3D solid: *[select the 3D solid on which you wish to make the impression]*
Select an object to imprint: *[select the object you wish to imprint]*
Delete the source object [Yes/No] <N>: *[AutoCAD allows you the opportunity to keep or delete the object you're using to create your impression]*
Select an object to imprint: *[you can continue to imprint objects if you wish]*

This will become clearer with an exercise. Let's see what we can do with it.

Do This: 9.4.1.1 Imprinting Edges on a 3D Solid

I. Be sure you're still in the *SE-Box.dwg* file in the C:\Steps3D\Lesson09 folder. If not, please open it now.

II. Follow these steps.

TOOLS	COMMAND SEQUENCE	STEPS
	**Command: *l***	1. Draw a line between the midpoints of the vertical edges defining the western face of the 3D solid (Figure 9.015). We'll use this line to imprint a new edge.

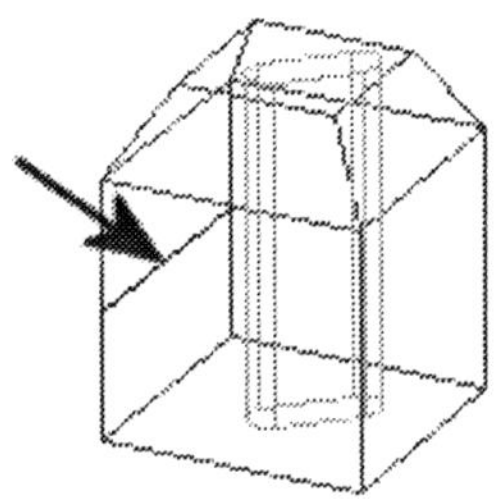

Figure 9.015

TOOLS	COMMAND SEQUENCE	STEPS
Imprint Button	**Command:** ***solidedit*** **Solids editing automatic checking: SOLIDCHECK=1** **Enter a solids editing option [Face/ Edge/Body/Undo/eXit] <eXit>:** ***b*** **Enter a body editing option** **[Imprint/seParate solids/Shell/ cLean/Check/Undo/eXit] <eXit>:** ***i***	2. Enter the command sequence shown to access the ***Imprint*** routine of the Body Category. Alternately, you can pick the **Imprint** button on the Solids Editing toolbar.
	Select a 3D solid:	3. Select the 3D solid.
	Select an object to imprint:	4. Select the line you created in Step 1.
	Delete the source object [Yes/No] <N>: ***y***	5. We won't need the line after we make the impression, so allow AutoCAD to delete it.
	Select an object to imprint: ***[enter]*** **Enter a body editing option** **[Imprint/seParate solids/Shell/ cLean/Check/Undo/eXit] <eXit>:** ***[enter]*** **Solids editing automatic checking: SOLIDCHECK=1** **Enter a solids editing option [Face/ Edge/Body/Undo/eXit] <eXit>:** ***[enter]***	6. We can continue to imprint objects, but we won't need to do so now. Complete the command.
	Command: ***solidedit***	7. Follow the procedure outlined in Exercise 9.2.3.1 to rotate the new face 30°. The completed drawing will look like Figure 9.016.

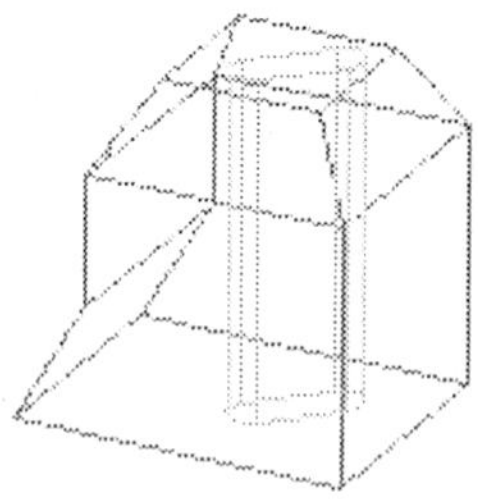

Figure 9.016

	Command: ***qsave***	6. Save the drawing.

You can probably see that **Imprint** will be one of the more useful of the ***SolidEdit*** command's options. You may find it easier to imprint and modify than to create a new 3D solid and join it (via the ***Union*** command) to an existing 3D solid.

9.4.2 Separating 3D Solids with the seParate Solids Routines

At first glance, the **seParate solids** option looked very promising – after all, it separates 3D solids into their constituencies. Simply put, this means that, if you created a 3D solid from a box and a cylinder, it would separate the 3D solid into the box and cylinder again. The way it works, however, *you can only separate the constituent objects when they don't actually touch.*

Still, there will be times when you find the **seParate solids** option quite handy. You'll see this in our exercise.

The command sequence is one of the simplest:

> **Command:** ***solidedit***
>
> **Solids editing automatic checking: SOLIDCHECK=1**
>
> **Enter a solids editing option [Face/Edge/Body/Undo/eXit] <eXit>:** ***b***
>
> **Enter a body editing option**
>
> **[Imprint/seParate solids/Shell/cLean/Check/Undo/eXit] <eXit>:** ***p***
>
> **Select a 3D solid:** ***[select the solid you wish to separate]***

Let's take a look.

Do This: 9.4.2.1 Separating Parts of a 3D Solid

I. Open the *SE-Box-2.dwg* file in the C:\Steps3D\Lesson09 folder. The drawing looks like Figure 9.4.2.1a.

II. Use the ***List*** command to verify that all objects shown are part of a single 3D solid.

III. Follow these steps.

TOOLS	COMMAND SEQUENCE	STEPS
 Separate Button	**Command:** ***solidedit*** **Solids editing automatic checking: SOLIDCHECK=1** **Enter a solids editing option [Face/ Edge/Body/Undo/eXit] <eXit>:** ***b*** **Enter a body editing option** **[Imprint/seParate solids/Shell/ cLean/Check/Undo/eXit] <eXit>:** ***p***	1. Enter the command sequence shown to access the ***seParate solids*** routine of the Body Category. Alternately, you can pick the **Separate** button on the Solids Editing toolbar.
	Select a 3D solid:	2. Select either of the objects on the screen (as you've seen, although they don't touch, they're both part of a single 3D solid).
	Enter a body editing option **[Imprint/seParate solids/Shell/ cLean/Check/ Undo/eXit] <eXit>:** ***[enter]*** **Solids editing automatic checking: SOLIDCHECK=1** **Enter a solids editing option [Face/**	3. Complete the command.

TOOLS	COMMAND SEQUENCE	STEPS
	Edge/Body/Undo/eXit] <eXit>: ***[enter]***	
	Command: ***e***	4. Erase the object on the left to verify that the 3D solids have separated.
	Command: ***qsave***	5. Save the drawing, but don't exit.

9.4.3 Clean

The **Clean** option of the Body Category removes extra (redundant) edges and vertices – including imprinted and unused edges – from a 3D solid. Use it as a final cleanup tool once you've completed your 3D solid.

The command sequence is identical to that of the **seParate solids** option – and is, therefore, one of AutoCAD's simplest.

Did you notice the extra circular edge at the top of the new 3D solid? We don't need this one, so we'll use the **Clean** option to remove it.

Let's proceed.

Do This: 9.4.3.1 Cleaning Up a 3D Solid

I. Be sure you're still in the *SE-Box-2.dwg* file in the C:\Steps3D\Lesson09 folder. If not, please open it now.

II. Follow these steps.

TOOLS	COMMAND SEQUENCE	STEPS
Clean Button	**Command:** ***solidedit*** **Solids editing automatic checking: SOLIDCHECK=1** **Enter a solids editing option [Face/ Edge/Body/Undo/eXit] <eXit>:** ***b*** **Enter a body editing option** **[Imprint/seParate solids/Shell/ cLean/Check/Undo/eXit] <eXit>:** ***l***	1. Enter the command sequence shown to access the ***Clean*** routine of the Body Category. Alternately, you can pick the **Clean** button on the Solids Editing toolbar.
	Select a 3D solid:	2. Select the 3D solid.
	Enter a body editing option **[Imprint/seParate solids/Shell/ cLean/Check/ Undo/eXit] <eXit>:** ***[enter]*** **Solids editing automatic checking: SOLIDCHECK=1** **Enter a solids editing option [Face/ Edge/Body/Undo/eXit] <eXit>:** ***[enter]***	3. Complete the command. The object looks like Figure 9.017. Notice that AutoCAD has removed the extra circular edge on the top.

TOOLS	COMMAND SEQUENCE	STEPS
	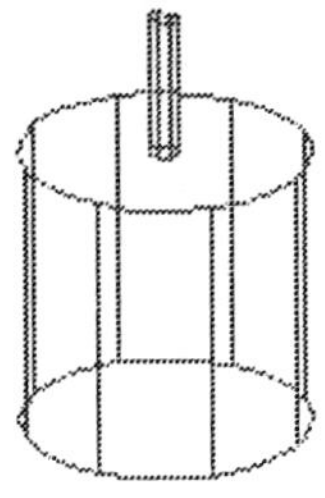 Figure 9.017	
	Command: ***qsave***	4. Save the drawing, but don't exit.

9.4.4	Shell

The **Shell** routine is one of the niftiest in the ***SolidEdit*** stable. With it, you can convert a 3D solid into a solid object similar to a surface model. In other words, you can convert a 3D solid into a hollow "shell" made up of a single 3D solid object.

To better understand this, consider your computer's monitor. Imagine the monitor with all the "guts" taken out – leaving just the plastic shell. The shell is a single object. The programmers at AutoCAD designed the **Shell** routine to create such objects!

The command sequence resembles those in the Face Category:

> **Command:** ***solidedit***
>
> **Solids editing automatic checking: SOLIDCHECK=1**
>
> **Enter a solids editing option [Face/Edge/Body/Undo/eXit] <eXit>:** ***b***
>
> **Enter a body editing option**
>
> **[Imprint/seParate solids/Shell/cLean/Check/Undo/eXit] <eXit>:** ***s***
>
> **Select a 3D solid:** ***[select the 3D solid you wish to shell]***
>
> **Remove faces or [Undo/Add/ALL]:** ***[select an edge of the face(s) you wish to remove]***
>
> **Remove faces or [Undo/Add/ALL]:** ***[complete the selection]***
>
> **Enter the shell offset distance:** ***[this figure defines the thickness of your shell]***

Let's create a shell from our original object.

Do This: 9.4.4.1	Shelling a 3D Solid

I. Be sure you're still in the *SE-Box-2.dwg* file in the C:\Steps3D\Lesson09 folder. If not, please open it now.

II. Change the viewpoint to 1,-2,-1.

III. Follow these steps.

TOOLS	COMMAND SEQUENCE	STEPS
Shell Button	**Command:** ***solidedit*** **Solids editing automatic checking: SOLIDCHECK=1** **Enter a solids editing option [Face/**	1. Enter the command sequence shown to access the ***Shell*** routine of the Body Category. Alternately, you can pick the **Shell** button on the Solids

Tools	Command Sequence	Steps
	Edge/Body/Undo/eXit] <eXit>: ***b*** **Enter a body editing option** **[Imprint/seParate solids/Shell/ cLean/Check/ Undo/eXit] <eXit>:** ***s***	Editing toolbar.
	Select a 3D solid:	2. Select the 3D solid.
	Remove faces or [Undo/Add/ALL]: **Remove faces or [Undo/Add/ALL]:** ***[enter]***	3. Remove the bottommost face (pick in the middle of the face – not on an edge).
	Enter the shell offset distance: ***.125***	4. Make the shell thickness 1/8".
	Solid validation started. **Solid validation completed.** **Enter a body editing option** **[Imprint/seParate solids/Shell/ cLean/Check/Undo/eXit] <eXit>:** ***[enter]*** **Solids editing automatic checking: SOLIDCHECK=1** **Enter a solids editing option [Face/ Edge/Body/Undo/eXit] <eXit>:** ***[enter]***	5. Complete the command. Your drawing looks like Figure 9.018 (shown with a **Gouraud+edges** Shademode).

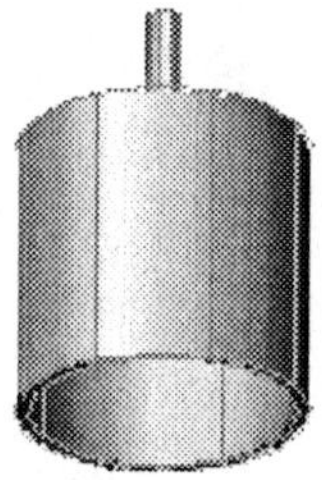

Figure 9.018

	Command: ***qsave***	4. Save the drawing.

9.4.5 Checking to Be Certain You Have an ACIS Solid

Have you noticed this note at the beginning of each of the options you've used in the ***SolidEdit*** command?

Solids editing automatic checking: SOLIDCHECK=1

This is telling you that the **SolidCheck** system variable has been set to **1** (that is, it's been activated). This means that AutoCAD will automatically check any 3D solid objects selected for editing to verify that they're valid ACIS solids.

To keep it simple, this just means that other software than uses ACIS can use 3D solids created in AutoCAD. If the object fails the verification, it's a good idea to redraw it.

9.5 Extra Steps

Just for fun, use the ***ACISOut*** command to create an ACIS text file (AutoCAD will automatically export the file with a .sat extension). Select one of the three-dimensional objects in the last drawing you had open.

Once you've created the file, open it with **Notepad**. Scan the text; it sure takes a lot to create a three-dimensional object!

9.6 What Have We Learned?

Items covered in this lesson include

- *Tools used to edit 3D Solid Faces, including*
  - ***Extrude***
  - ***Move***
  - ***Rotate***
  - ***Offset***
  - ***Taper***
  - ***Delete***
  - ***Copy***
  - ***coLor***
- *Tools used to edit 3D Solid Edges, including*
  - ***Copy***
  - ***coLor***
- *Tools used to edit 3D Solid Bodies, including*
  - ***Imprint***
  - ***seParate solids***
  - ***Clean***
  - ***Shell***
- *The SolidCheck system variable*

Wow! What a lesson! Did you ever imagine a single command could have so many different options?!

You've learned many ways to create and modify 3D solids. As I promised, you're no longer simply a CAD draftsman. By experience and training, you've become a CAD operator. You no longer draw. Now you create actual objects in the computer. There's very little left in the three-dimensional world for you to learn! (Okay. Don't get too excited – we still have two more chapters!)

In our next lesson, I'll show you how to handle blocks in Z-Space. I'll also show you some tools to help you plot the objects you've learned to create. It'll be an easier and less involved lesson than this one, so you can relax a bit. After that, we'll look at rendering the objects you create (making them look "real" by assigning materials to them).

But first, as always, let's practice what we've learned.

9.7 Exercises

1. Open the *Slotted Guide #1.dwg* file in the C:\Steps3D\Lesson09 folder. Using only the procedures discussed in this lesson, create the drawing in Figure 9.019.
   1.1. Remember that a negative number entered as an extrusion height will extrude *into* the object.
   1.2. Remember that a negative number entered as a tapering angle will angle outward.
   1.3. Add a 8½" x 11" title block.
   1.4. Save the drawing as *MySG#1.dwg* in the C:\Steps3D\Lesson09 folder.

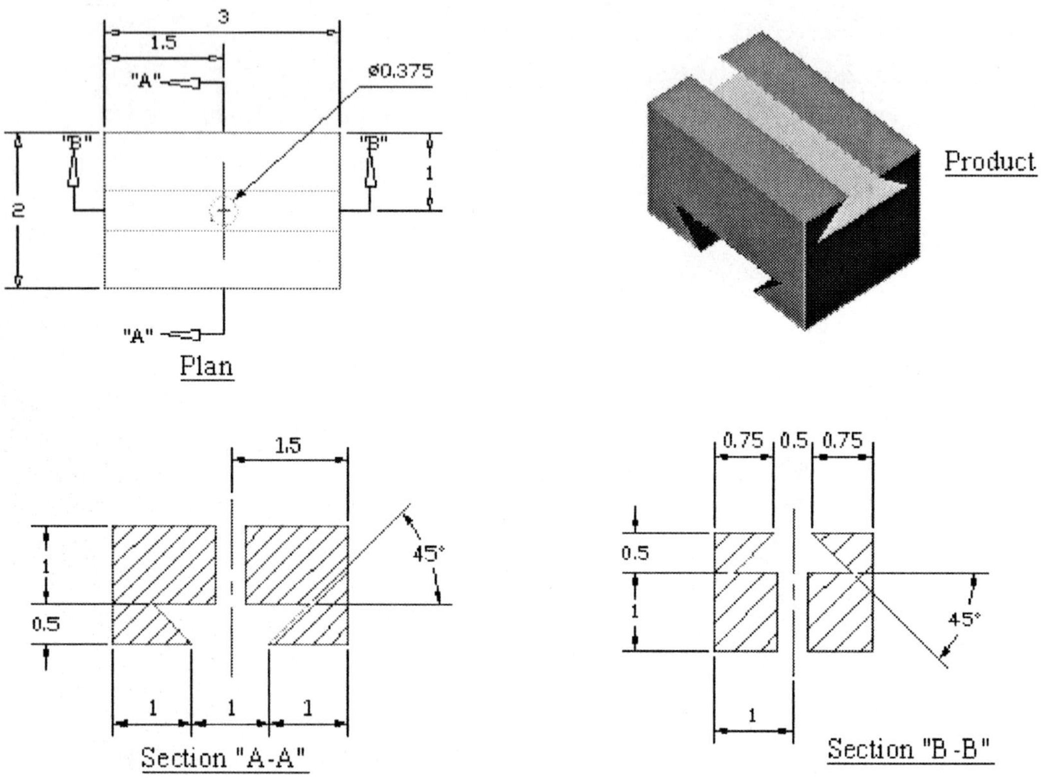

Figure 9.019

2. Open the *Slotted Guide #2.dwg* file in the C:\Steps3D\Lesson09 folder. Using only the procedures discussed in this lesson, create the drawing in Figure 9.020.
   2.1. Add a 8½" x 11" title block.
   2.2. Save the drawing as *MySG#2.dwg* in the C:\Steps3D\Lesson09 folder.

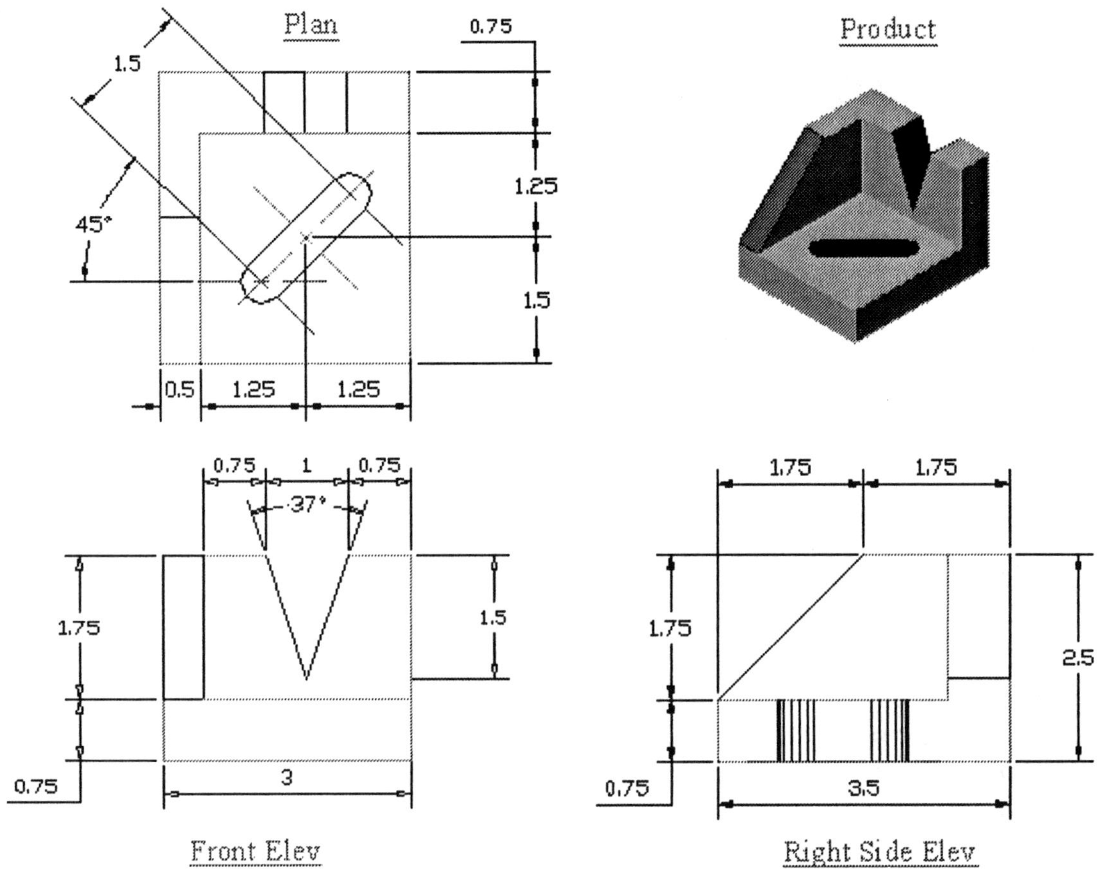

Figure 9.020

3. Create the corner bracket drawing in Figure 9.021.
   3.1. Use a 1/16" fillet along the edges.
   3.2. The rounded indentations are visible front and back.
   3.3. Add a 8½" x 11" title block.
   3.4. Save the drawing as *MyCB.dwg* in the C:\Steps3D\Lesson09 folder.

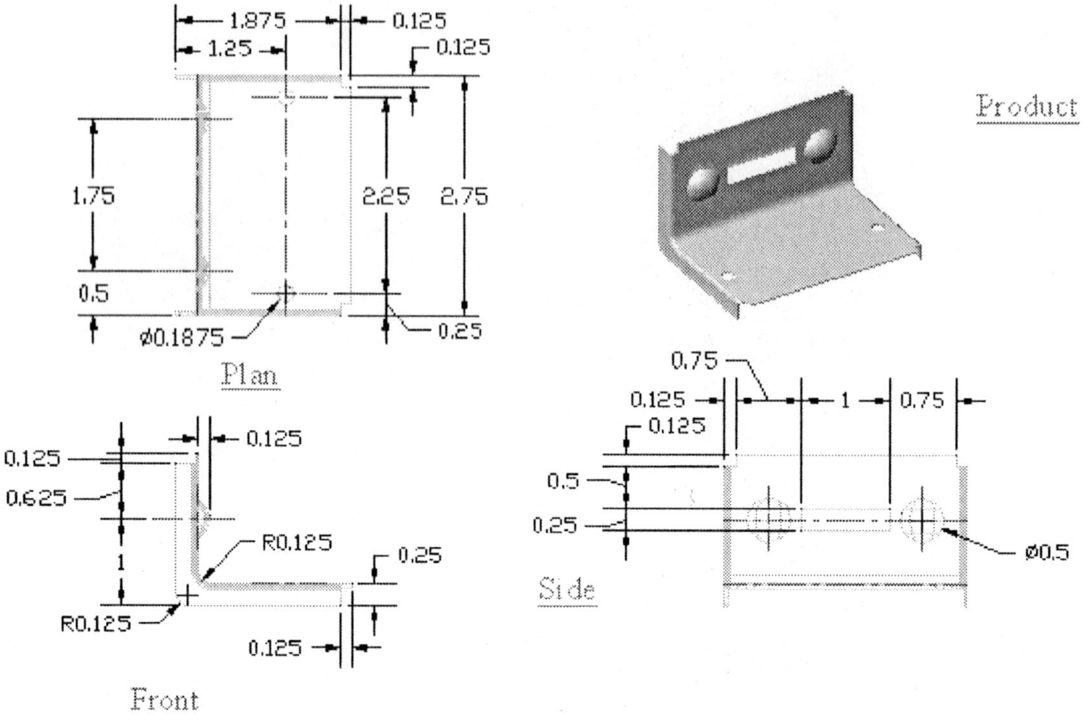

Figure 9.021

4. Create the lid drawing in Figure 9.022.
   4.1. Add a 8½" x 11" title block.
   4.2. Save the drawing as MyLid.dwg in the C:\Steps3D\Lesson09 folder.

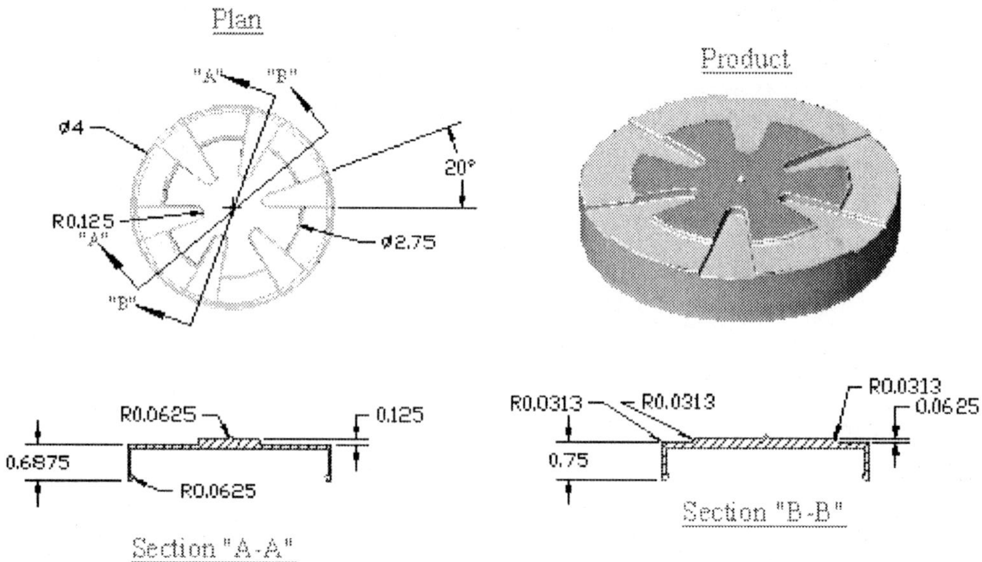

Figure 9.022

5. Create the plug drawing in Figure 9.023.
    5.1. Add a 8½" x 11" title block.
    5.2. Save the drawing as *MyPlug.dwg* in the C:\Steps3D\Lesson09 folder.

Thanks to George Gilbert of G. Gilbert EngineeringServices Ltd. for permission to use this drawing. For more on G. Gilbert Engineering Services Ltd., and more of George's drawings, visit his web site at: http://ourworld.compuserve.com/homepages/george_gilbert/

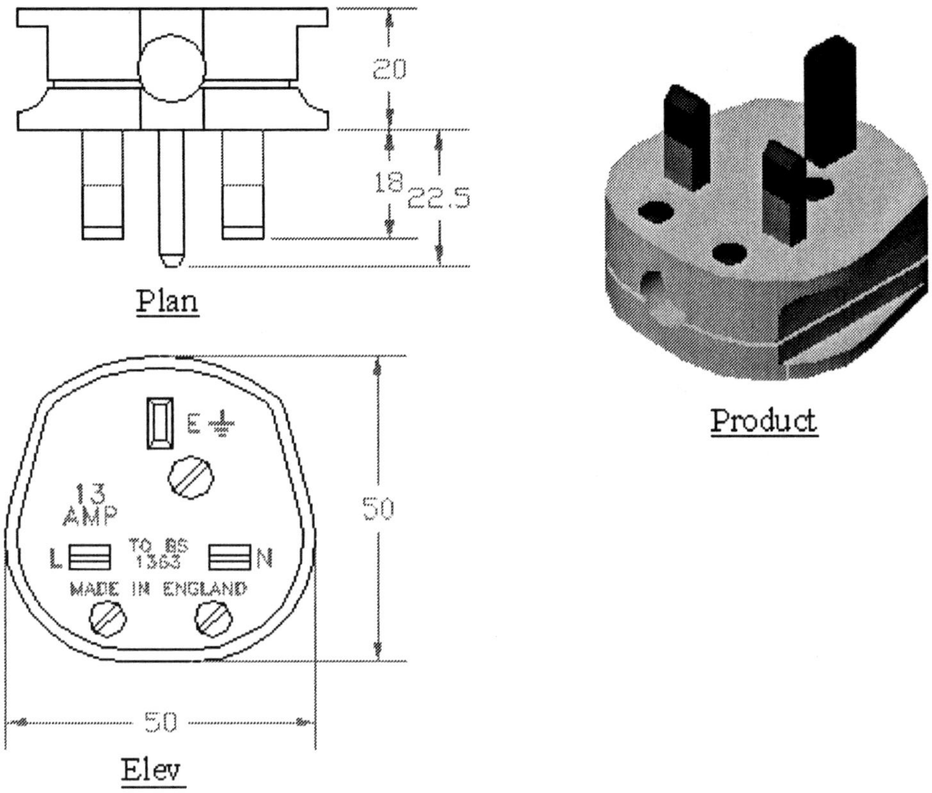

Figure 9.023

6. Create the switch drawing in Figure 9.024.
   6.1. Add a 8½" x 11" title block.
   6.2. Save the drawing as Switch.dwg in the C:\Steps3D\Lesson09 folder.

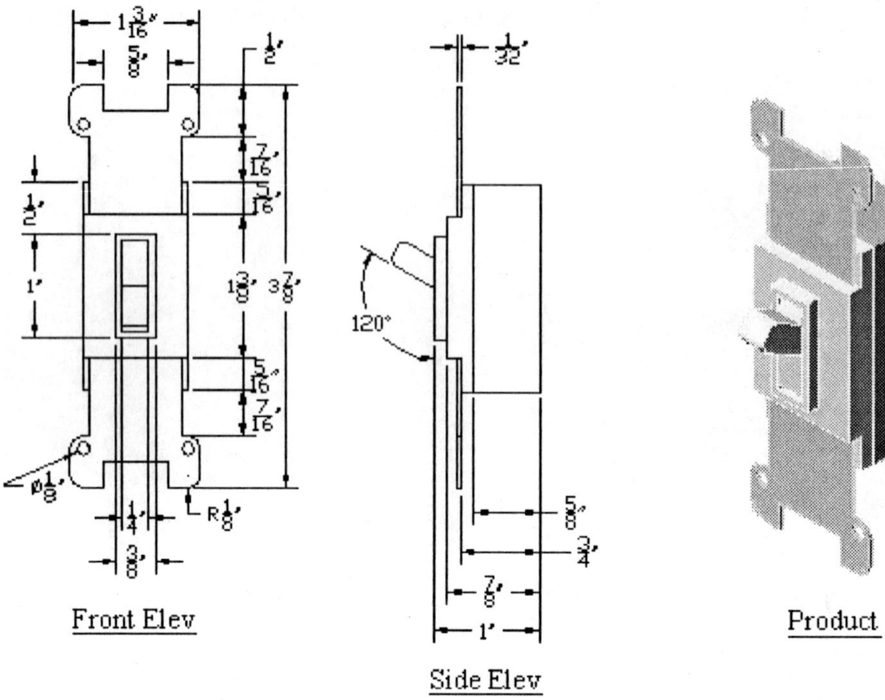

Figure 9.024

7. Create the wheel drawing in Figure 9.025.
   7.1. Place this drawing on a C-size (22" x 17") sheet of paper. Use a title block of your choice.
   7.2. Fillet the rim and hub with a ¼" fillet.
   7.3. Save the drawing as *MyWheel.dwg* in the C:\Steps3D\Lesson09 folder.

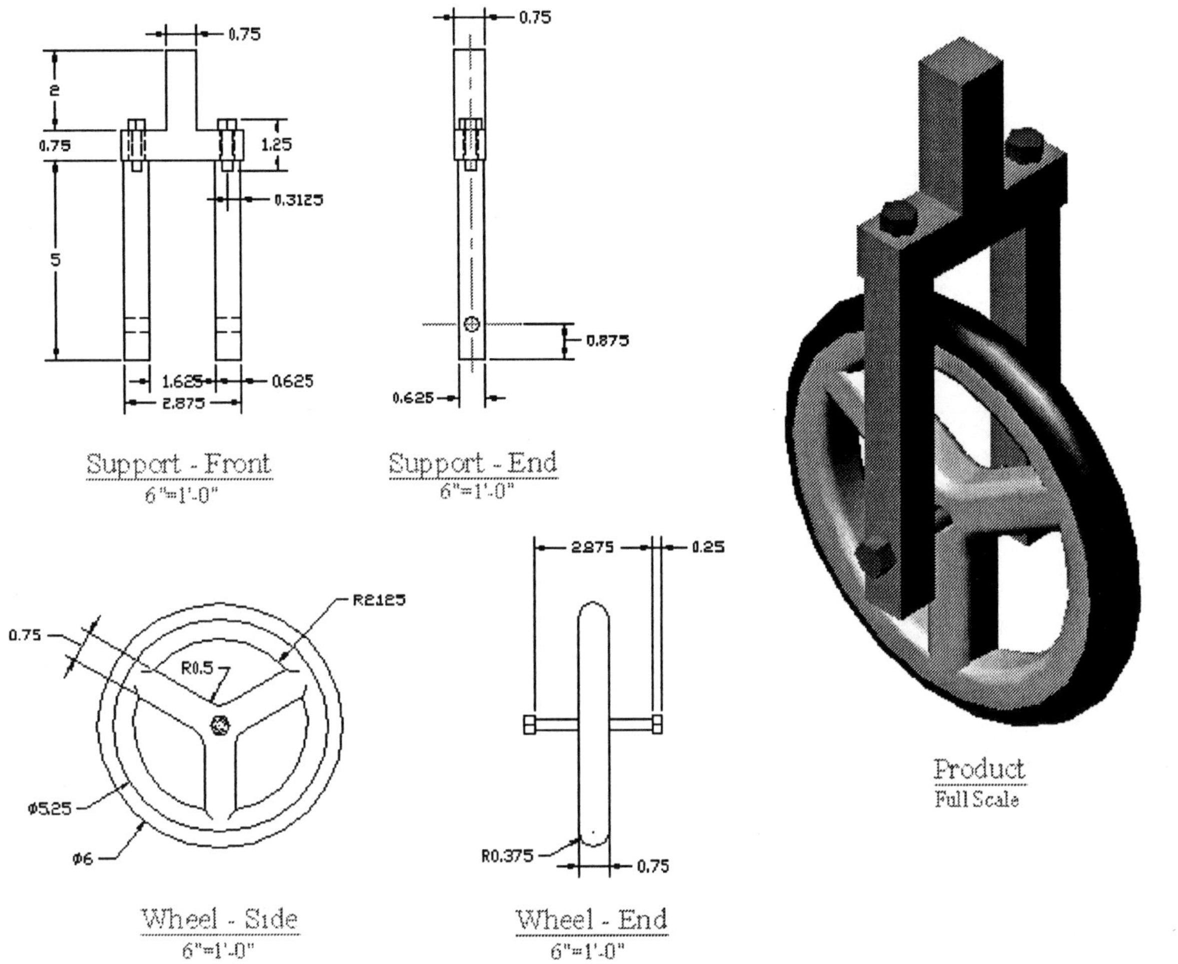

Figure 9.025

8. Create the level drawing in Figure 9.026.
   8.1. Don't join the glass pieces to the body of the level.
   8.2. Add a 8½" x 11" title block.
   8.3. Save the drawing as *MyLevel.dwg* in the C:\Steps3D\Lesson09 folder.

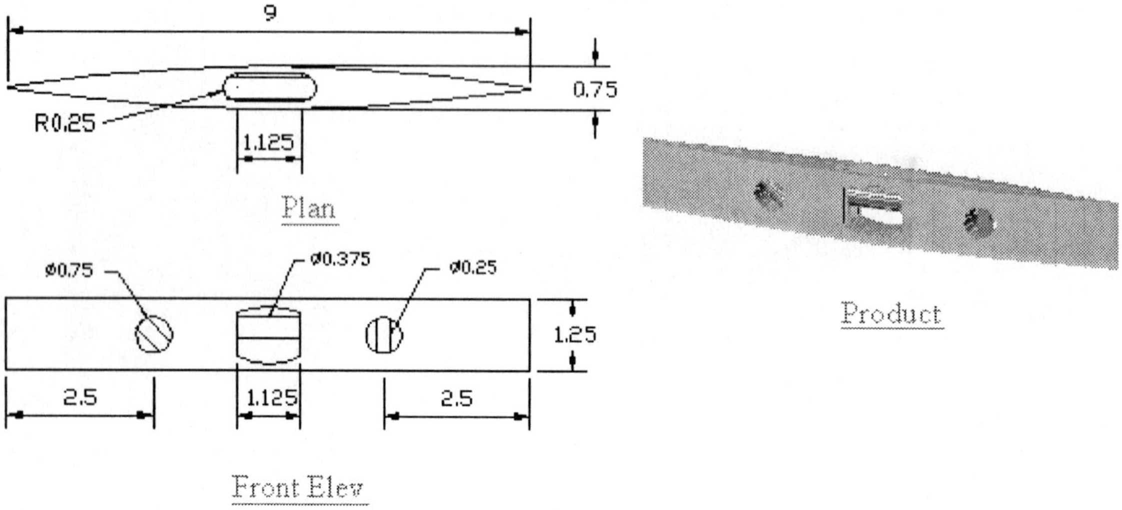

Figure 9.026

9.8 For Web-Based Review Questions, visit: http://www.uneedcad.com/2006/Files/06R9-3D.pdf

Lesson 10

Following this lesson, you will:

- ✓ *Know how to use blocks in Z-Space*
  - o *Creating three-dimensional blocks*
  - o *Inserting three-dimensional blocks*
- ✓ *Know how to use the Solid Plotting Tools*
  - o ***Solview***
  - o ***Soldraw***
  - o ***Solprof***

Three-Dimensional Blocks and Three-Dimensional Plotting Tools

As I mentioned when concluding the last lesson, there's very little left for you to learn with respect to creating three-dimensional objects in AutoCAD. This lesson, then, will serve as a wrap-up of Z-Space methods and techniques before we move on to rendering.

In Lesson 10, we'll first consider the behavior of blocks in a three-dimensional world. While the differences between two-dimensional blocks and three-dimensional blocks can be dramatic, they don't necessarily have to be difficult. We'll consider the effects of Z-Space and working planes on creation and insertion of blocks as well as the use of attributes on a three-dimensional block.

Then we'll discuss three special tools designed to help you set up and plot 3D solids with considerably less difficulty than you might have had previously.

Let's begin.

10.1 Using Blocks in Z-Space

As with two-dimensional blocks, three-dimensional blocks can save you a tremendous amount of time and effort. But there are a few things the three-dimensional operator must consider.

10.1.1 Three-Dimensional Blocks and the UCS

First among these considerations is the working plane (the current UCS). *AutoCAD creates blocks against the plane of the current UCS* (Figure 10.001) – *not* the WCS. In other words, the XYZ values of the current UCS (the X-axis, Y-axis, and Z-axis values of the object) become part of the block definition.

Likewise, AutoCAD inserts bocks by matching the axis values of the block to the current UCS (Figure 10.002).

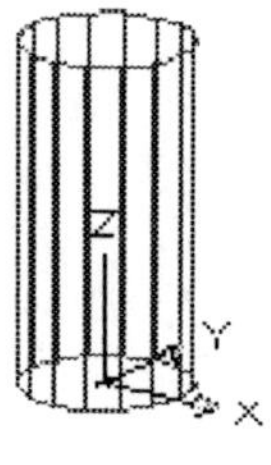

Figure 10.001

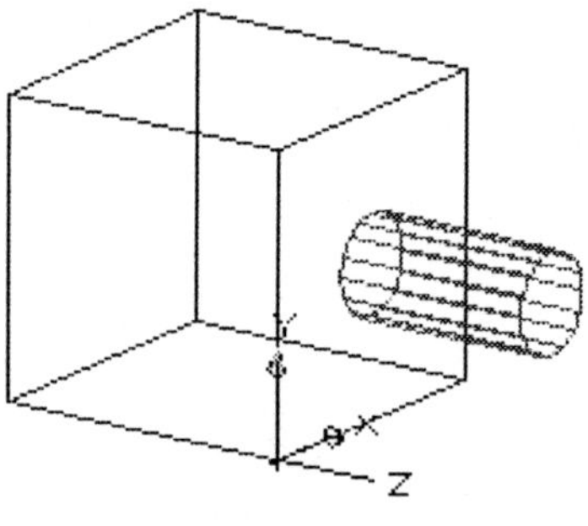

Figure 10.002

We'll see this in a series of exercises. First, we'll create two blocks – an elbow and some pipe. Our blocks will have two simple attributes defining what they are and their size. Second, we'll utilize our UCS and insertion scale factors to insert the blocks in a simple piping configuration. Third, we'll extract attribute data that'll give us a running total of the amount of pipe we've used.

Let's begin.

Do This: 10.1.1.1	Creating 3D Blocks

I. Open the *Blocks.dwg* file in the C:\Steps3D\Lesson10 folder. The drawing looks like Figure 10.003.

II. Notice that the elbow is open in the +X and +Y directions. Note also that the elbow and the pipe are drawn to scale as 4" fittings, and that the height of the pipe is 1".

III. Follow these steps.

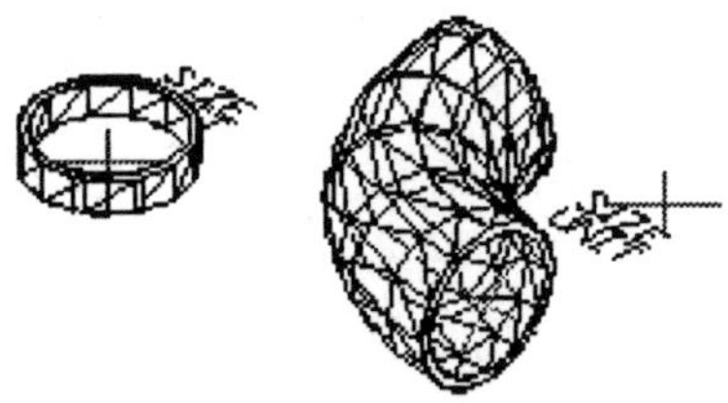

Figure 10.003

TOOLS	COMMAND SEQUENCE	STEPS
No Button Available	**Command:** ***w***	1. Make a note of the current UCS. Then make blocks (use the ***WBlock*** command) from the two objects you see. Call the block on the left *Pipe*. Use the node as the insertion point (include the node and the two attributes in the block). Call the block on the right *Ell*. Use the node inside the eastern opening of the elbow as the insertion point (include all three nodes and the two attributes in the block). Be sure to write both blocks to the C:\Steps3D\Lesson10 folder.
		2. Close the drawing.

Now we'll look at some new variations of the ***Insert*** command.

10.1.2 Inserting Three-Dimensional Blocks

The second three-dimensional block consideration involves the insertion scale of the block. In two-dimensional drafting, you could scale a block along the X- or Y-axis. Now you'll have an additional axis along which you can scale the block. You should give special attention to the use of a three-dimensional block because of the effect scale might have. Consider the images in Figure 10.004, 10.005, 10.006, and 10.007. We'll have an opportunity to use the scale options and see how the UCS affects the insertion when we insert our new blocks into a drawing.

Let's get started.

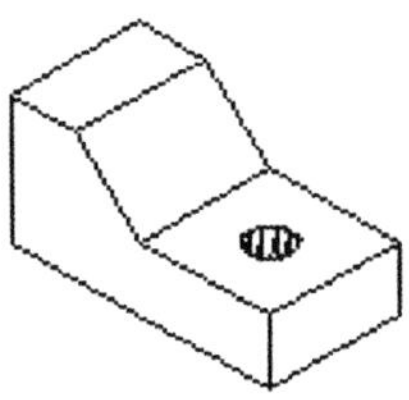

Figure 10.004

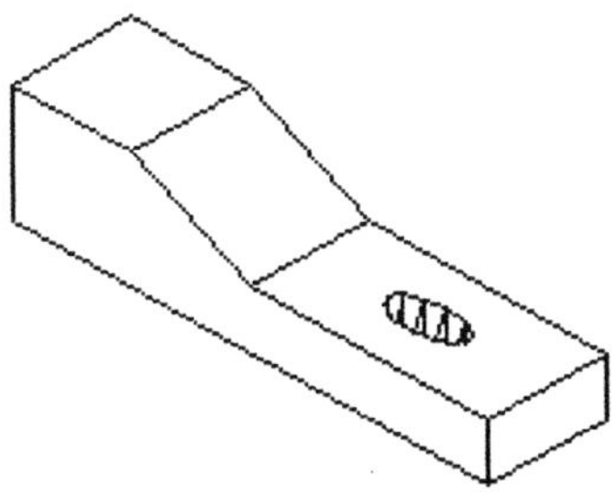

Figure 10.005

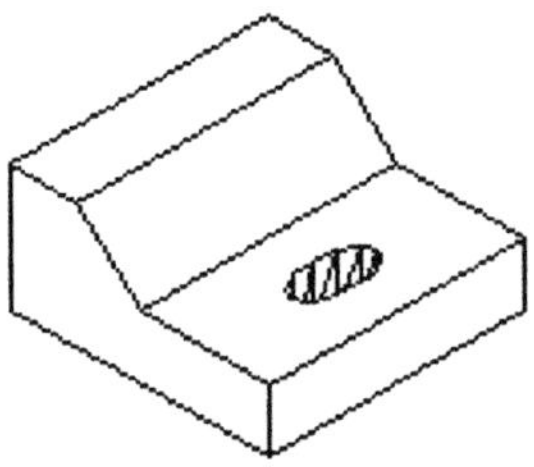

Figure 10.006

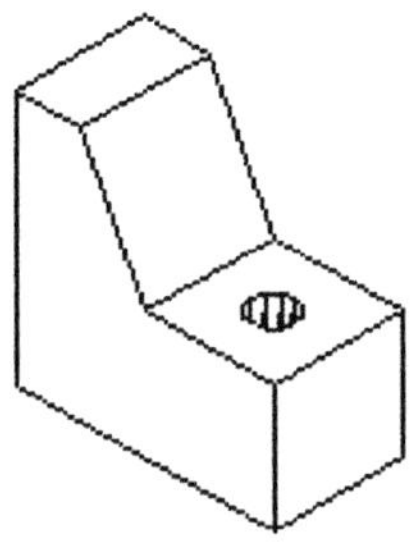

Figure 10.007

Do This: 10.1.2.1	Inserting 3D Blocks

I. Open the *Piping Configuration.dwg* file in the C:\Steps3D\Lesson10 folder. The drawing looks like Figure 10.008. (I set the nodes to help guide you through the block insertions. The **UCSIcon** system variable has been set to **ORigin**.)

II. Set the running OSNAP to **Node**. Clear all other settings.

III. Set the current layer to **Pipe**.

IV. Follow these steps.

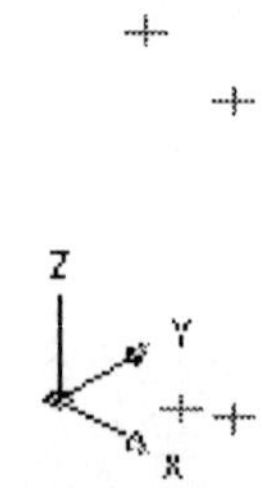

Figure 10.008

TOOLS	COMMAND SEQUENCE	STEPS
	Command: ***ucs***	1. Change the UCS as indicated in Figure 10.009. (Hint: Rotate the UCS 90°on the X-axis and the 270° on the Z-axis.)

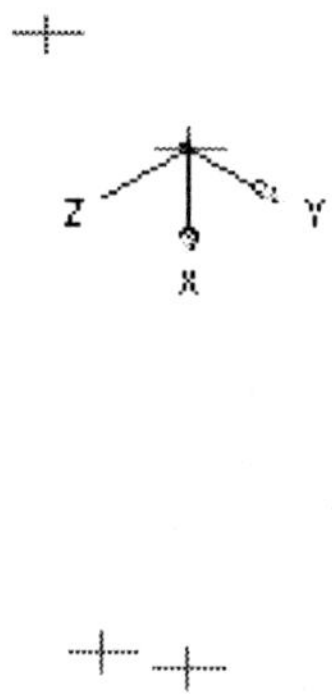

Figure 10.009

TOOLS	COMMAND SEQUENCE	STEPS
		2. Insert the *Ell* block at the 0,0,0 coordinate of the current UCS. (You created this block and placed it in the C:\Steps3D\Lesson10 folder in our last exercise.) Use a scale of 1 for each axis. Accept the default 4" size. The elbow looks like Figure 10.010 (hidden lines removed).

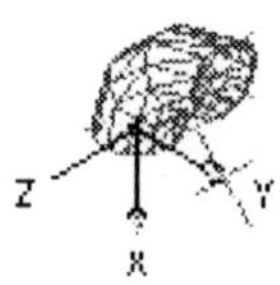

Figure 10.010

Tools	Command Sequence	Steps
		3. Add the rest of the elbows. (Place the UCS at each node and insert the elbows at 0,0,0.) Your drawing looks like Figure 10.011 (hidden lines and removed). (Your UCS icon may be in a different location.)

Figure 10.011

Tools	Command Sequence	Steps
	Command: ***ucs***	4. Reset the UCS to the WCS.
		5. Now we'll insert the pipe using a scale factor to make it fit between the elbows. Tell AutoCAD you wish to insert the *Pipe* block. Use the settings shown (Figure 10.012). (The distance between the elbows we'll use is 8' – or 96".)

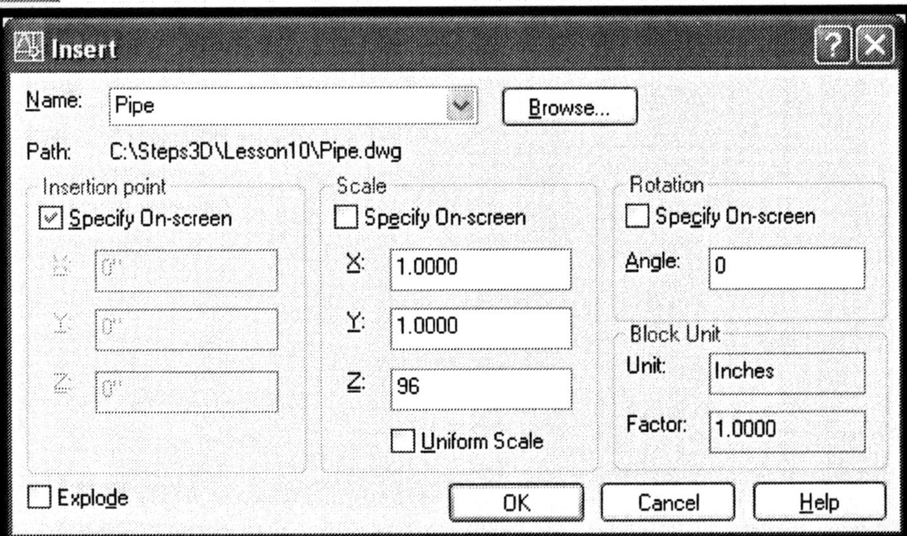

Figure 10.012

Tools	Command Sequence	Steps
	Specify insertion point or [Basepoint/Scale/X/Y/Z/Rotate/ PScale/PX/PY/PZ/ PRotate]:	6. Insert the block at the upper node of the lower-right elbow.
	Enter attribute values **What is the size of the unit? <4">:**	7. Accept the default size of the pipe.
		8. Repeat Steps 5 through 7 for the other vertical run of pipe. Your drawing looks like Figure 10.013 (hidden lines removed).

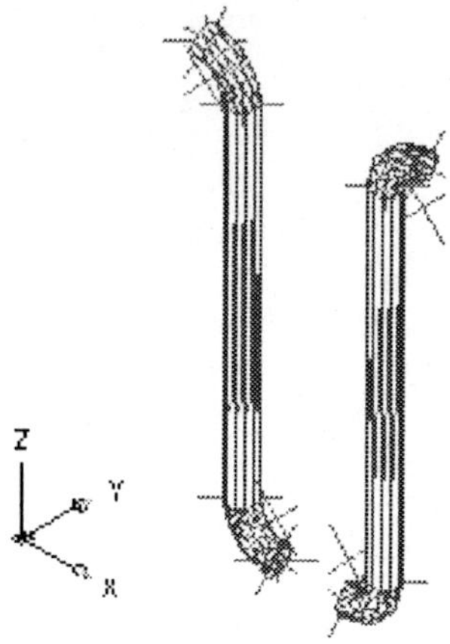

Figure 10.013

Tools	Command Sequence	Steps
		9. Adjust the UCS as needed to add the final run of pipe. The distance between the two elbows is 12". Your drawing looks like Figure 10.014 (hidden lines removed).

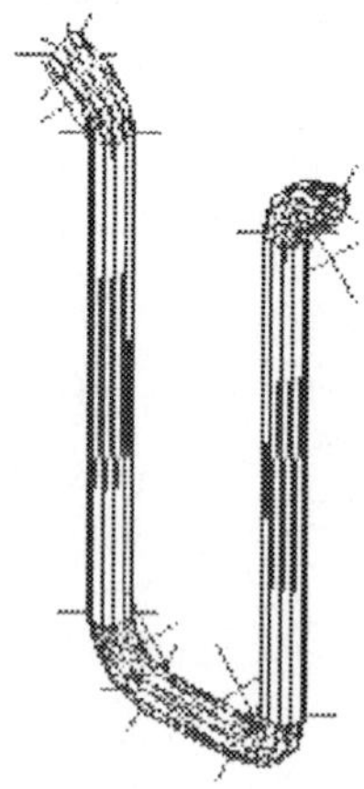
Figure 10.014

Tools	Command Sequence	Steps
	Command: ***qsave***	10. Save the drawing, but don't close it.

10.1.3	Making Good Use of Attributes

The last (and possibly most rewarding) things we should consider when using three-dimensional blocks is the possible use of attributes. When combining the abilities you've already seen in this lesson with a clever use of attributes, you'll discover a remarkably useful way to create a bill of materials for most projects.

We'll extract the attribute data to our blocks to see that the *Pipe* blocks contain data about the length of each piece of pipe. This data is accurate enough to be used on a cutting list!

Sound nifty? Let's try it!

Do This: 10.1.3.1	**Extracting the Attributes**

I. Be sure you're still in the *Piping Configuration.dwg* file in the C:\Steps3D\Lesson10 folder. If not, please open it now.

II. Set the UCS = World and restore the plan view (0,0,1).

III. Zoom out to about .25x and set the **Text** layer current.

IV. Follow these steps.

TOOLS	COMMAND SEQUENCE	STEPS
No Button Available	**Command: *Eattext***	1. Begin the attribute extraction wizard.
Next >		2. Accept the defaults on page 1 (Begin) and page 2 (Select Drawings) of the wizard.
		3. (Refer to Figure 10.015.) • Deselect **Ell** in the **Blocks** list box. • Clear the check box next to **Exclude general block properties**. (AutoCAD will list the blocks' general properties – such as scale and position.) • Clear the check box next to **UNIT** in the **Properties for checked blocks** list box. • Place a check next to **Scale Z** in the **Properties for checked blocks** list box. (This will give us our pipe lengths.)

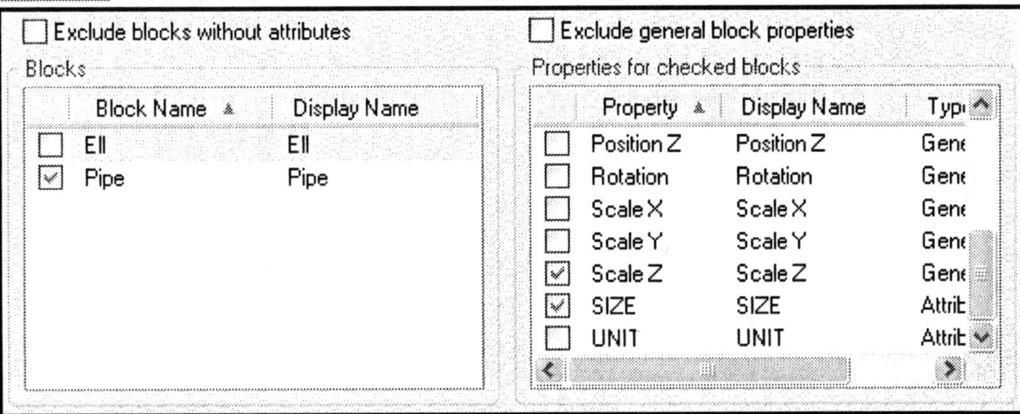

Figure 10.015

TOOLS	COMMAND SEQUENCE	STEPS
		4. Go on to the next page, accept the defaults on Page 4 (Finalizing Output), and go to Page 5 (Table Style).
		5. (Refer to Figure 10.016.) • Rename the **Scale Z** column to **Length**. • Rearrange the columns as shown.

Quantity	Name	SIZE	Length
1	Pipe	4"	12.0000
2	Pipe	4"	96.0000

Figure 10.016

TOOLS	COMMAND SEQUENCE	STEPS
		6. Go on to Page 5 and (referring to Figure 10.017): • Enter a title for your table as shown. • Use the Pipe Length Table (already created for you).

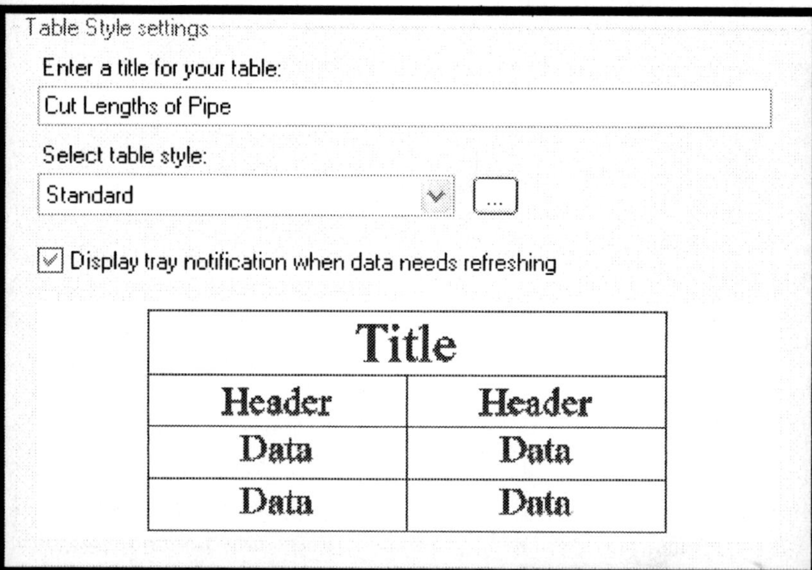

Figure 10.017

TOOLS	COMMAND SEQUENCE	STEPS
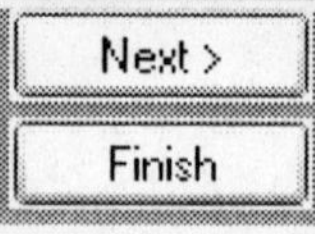		7. Go to Page 6 and complete the wizard.
		8. Place the table in a convenient location. It looks like Figure 10.018.

TOOLS	COMMAND SEQUENCE	STEPS

Cut Lengths of Pipe

Quantity	Name	SIZE	Length
1	Pipe	4"	12.0000
2	Pipe	4"	96.0000

Figure 10.018

		9. Save and exit the drawing.

How's that for convenience?

> Although we used pipe in our exercise, this method works as well when tracking board feet or lengths of steel. How might you use it in your profession?

10.2 Plotting a 3D Solid

We'll conclude our study of 3D solids with a group of special commands that AutoCAD designed to make plotting solid models easier. The "Sol Group" consists of three commands: ***Solview***, ***Soldraw***, and ***Solprof***. (You won't find a finer example of teamwork in the CAD world.) You'll use the first command – ***Solview*** – to set up the layout (the Paper Space viewports). Then you'll use the other two commands – ***Soldraw*** and ***Solprof*** – to create the actual drawings that go into the viewports.

Let's look at each.

10.2.1 Setting Up the Plot – the *Solview* Command

Of the three, ***Solview*** is the most complex command. It creates viewports according to your input. You can define the viewports by the XY-plane or a user-defined UCS, or by calculating orthographic projections, auxiliary projections, and cross sections from a UCS viewport. ***Solview*** places the viewports on the **VPorts** layer (which it creates if necessary). It also creates viewport-specific layers for visible lines (*Viewname-vis*), hidden lines (*Viewname-hid*), dimensions (*Viewname-dim*), and hatching (*Viewname-hat*).

You must enter the ***Solview*** command while a **Layout** tab is active. AutoCAD responds with the initial ***Solview*** prompt:

> **Command:** ***solview***
>
> **Enter an option [Ucs/Ortho/Auxiliary/Section]:**

Let's consider each option.

- The **UCS** option creates a two-dimensional profile view of the object. It uses the XY-plane of a user-specified UCS to define the profile. AutoCAD responds to selection of the UCS option with the prompt:

  > **Enter an option [Named/World/?/Current] <Current>:** ***[enter]***

Enter view scale <1.0000>: ***[enter the scale for the view (if incorrect, you can rescale the view later using Zoom XP or the Scale control box on the Viewports toolbar)]***

Specify view center: ***[pick a point where you'd like to place the center point of the view]***

Specify view center <specify viewport>: ***[reposition the view, if necessary, or hit* enter *to continue]***

Specify first corner of viewport: ***[define the viewport by specifying opposite corners]***

Specify opposite corner of viewport:

Enter view name: ***[give the viewport a unique name]***

The same sequence applies to the **Named** and **World** options, except that AutoCAD will precede the **Named** option sequence with a request for the name of the UCS to use. Use the **?** option to list the named UCSs available for use. Use the **UCS** option to create the first viewport – usually a front view or plan view of the 3D solid.

- Use the **Ortho** option (back at the first ***Solview*** prompt) to create orthographic projections from an existing view. The command sequence is the same as with the **UCS** option.
- Creating an auxiliary view – a view perpendicular to an inclined face – is as easy as drawing an orthographic projection for AutoCAD. AutoCAD responds to selection of the **Auxiliary** option with

  Specify first point of inclined plane: ***[specify two points that define the inclined plane]***

  Specify second point of inclined plane:

  Specify side to view from: ***[pick a point from where you wish to see the inclined plane]***

  AutoCAD continues with the options to size and locate the viewport.
- The **Section** option creates a cross section complete with section lines. (It takes the ***Soldraw*** command to actually create the section.) The sequence is

  Specify first point of cutting plane:

  Specify second point of cutting plane:

  Specify side to view from:

  Again, AutoCAD continues with the options to size and locate the viewport.

All of these options will become much clearer with an exercise.

> You'll notice in our next exercise that, in fact, none of these options creates a drawing. What they do is *set up* a viewport for the orthographic, auxiliary, or cross sectional drawings that you'll create later with the ***Soldraw*** command. You'll see the actually 3D solid in each viewport until you use the ***Soldraw*** command.

Let's begin.

> You can access all of the ***Sol...*** commands through the Draw pull-down menu. Follow this path:
>
> *Draw – Solids – Setup – (Drawing, View, or Proflie)*

Do This: 10.2.1.1	Using *Solview* to Set Up a Layout

I. Open the *Sol1.dwg* file in the C:\Steps3D\Lesson10 folder. It looks like Figure 10.019.

II. Activate the **Layout1** tab. Erase any viewports that appear.

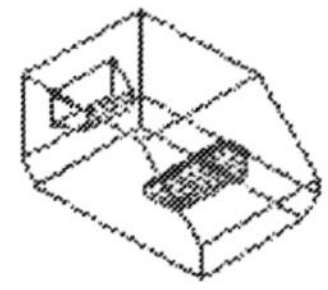

Figure 10.019

TOOLS	COMMAND SEQUENCE	STEPS
Setup View Button	**Command:** ***solview***	1. Enter the ***Solview*** command. Alternately, you can pick the **Setup View** button on the Solids toolbar.
Ucs Ortho Auxiliary Section	**Enter an option [Ucs/Ortho/ Auxiliary/Section]:** ***U***	2. Select the **UCS** option.
Named	**Enter an option [Named/World/ ?/Current] <Current>:** ***N*** **Enter name of UCS to restore:** ***front***	3. Tell AutoCAD to use the **Named** UCS option, and use the UCS called ***Front***.
	Enter view scale <1.0000>: ***.75***	4. We'll use a three-quarter scale for our viewport.
	Specify view center: ***3,3.5*** **Specify view center <specify viewport>:** ***[enter]***	5. Center the viewport on Paper Space coordinate 3,3.5.
	Specify first corner of viewport: ***1,4.5*** **Specify opposite corner of viewport:** ***@4,-2***	6. Size the viewport as indicated (you can pick approximate coordinates).
	Enter view name: ***Front***	7. Call the view ***Front***. Your drawing looks like Figure 10.020.

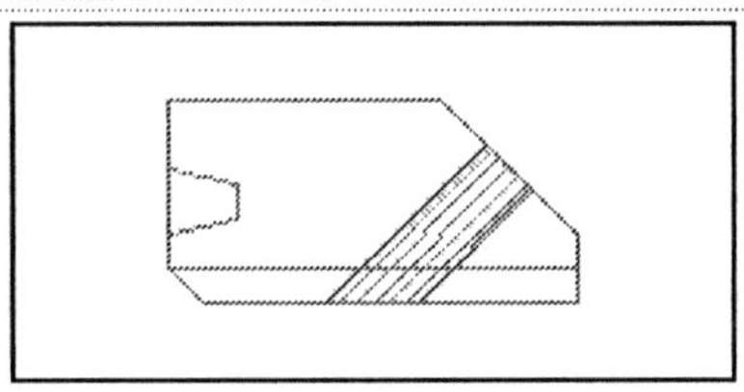

Figure 10.020

	Enter an option [Ucs/Ortho/Auxiliary/Section]: ***[enter]***	8. Complete the command.
	Command: ***[enter]***	9. Repeat the command.

TOOLS	COMMAND SEQUENCE	STEPS
Ortho	**Enter an option [Ucs/Ortho/ Auxiliary/Section]: *O***	10. Now tell AutoCAD to create an orthographic projection.
	Specify side of viewport to project:	11. Pick a point on the right side of the existing viewport (notice that AutoCAD automatically uses the midpoint OSNAP) …
	Specify view center: *@3<0* **Specify view center <specify viewport>: *[enter]***	12. … and center the viewport about 3 units to the right of the UCS view.
	Specify first corner of viewport: *6.5,4.5* **Specify opposite corner of viewport: *@3,-2***	13. Locate the viewport at about the coordinates indicated.
	Enter view name: *Right* **Enter an option [Ucs/Ortho/ Auxiliary/Section]: *[enter]***	14. Call the viewport ***Right***, and complete the command.
	Command: *[enter]*	15. Repeat Steps 10 through 14 to create a Top viewport as shown in Figure 10.021.

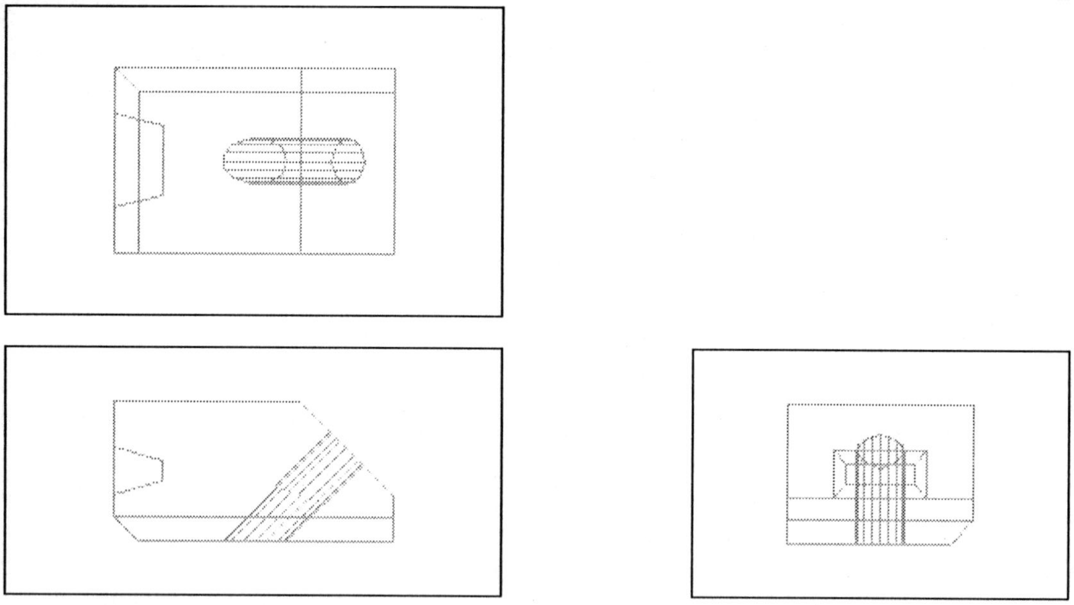

Figure 10.021

TOOLS	COMMAND SEQUENCE	STEPS
Auxiliary	**Enter an option [Ucs/Ortho/ Auxiliary/Section]: *a***	16. Now we'll create an **Auxiliary** view of the inclined surface.
	Specify first point of inclined plane: **Specify second point of inclined plane:**	17. In the original viewport (*Front*), pick the endpoints of the inclined surface. (Pick anywhere in the Front viewport to activate it.)

TOOLS	COMMAND SEQUENCE	STEPS
	Specify side to view from:	18. Tell AutoCAD that you wish to view the surface from the upper-right corner of the viewport.
	Specify view center: **Specify view center <specify viewport>:** ***[enter]***	19. Pick a point about even with the center of the upper viewport (*Top*).
	Specify first corner of viewport: **Specify opposite corner of viewport:** **Enter view name:** ***Aux*** **UCSVIEW = 1 UCS will be saved with view** **Enter an option [Ucs/Ortho/ Auxiliary/Section]:** ***[enter]***	20. Place the viewport around the auxiliary image (don't worry that the viewports overlap), and name the view ***Aux***. Complete the command.
	Command: ***m***	21. Move the new viewport to the position shown in Figure 10.022.

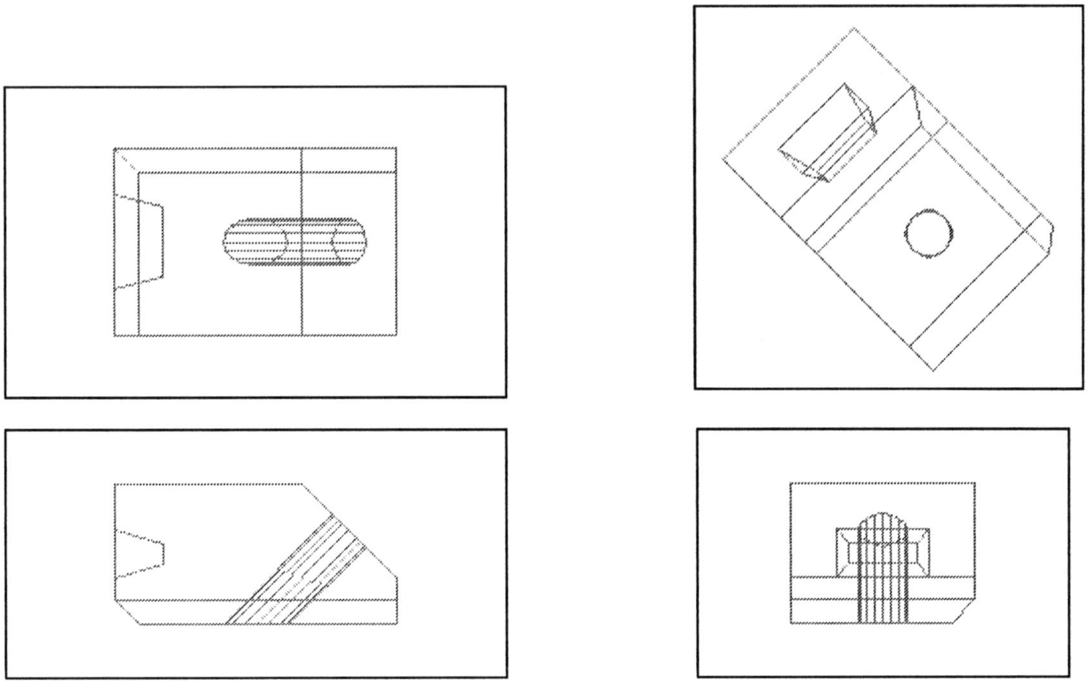

Figure 10.022

TOOLS	COMMAND SEQUENCE	STEPS
	Command: ***qsave***	22. Save the drawing.
	Command: ***solview***	23. Repeat the ***Solview*** command.
Section	**Enter an option [Ucs/Ortho/ Auxiliary/Section]:** ***S***	24. Select the **Section** option.

TOOLS	COMMAND SEQUENCE	STEPS
	Specify first point of cutting plane: **Specify second point of cutting plane:**	25. In the upper-left viewport (*Top*), specify the cutting plane as shown in Figure 10.023 (use Ortho).

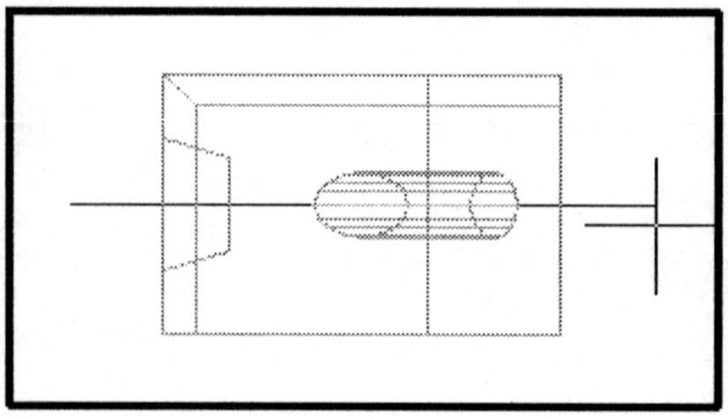

Figure 10.023

TOOLS	COMMAND SEQUENCE	STEPS
	Specify side to view from: **Enter view scale <0.7500>: *[enter]***	26. View the object from the lower half of the viewport, and accept the default view scale.
	Specify view center: **Specify view center <specify viewport>: *[enter]***	27. Center the sectional viewport below the original (*Front*) viewport.
	Specify first corner of viewport: **Specify opposite corner of viewport:** **Enter view name: *Sect***	28. Place the viewport around the image, and call the view ***Sect***. Your drawing looks like Figure 10.024.

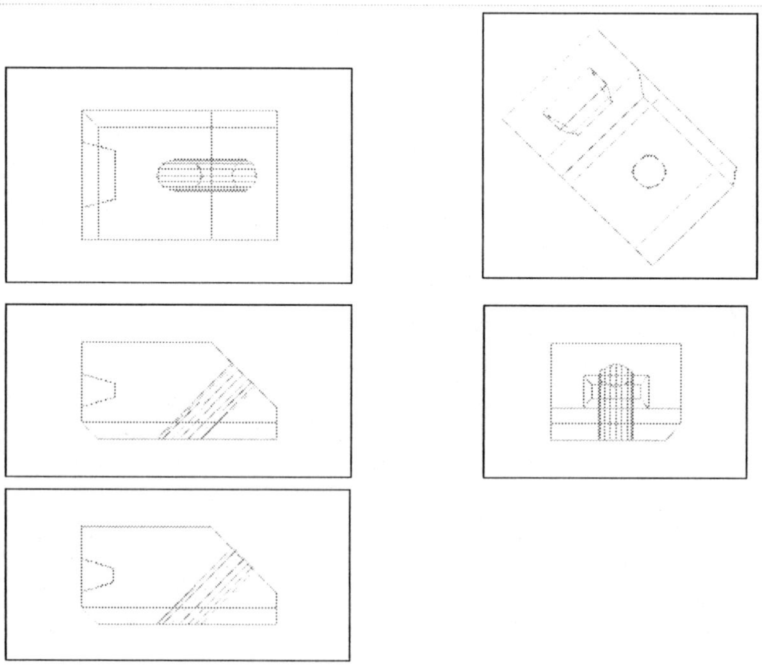

Figure 10.024

TOOLS	COMMAND SEQUENCE	STEPS
	UCSVIEW = 1 UCS will be saved with view **Enter an option [Ucs/Ortho/ Auxiliary/Section]:** ***[enter]***	29. Complete the command.
	**Command:** ***qsave***	30. Save the drawing, but don't exit.

How's that for a quick way to set up several viewports? If you look a little further than your immediate screen, you'll find that AutoCAD has also set up layers specific to each viewport.

Obviously, however, the drawing isn't yet ready to plot – each viewport still shows the full 3D solid. To complete the drawing, we need to show profiles in each viewport. Let's look at the ***Soldraw*** and ***Solprof*** commands next.

10.2.2 Creating the Plot Images – The *Soldraw* and *Solprof* Commands

Both ***Soldraw*** and ***Solprof*** create profiles in a viewport. The biggest difference is that the programmers designed ***Soldraw*** to work specifically with viewports created by the ***Solview*** command. ***Solprof*** will create a profile (see insert) in a viewport created by the ***MView*** or ***MVSetup*** commands. Additionally, ***Soldraw*** will create cross sections where they were set up with the ***Solview*** command.

> A profile shows only those edges and/or silhouettes of a 3D solid that are visible in the specified viewport when hidden lines are removed.
>
> An important thing to remember about the Sol… commands is that they were designed to work only with solids. They won't work with surfaces or blocks.

The command sequence for ***Soldraw*** is one of AutoCAD's simplest:

Command: ***soldraw***

Select viewports to draw..

Select objects:

AutoCAD does the rest automatically. Try it.

Do This: 10.2.2.1 Using *Solview* to Set Up a Layout

I. Be sure you're still in the *Sol1.dwg* file in the C:\Steps3D\Lesson10 folder. If not, please open it now.

II. Follow these steps.

TOOLS	COMMAND SEQUENCE	STEPS
 Setup Drawing Button	**Command:** ***soldraw***	1. Enter the ***Soldraw*** command. Alternately, you can pick the **Setup Drawing** button on the Solids toolbar.
	Select viewports to draw.. **Select objects:** **Select objects:** ***[enter]***	2. Select each of the viewports. AutoCAD creates the profiles and sections. Your drawing looks like Figure 10.025.

Tools	Command Sequence	Steps

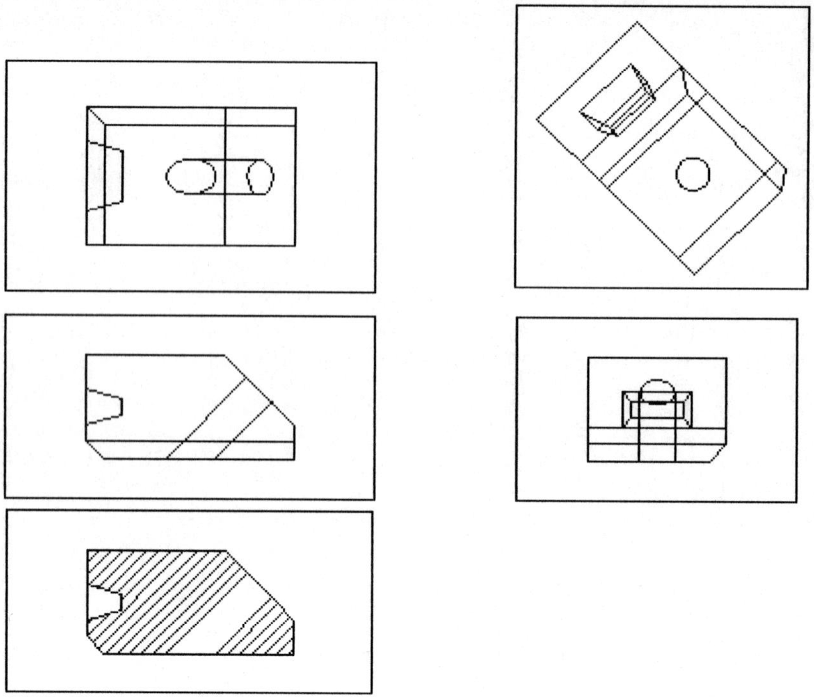

Figure 10.025

Tools	Command Sequence	Steps
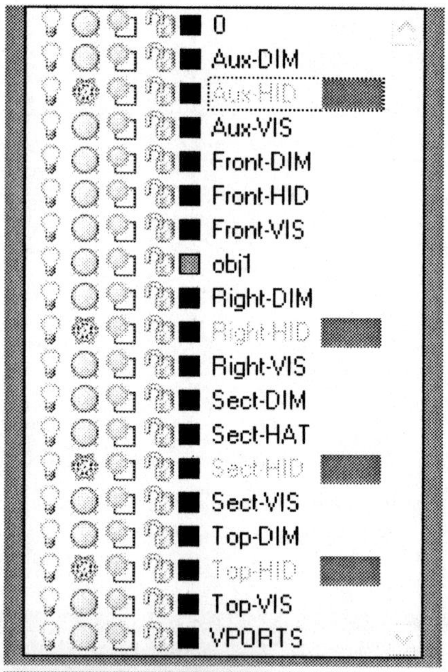	**Command:** ***la***	3. Not quite satisfied? Freeze all of the **[Name]-hid** layers except **Front-hid**.
		4. Load the **Hidden** linetype and assign it to the **Front-hid** layer. Your drawing lows like Figure 10.026.

TOOLS	COMMAND SEQUENCE	STEPS

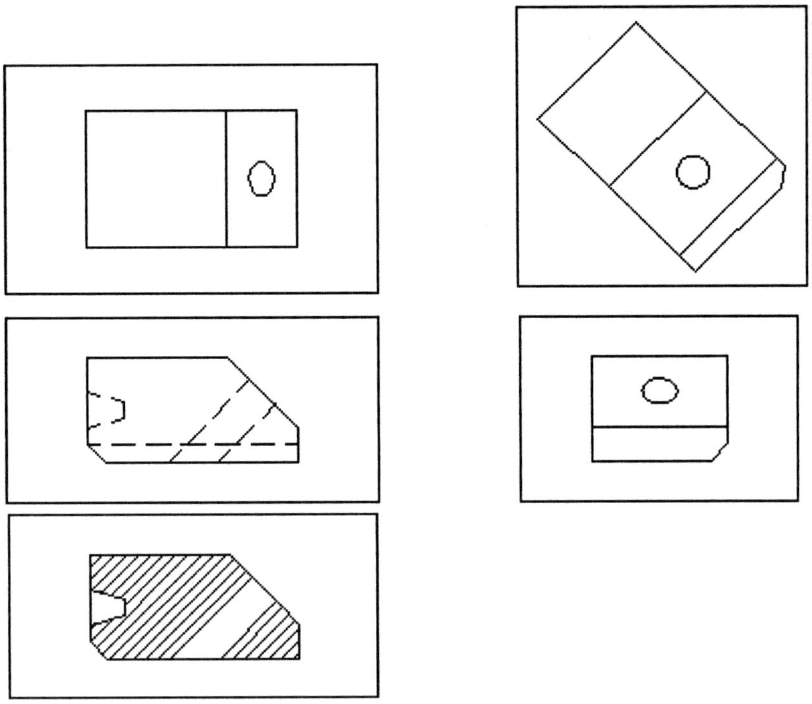

Figure 10.026

TOOLS	COMMAND SEQUENCE	STEPS
	Command: ***qsave***	5. Save the drawing, but don't exit.

Solprof works almost as easily, but the profiles it creates are actually blocks. Here's the command sequence:

> **Command:** ***solprof***
>
> **Select objects:**
>
> **Select objects:**
>
> **Display hidden profile lines on separate layer? [Yes/No] <Y>:**
>
> **Project profile lines onto a plane? [Yes/No] <Y>:**
>
> **Delete tangential edges? [Yes/No] <Y>:**

Let's look at the options.

- The first option – **Display hidden profile lines on separate layer** – asks if you'd like to place all profile lines on one layer or place the hidden lines on a separate layer. The default is to place hidden lines on a separate layer.

  When you accept the default (generally a good idea), AutoCAD places visible lines on layer *PV-[viewport handle]* and hidden lines on layer *PH-[viewport handle]*. By using the AutoCAD-assigned viewport handle as part of the layer, AutoCAD assures you of a unique layer name. This way, you can freeze the layer or change the linetype of the hidden lines.

- The next option – **Project profile lines onto a plane** – allows you to create a two-dimensional profile by projecting the lines onto the view plane (the default), or to create three-dimensional lines.

- The last option – **Delete tangential edges** – allows you to remove tangential lines. These are objects (lines) that show the transition between arcs or circles. They're essentially the same things as isolines, except that they're actual objects.

Let's use the ***Solprof*** command to create an isometric view of our object.

Do This: 10.2.2.2	Using *Solview* to Set Up a Layout

I. Be sure you're still in the *Sol1.dwg* file in the C:\Steps3D\Lesson10 folder. If not, please open it now.

II. Create a new viewport in the lower-right corner of the layout. (Use the ***MView*** command.)

III. Activate the new viewport and set up an isometric view (1,1,1),

IV. Follow these steps.

Tools	Command Sequence	Steps
 Setup Profile Button	**Command: *solprof***	1. Enter the ***Solprof*** command. Alternately, you can pick the **Setup Profile** button on the Solids toolbar.
	Select objects: **Select objects: *[enter]***	2. Select the 3D solid.
	Display hidden profile lines on separate layer? [Yes/No] <Y>: *[enter]* **Project profile lines onto a plane? [Yes/No] <Y>: *[enter]*** **Delete tangential edges? [Yes/No] <Y>: *[enter]***	3. Accept the defaults for the next three prompts.
 	**Command: *la***	4. Freeze the **PH-[viewport handle]** layer. (The viewport handle will vary.)
 		5. Freeze the **obj1** layer in the active viewport. Your drawing looks like Figure 10.027.

TOOLS	COMMAND SEQUENCE	STEPS

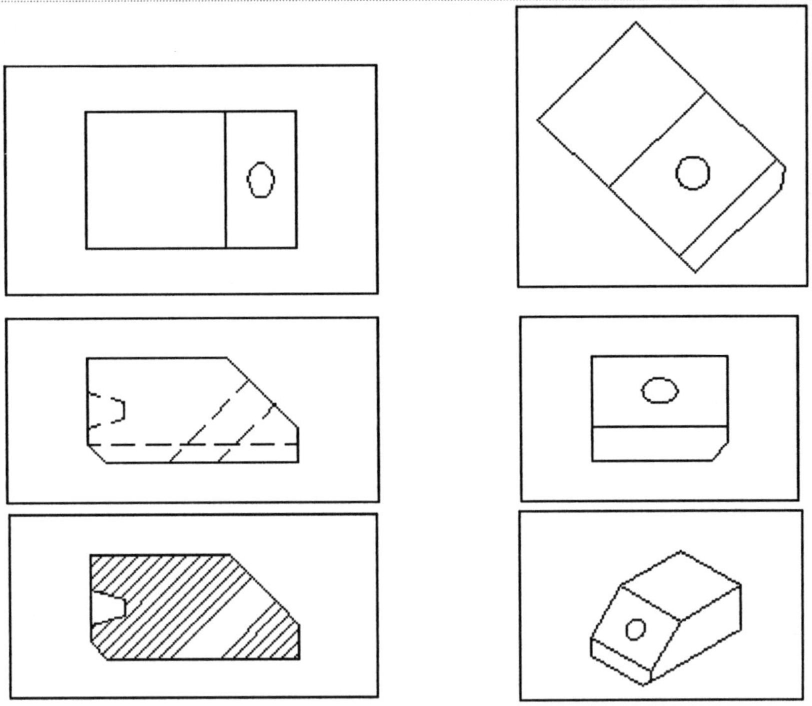

Figure 10.027

	Command: ***qsave***	6. Save the drawing.

Now you can use the other techniques you've learned to dimension each view, add appropriate text, and otherwise complete the drawing.

10.3 Extra Steps

Return to any of the drawings you created in the exercises at the end of Lessons 8 and 9. Create the plotting layouts shown in the exercises (if you created layouts, open **Layout2** and recreate them). This will work best if you use drawings for which you've already created layouts – it'll help you compare the method you used previously with the ***Sol...*** commands.

10.4 What Have We Learned?

Items covered in this lesson include:

- *Three-dimensional uses and techniques for blocks*
- *The* ***Sol...*** *tools used to set up plots for 3D solid*
  - ***Solview***
  - ***Soldraw***
  - ***Solprof***

In this lesson, you discovered some easier ways to set up a Paper Space plot and some new (and useful) techniques for working with blocks. And you wrapped up your study of three-dimensional drafting and modeling techniques. You can relax for two minutes and pat yourself on the back for having accomplished quite a lot of often-difficult material. Then tackle the exercises at the end of the lesson.

In Lesson 11, you'll see how to render a drawing. While not always useful as a drafting tool, rendering takes you one step further – to adding material qualities to your objects. This means having a table that shows wood grain or a glass lamp that appears transparent. You'll see how to show your drawing in perspective rather than isometric mode. You'll create photographic-quality images suitable for brochures or posters, and much more!

So, complete the exercises and hurry into that place where AutoCAD meets computer graphics!

10.5 Exercises

1. Open the *slotted guide.dwg* file in the C:\Steps3D\Lesson10 folder. Create the layout shown in Figure 10.028.
   1.1. Hint: The ***Sol...*** commands work best when the object is viewed through a 2D Shademode.
   1.2. Save the drawing as *MySG.dwg* in the C:\Steps3D\Lesson10 folder.

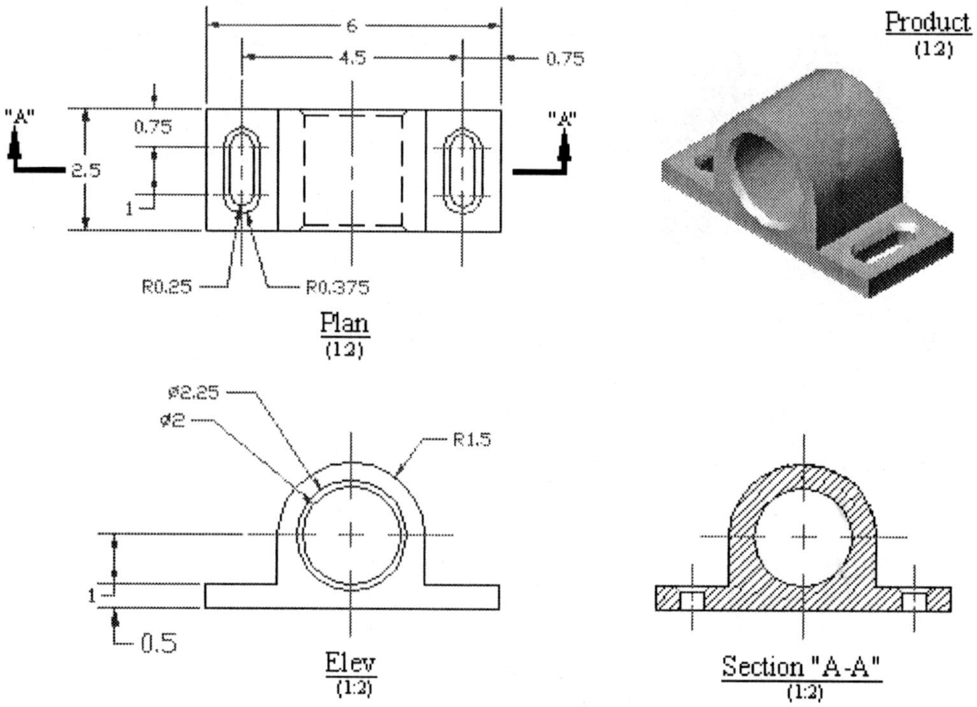

Figure 10.028

2. Open the *My Flange 10.dwg* file you created in Lesson 8 (it should be in the C:\Steps3D\Lesson10 folder). If that one isn't available, open the *flange10.dwg* file instead. Create the layout shown in Figure 10.029.
   2.1. Most of the centerlines already exist on layer **Cl**.
   2.2. Save the drawing as *MyFlg.dwg* in the C:\Steps3D\Lesson10 folder.

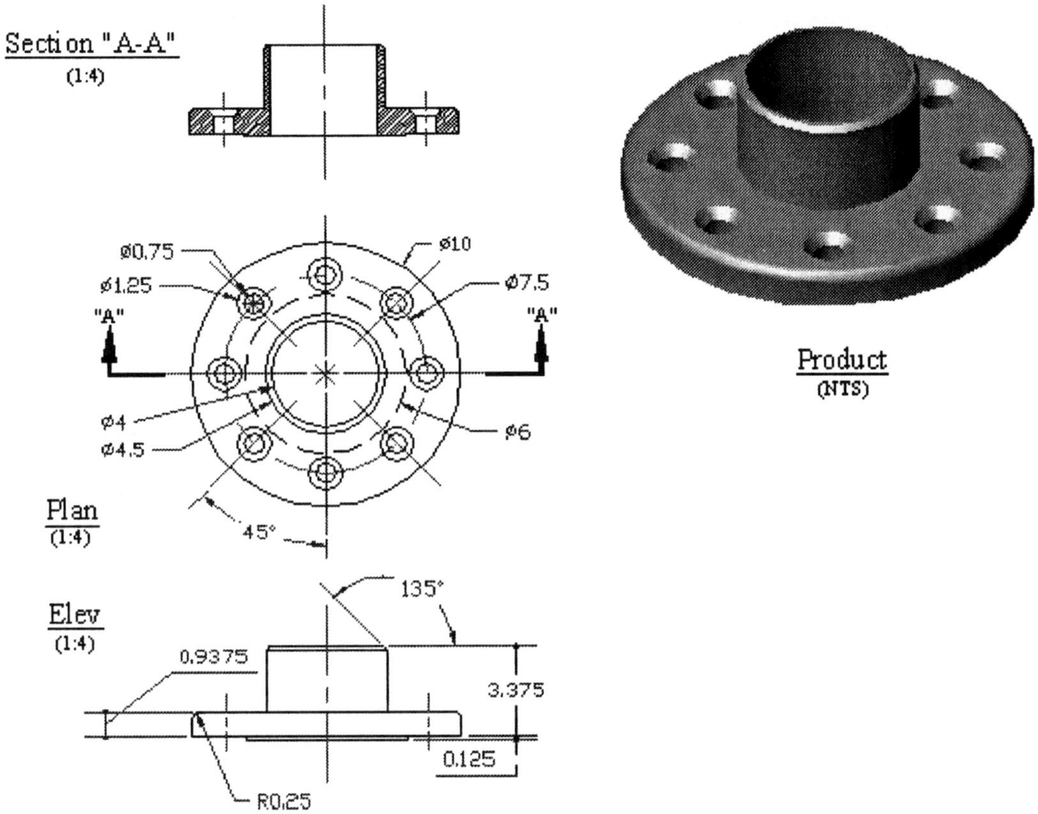

Figure 10.029

3. Open the *Jig.dwg* in the C:\Steps3D\Lesson10. Create the layout shown in Figure 10.030.
   3.1. Set up the drawing on a 17" x 11" sheet of paper.
   3.2. Save the drawing as *MyJig.dwg* in the C:\Steps3D\Lesson10 folder.

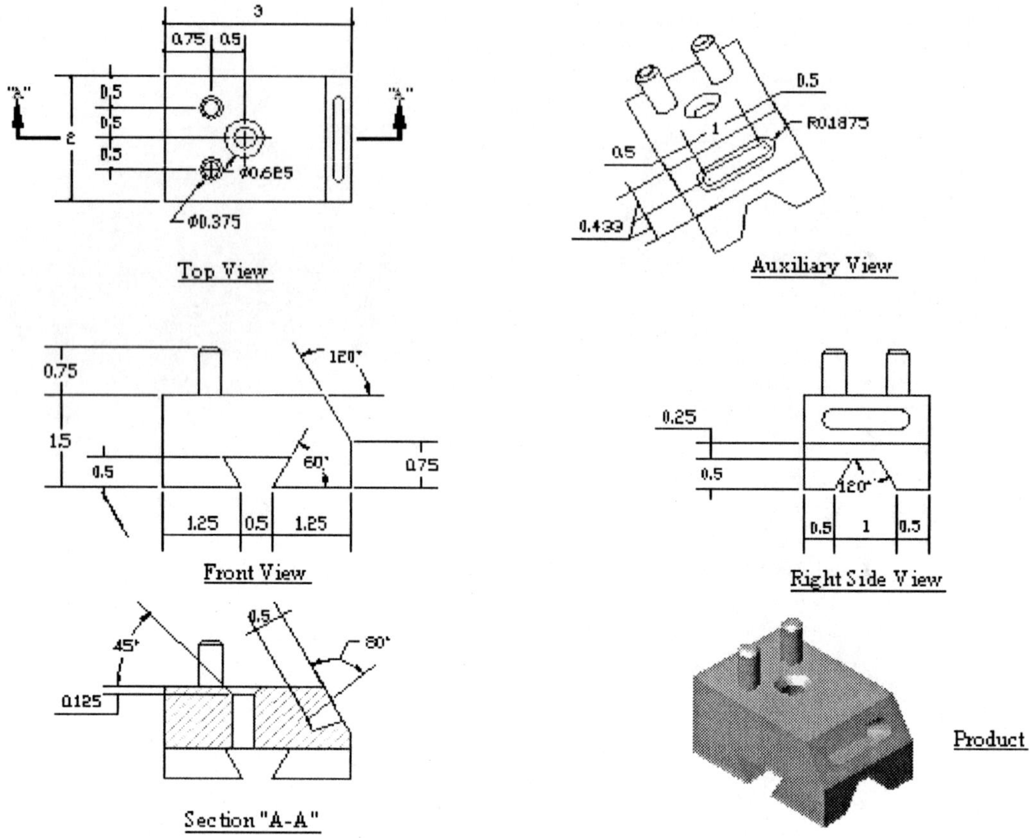

Figure 10.030

4. Open the *Thermometer.dwg* file in the C:\Steps3D\Lesson10 folder. Create the layout shown in Figure 10.031.
    4.1. Set up the drawing on an 11" x 17" sheet of paper.
    4.2. Save the drawing as *MyThermometer.dwg* in the C:\Steps3D\Lesson10 folder.

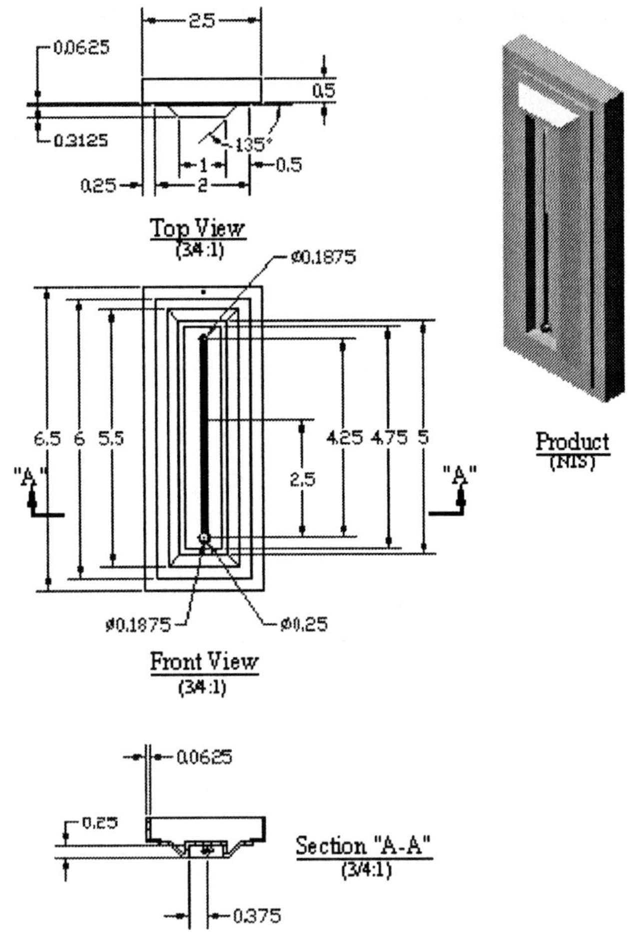

Figure 10.031

5. Create the service cart drawing shown in Figure 10.032.
   5.1. Create the layout on an 11" x 17" sheet of paper.
   5.2. Adjust the Z-Space and UCS as needed to use a single 2" x 2" block to build the frame.
   5.3. Use the caster you created in Lesson 4 (or the *Caster.dwg* file in the C:\Steps3D\Lesson10 folder) for the caster block.
   5.4. Use attributes and the ***Attext*** command to create the cutting list.
   5.5. You'll notice that the ***Sol...*** commands won't work properly on blocks, so you'll have to use the ***MView*** or ***MVSetup*** command to create your viewports.
   5.6. Save the drawing as *MyCart.dwg* in the C:\Steps3D\Lesson10 folder.

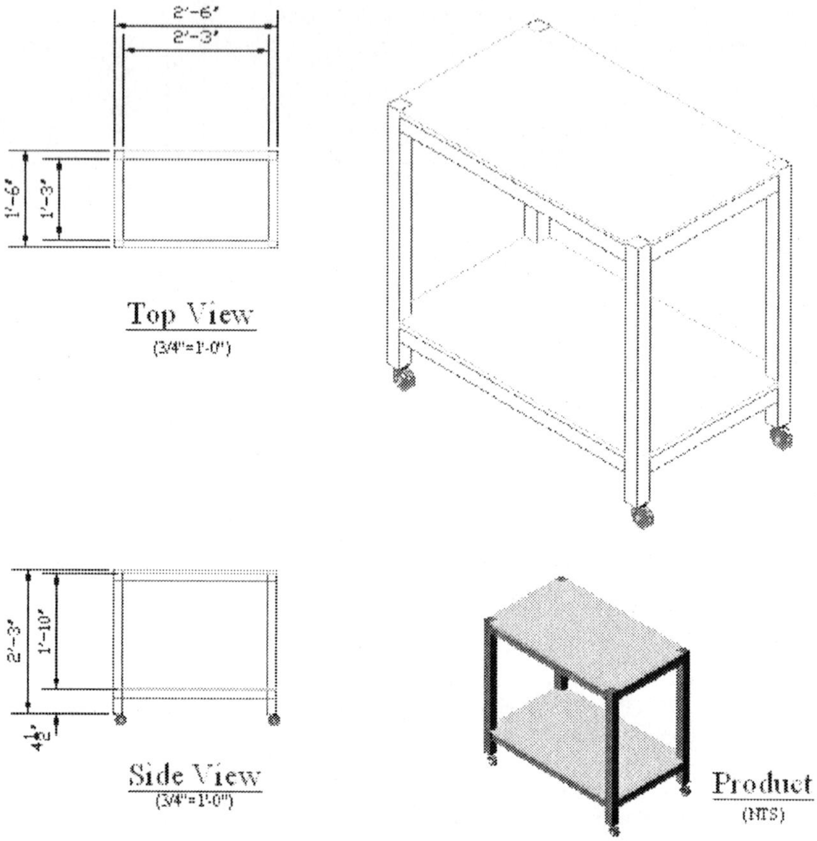

Cutting List		
Quantity	Item	Scale Z
4	2x2	15
8	2x2	27
4	caster	1

Figure 10.032

6. Create the *1_2Ell* drawing shown in Figure 10.033.
   6.1. Make sure the base point of the elbow is as indicated. (Use the ***Base*** command to move it, if necessary).
   6.2. Be sure the UCS = WCS when you finish.
   6.3. Save the drawing as *My1_2Ell.dwg* in the C:\Steps3D\Lesson10 folder.

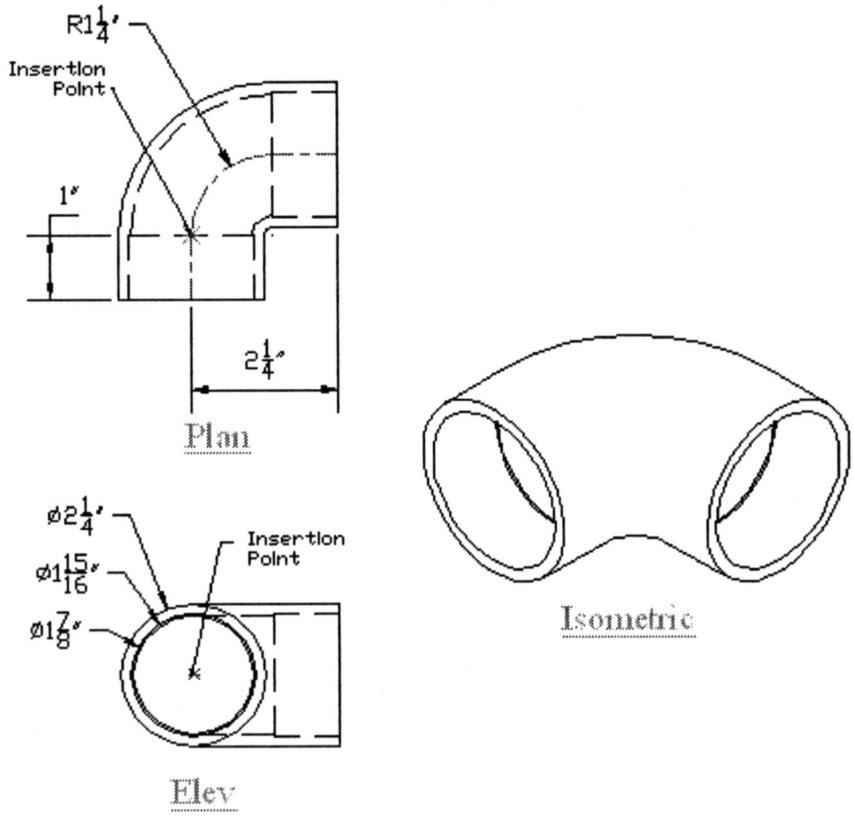

Figure 10.033

7. Create the *1_2Tee* drawing shown in Figure 10.034.
   7.1. Make sure the base point of the tee is as indicated.
   7.2. Be sure the UCS = WCS when you finish.
   7.3. Save the drawing as *My1_2Tee.dwg* in the C:\Steps3D\Lesson10 folder.

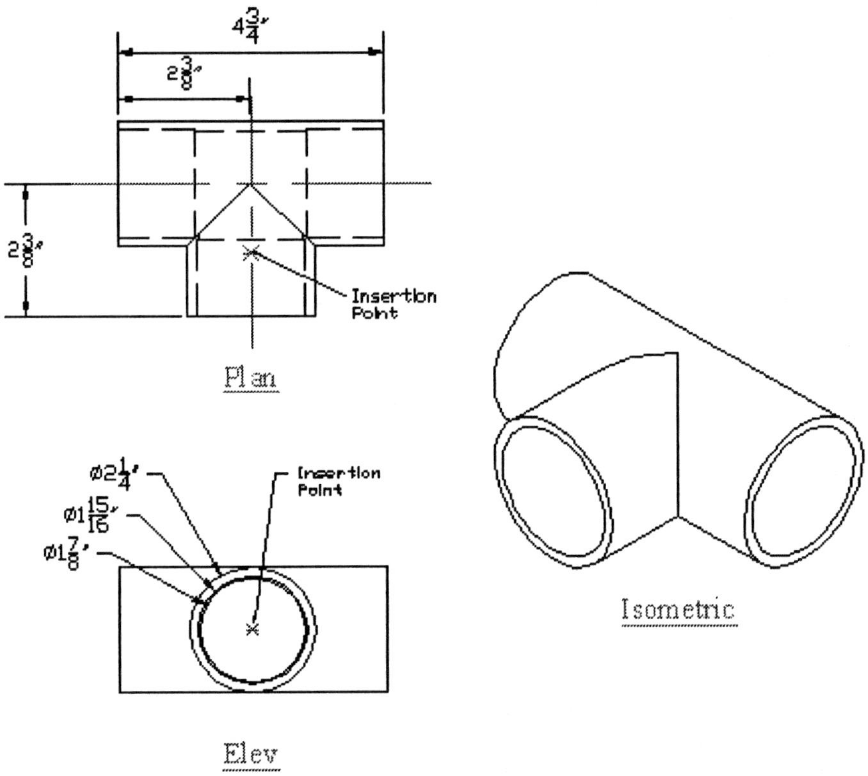

Figure 10.034

8. Create the bike rack drawing in Figure 10.035.
   8.1. Use the elbow you created in Exercise 10.5.6. (If this isn't available, use the *1_2Ell* drawing found in the C:\Steps3D\Lesson10 folder.)
   8.2. Use the *1_2Pipe* drawing to provide the pipe between the elbows (just as you did in Exercise 10.1.2.1).
   8.3. Be sure the UCS = WCS when you finish.
   8.4. Save the drawing as *MyBikeRack.dwg* in the C:\Steps3D\Lesson10 folder.

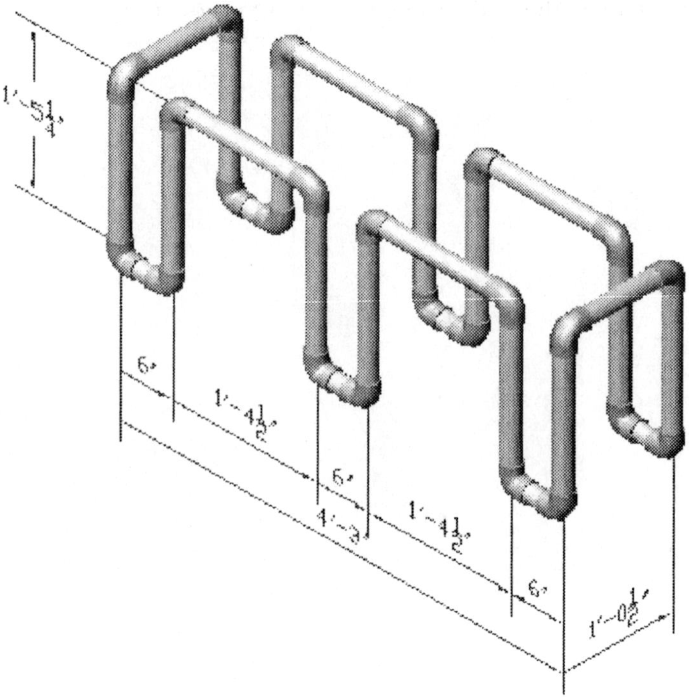
Figure 10.035

9. Create the lawn chair shown in Figure 10.036.
   9.1. Use the elbow you created in Exercise 10.5.6. (If this isn't available, use the *1_2Ell* drawing found in the C:\Steps3D\Lesson10 folder.)
   9.2. Use the tee you created in Exercise 10.5.7. (If this isn't available, use the *1_2Tee* drawing found in the C:\Steps3D\Lesson10 folder.)
   9.3. Use the *1_2Pipe* drawing to provide the pipe between the elbows (just as you did in Exercise 10.1.2.1).
   9.4. Be sure the UCS = WCS when you finish.
   9.5. Save the drawing as *MyLawnChair.dwg* in the C:\Steps3D\Lesson10 folder.

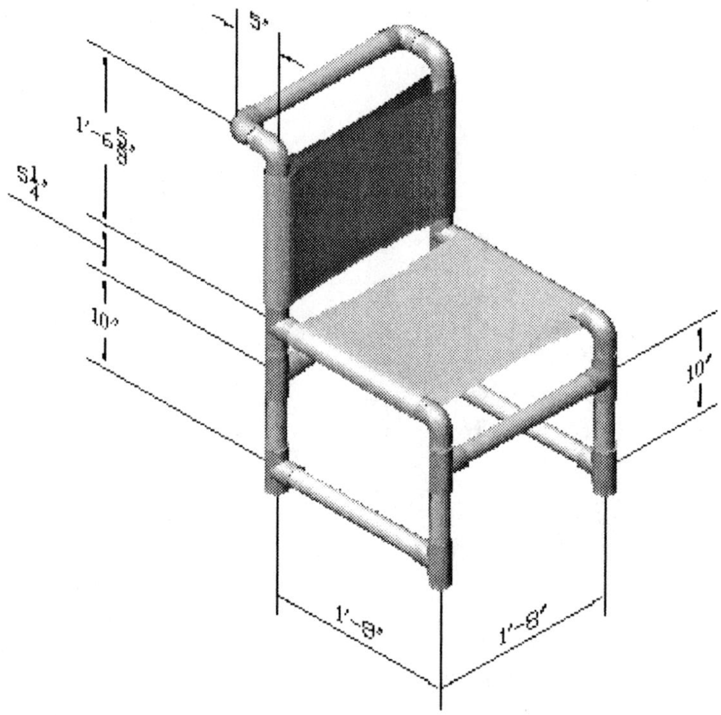
Figure 10.036

10. Create the patio scene shown in Figure 10.037.
    10.1. Use the equipment you created in Exercises 10.5.8a and 10.5.9a. (If these aren't available, use the corresponding drawing found in the C:\Steps3D\Lesson10 folder.)
    10.2. You created the garden fence in Lesson 7 and the fountain in Lesson 8. (Both are provided in the C:\Steps3D\Lesson10 folder if you didn't save your drawings.)
    10.3. Save the drawing as *MyPatioScene* in the C:\Steps3D\Lesson10 folder.

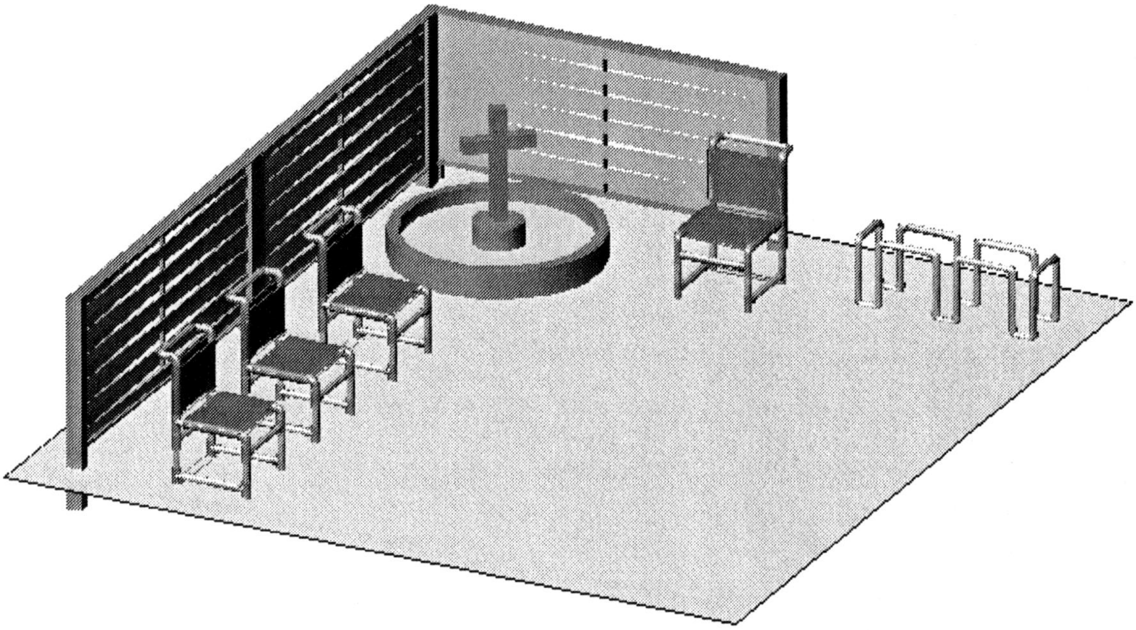

Figure 10.037

10.6 For Web-Based Review Questions, visit: http://www.uneedcad.com/2006/Files/06R10-3D.pdf

Lesson

11

Following this lesson, you will:

- ✓ *Know how to render an AutoCAD drawing*
    - *Rendering*
    - *Assigning materials*
    - *Adding graphics*
    - *Adding lights*
    - *Creating scenes*

Is It Real or Is It Rendered

From childhood's hour I have not been
As others were – I have not seen
As other saw.
Alone – Edgar Allan Poe

Far better it is to dare mighty things, to win glorious triumphs, even though checkered by failure, than to take rank with those poor spirits who neither enjoy much nor suffer much, because they live in the gray twilight that knows not victory nor defeat.
Theodore Roosevelt

The measure of one's soul is calculated
not in successes or failures, but in the number of
attempts one is willing to make.
Anonymous

As you might guess from the three preceding quotes, you now face the most challenging of the lessons you'll undertake in our One Step at a Time series. So before you start, think back to what you knew when you began Lesson 1 of AutoCAD 2006: One Step at a Time. You began each lesson with anticipation and a touch of anxiety, but you finished each knowing more than you did when you started. It hasn't always been easy, but you've persevered (or else you wouldn't be here). Consider your accomplishments. And take Teddy's advice and "dare mighty things" in Lesson 11.

11.1 What Is Rendering and Why Is It So Challenging?

Rendering is a procedure that takes the objects you've created and gives them properties to make them appear "real." The degree to which they appear real depends on a host of user-defined settings and assignments, including materials, types and positions of lights, and light intensity.

Why is rendering so challenging? Consider what Edgar Allan Poe said in the quote that began this lesson. Every individual will "see" a scene in a different way. Translating what your mind sees to what appears on the screen involves often subtle manipulation of several variables.

Remember Lesson 5 – I told you that we'd reached the edge between CAD operating and CAD programming. Well, in Lesson 11, you've reached the edge between CAD operating and art. Just as not every whittler is a sculptor, not every draftsman is an artist. (A fact I found myself repeating … and repeating … to my employer back when I designed those nifty little houses that Santa sits in down at the mall.) This is where you face the challenge.

11.2 Beyond Shademode – The *Render* Command

It may fortify you to know that you've been using a rudimentary form of rendering all along when you used the **Shademode** system variable. But here again, consider the whittler and the sculptor. Whereas the **Shademode** system variable has a few settings from which to choose, the ***Render*** command presents (quite literally) infinite possibilities. Fortunately, we'll navigate the possibilities using dialog boxes. (This should make you appreciate the fact that you're not using one of the earlier – primitive – releases of AutoCAD!)

Let's begin with the basic Render dialog box (Figure 11.001). Access the dialog box by entering the ***Render*** command (or its hotkeys – ***rr***) at the command prompt or by picking the **Render** button on the Render toolbar. The first thing you'll notice is that the dialog box has been sectioned into four frames and several spaces between the frames.

Let's take a look.

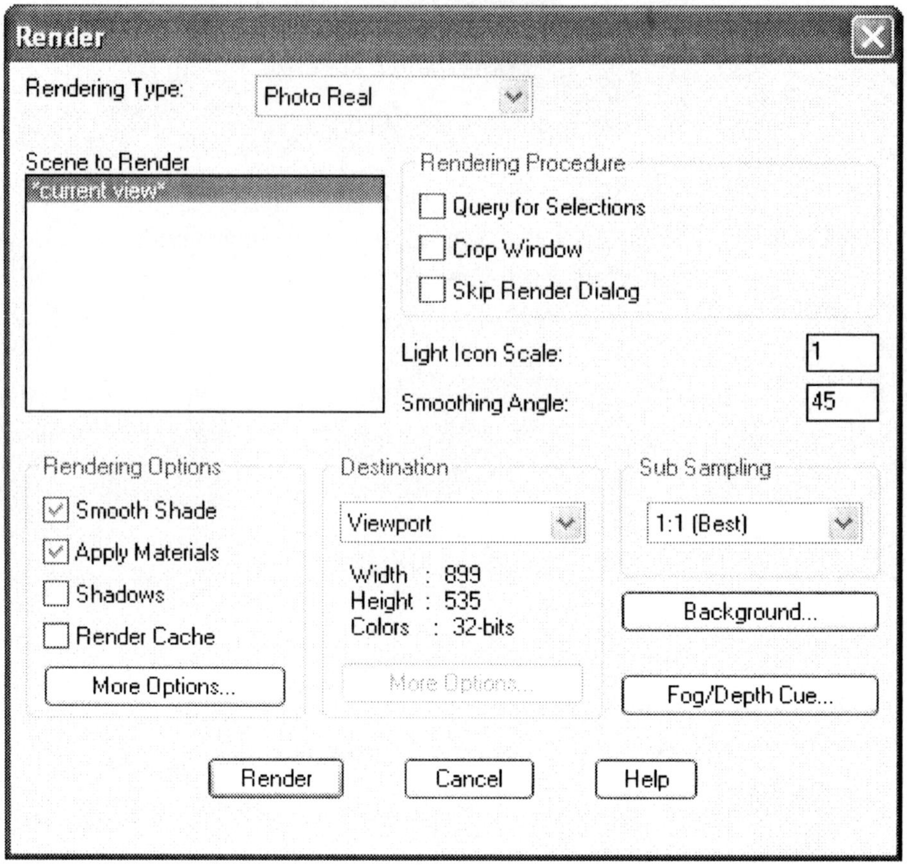

Figure 11.001

- AutoCAD makes three types of rendering available in the **Rendering Type** control box. These are (in order of quality from lowest to highest) **Render**, **Photo Real**, and **Photo Raytrace**. Each has its own options (more on these in a few moments) and each has it uses. To make it simple, remember that the lowest-quality renderer (**Render**) is fastest and that the highest quality (**Photo Raytrace**) is slowest. Even on a fast computer, rendering speeds remain an issue.
- The **Scene to Render** list box presents a selection list of available scenes to render. By default, AutoCAD will render the **current view**. But you'll see how to create additional scenes in Section 11.6.
- The three check box options in the **Rendering Procedure** frame can help save time when rendering.
    - **Query for Selections** allows you to render just those items you select when prompted.
    - **Crop Window** prompts you to **Pick crop window to render**. Place a window around the area you wish to render. AutoCAD renders that area and hides the rest of the view.
    - Use **Skip Render Dialog** to use the same settings for subsequent renderings. AutoCAD won't display the Render dialog box. To again display the dialog box, remove the check in the appropriate box in the Render Preferences dialog box (identical to the Render dialog box without the **Render** button). Access this dialog box with the ***RPref*** command.
- The two (unframed) options below the **Rendering Procedure** frame are unrelated.
    - When you add lights to your drawing, AutoCAD places an icon to mark the location of each. Use the **Light Icon Scale** text box to change the size of the icon if necessary for clarity. To determine the proper size, start with the drawing scale factor and adjust until you're satisfied.
    - The **Smoothing Angle** is the angle at which AutoCAD determines an edge when rendering. AutoCAD considers objects that form any angle greater than that shown in the

Smoothing Angle text box to have an edge between them. If the objects form less of an angle, AutoCAD will render as though they were a single object.

- Pay particular attention to the options listed in the **Rendering Options** frame in the lower-left corner of the Render dialog box. Changes in these settings can dramatically affect the appearance of the rendered objects.
  - **Smooth Shade** smoothes edges on multifaceted surfaces (like spheres).
  - A check in the **Apply Materials** box tells AutoCAD to apply (when rendering) any materials that you've assigned to specific objects.
  - **Shadows** are only available for the **Photo Real** and **Photo Raytrace** types of rendering. When selected, AutoCAD will calculate shadows cast by objects as a result of any lights you've defined. The use of shadows in a rendered scene goes a long way toward creating a "real" appearance.
  - **Render Cache** can be a tremendous timesaver. When checked, AutoCAD writes rendering information to a file on the computer's hard disk. It then uses this information in subsequent renderings.
  - **Rendering Type** determines which dialog box the **More Options** button presents.
    - When the rendering type is **Render**, AutoCAD presents the Render Options dialog box (Figure 11.002).

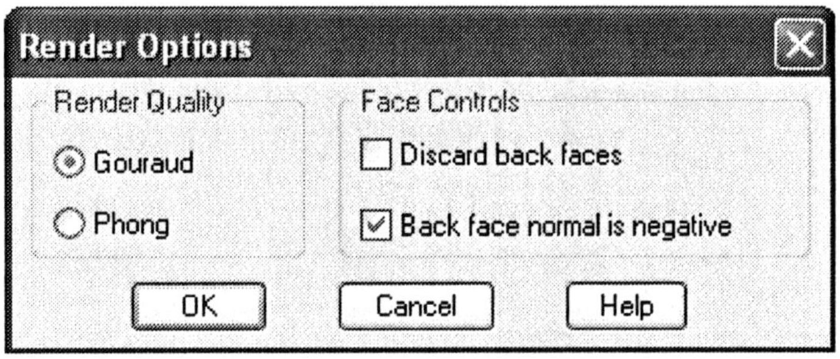

Figure 11.002

      - The **Render Quality** frame allows you to select between **Gouraud** rendering (this is what you're accustomed to seeing with the **Gouraud** setting of the **Shademode** system variable) and **Phong** rendering (this is more sophisticated than Gouraud – it produces a more realistic image).
      - The **Discard back faces** option of the **Face Controls** frame allows you another opportunity to increase rendering speed. A check here tells AutoCAD not to consider back faces (those normally hidden with the ***Hide*** command) when rendering.

        The **Back face normal is negative**

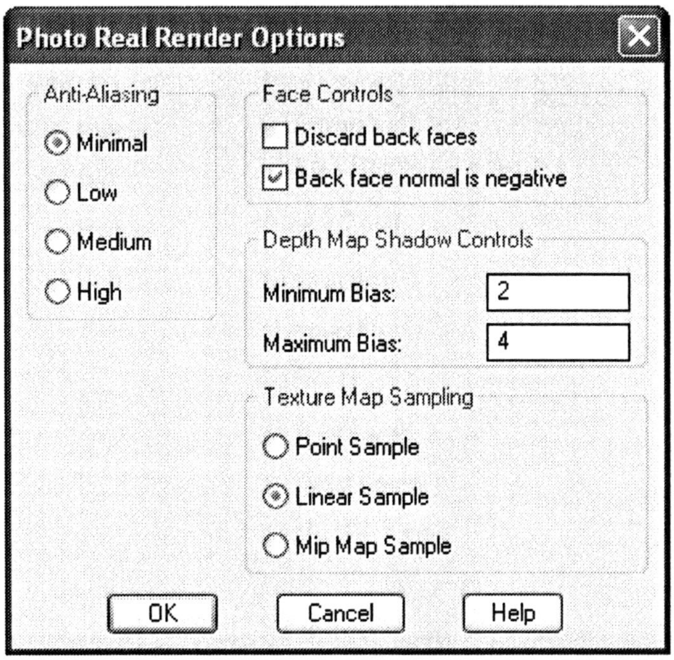

Figure 11.003

option sounds more complicated than it is. A check here tells AutoCAD that back is back and front is front. Remove the check and AutoCAD will consider the back faces to be the front (and vice versa). Then, if the **Discard back faces** option is checked, AutoCAD will render the back faces and hide the front ones!

- When the rendering type is **Photo Real**, AutoCAD presents the Photo Real Render Options dialog box (Figure 11.003) when you pick the **More Options** button. Here you have four frames of options.
  - The **Anti-Aliasing** frame allows you to control how jagged edges will appear in a rendered scene. **Minimal** anti-aliasing is faster but produces more jagged edges; **High** anti-aliasing has just the opposite effect.
  - The **Face Controls** frame is identical to the same frame on the Render Options dialog box previously discussed.
  - **Depth Map Shadow Controls** allow you to adjust how AutoCAD creates a shadow. These settings are best left to AutoCAD, but if you notice problems
  - with shadowing (i.e., shadows casting shadows), try adjusting these numbers. The **Minimum Bias** is generally between 2 and 20. The **Maximum Bias** value should be less than 10 above the **Minimum Bias** value.
  - **Texture Map Sampling** allows three choices of what to do when a texture map is larger than the object to which it has been attached.

    Point Sample produces a sharp image, but the lines may appear jagged.

    Linear Sample's image is less focused, but the lines are not as jagged.

    Mip Map Sample produces very soft edges, but lines may appear quite blurred.
- When the rendering type is **Photo Raytrace**, AutoCAD presents the Photo Raytrace Render Options dialog box (Figure 11.004) when you pick the **More Options** button. Here you have five frames of options. Fortunately, only two of the frames are new.
  - **Adaptive Sampling** is only available for the **Low**, **Medium**, and **High Anti-Aliasing** options. It provides a tool to speed up anti-aliasing. Enter a value between 0 and 1 in the **Contrast Threshold** box – the higher the number, the faster the rendering but the lower the quality.

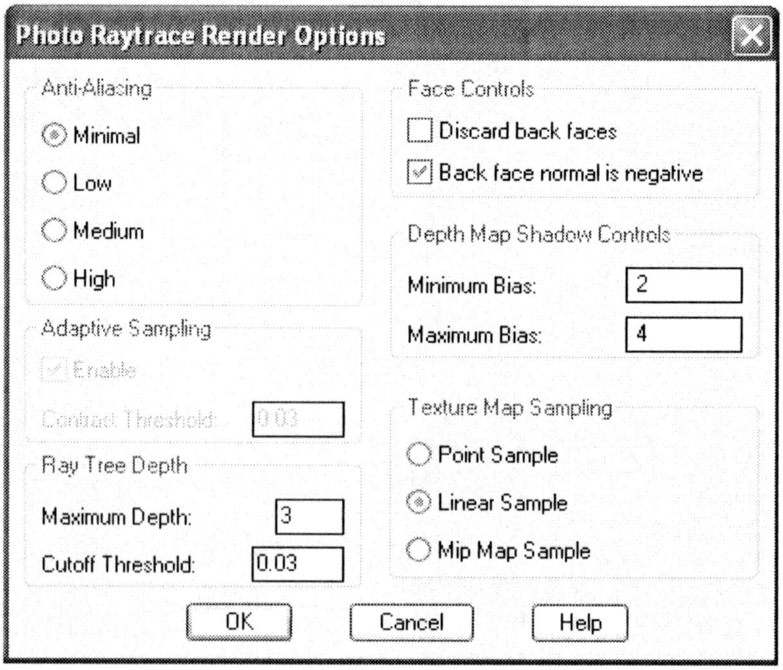

Figure 11.004

- The **Ray Tree Depth** (**Maximum Depth**) setting allows you to control the amount of reflected or refracted light rays in a rendering. Higher numbers yield better results but take longer. AutoCAD recommends a maximum value of 10.

  The **Cutoff Threshold** controls how far a ray of light will travel. Again, higher numbers produce better-quality renderings but take longer.

- Use the **Destination** frame of the Render dialog box (Figure 11.001) to tell AutoCAD where to place the rendering. Choices include:
  - Place the rendering in the current **Viewport**.
  - Open a separate **Render Window** in which to place the rendering.
  - Create a separate graphics **File** for the rendering.

    When you select the **File** option, AutoCAD makes the **More Options** button available. This button presents the File Output Configuration dialog box (Figure 11.005). Options available in the different frames depend on the type of file selected in the **File Type** frame.

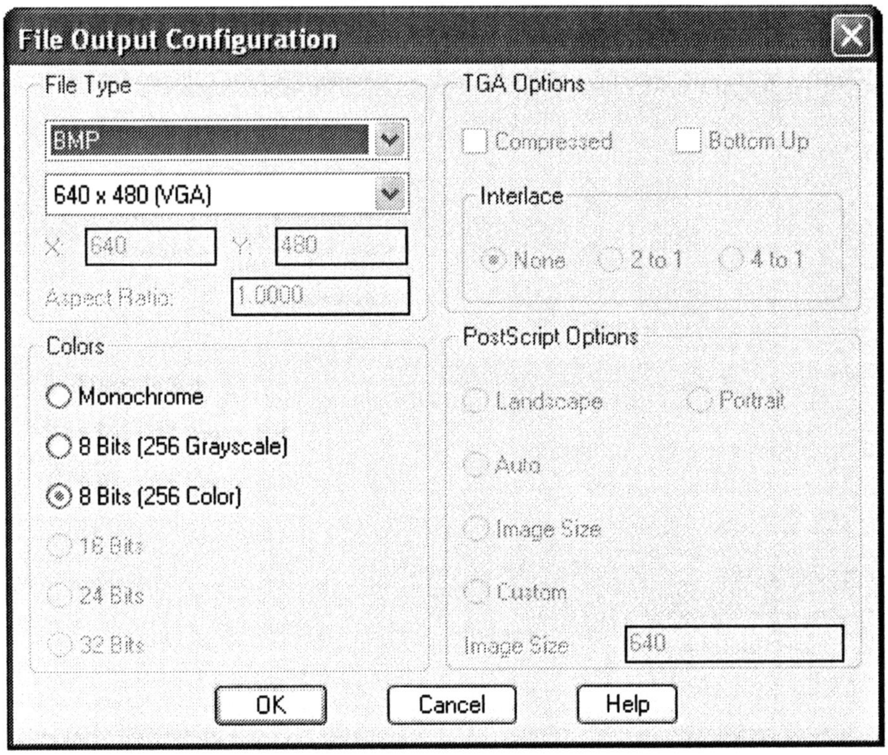

Figure 11.005

    - File types available in the control box include BMP, PCX, PostScript, TGA, and TIF.
    - Below the file type control box is a resolution control box. Use this to control the resolution of the output file (the higher the resolution, the finer the quality of the rendered file).

      The **Aspect Ratio** (width to height) is available when you select **User Defined** in the resolution control box.
  - **TGA Options** are available when you select the TGA file type.
    - **Compressed** makes the final file size smaller. I always select this for efficiency.

        - **Bottom Up** causes the file to be read from the lower-left corner rather than the upper-left corner.
        - **Interlace** controls how the file will be read by a graphics program. I normally set this to **2 to 1** for speed.
    - The **Colors** frame allows you to determine the number of colors used to create the file. The larger the number, the finer the quality. I usually set this to the highest quality available even though that tends to increase the size of the file.
    - All the options in the **PostScript** frame deal with size and orientation of the image.
        - **Portrait** stands the page up (i.e., 8½" x 11"); whereas **Landscape** lays it on its side (i.e. 11" x 8½").
        - **Auto** automatically scales the image for you.
        - **Image Size** makes the image the actual size.
        - **Custom** uses the value in the **Image Size** box to set the size of the image in pixels.
- **Sub Sampling** (back on the Render dialog box) speeds the rendering process by rendering only a ratio of the pixels. Highest quality but slowest rendering is 1:1 (render all pixels); lowest quality but fastest rendering is 8:1 (render every eighth pixel).
- The **Background** button calls the Background dialog box (Figure 11.006). Use this to control the background for the rendering. AutoCAD provides four options – these control what other options are available.

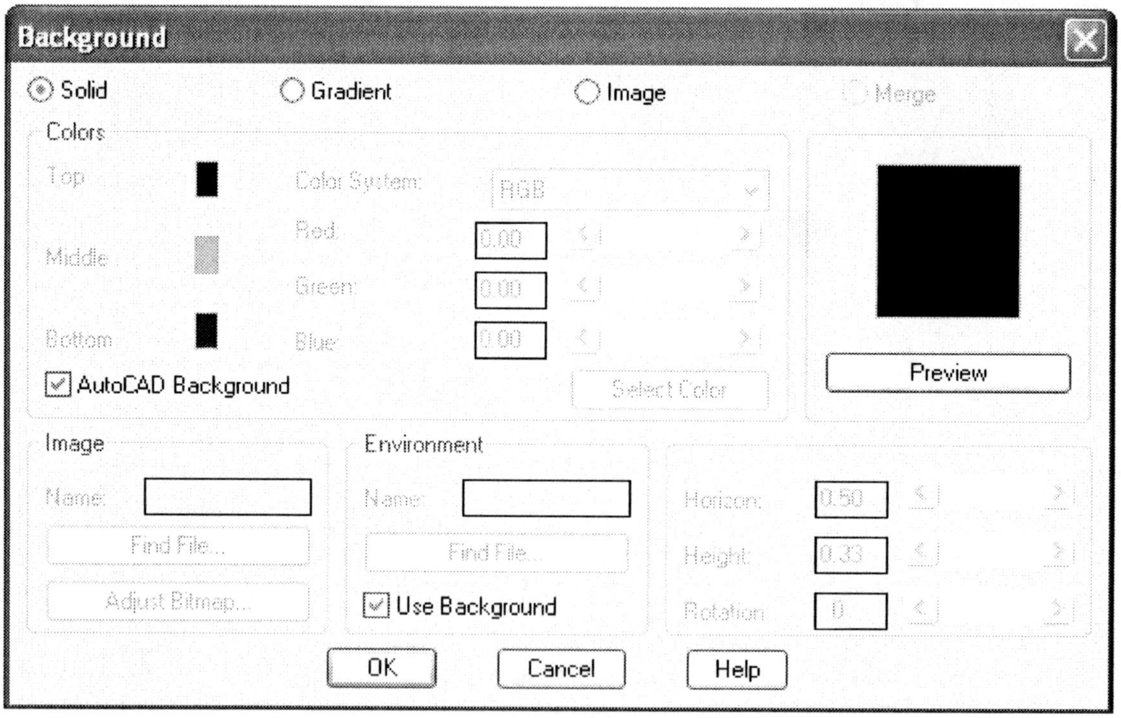

Figure 11.006

    - **Solid** allows you to use a single-color background. Use the current color (leave a check in the **AutoCAD Background** check box) or select/define a color in the **Colors** frame. (Preview your settings using the **Preview** button in the upper-right frame.)
    - **Gradiant** creates a two- or three-color graduated background. When selected, AutoCAD makes the **Colors** frame available as well as the lower-right frame. Use the lower-right

frame to define the size of the graduations. (Set the **Height** value to **0** for a two-color background.)

- The **Image** option, of course, allows you to use an image in the background. Possible file types you can use include BMP, PNG, JPG, TGA, TIF, GIF, and PCX. Once you've selected an image, the **Adjust Bitmap** button will present a dialog box that will allow you to adjust the location and tiling of the image.
- The **Merge** option uses the current background image as the background.
- The **Environment** option (frame), when used with a **Photo Real** type of rendering, causes a mirroring effect. When used with a **Photo Raytrace** rendering, AutoCAD uses the image to effect changes to light reflection and refraction in the rendering.

- The **Fog/Depth Cue** button on the Render dialog box calls the Fog/Depth Cue dialog box (Figure 11.007). Here you can fog (tint) the objects being rendered as well as the background of the rendering.

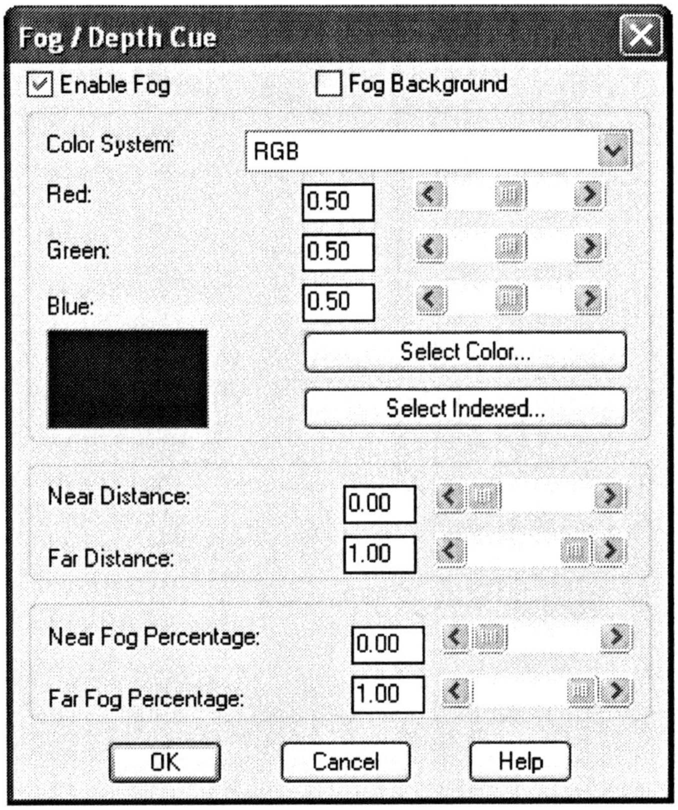

Figure 11.007

  - **Enable Fog**, when selected alone, causes the objects in the drawing to be tinted according to the color defined in the uppermost frame. When **Fog Background** is also selected, the entire image will be tinted.
  - Use the uppermost frame of the Fog/Depth Cue dialog box to define the color for the fogging.

    The **Color System** control box provides two methods for selecting colors – **RGB** (Red, Green, Blue) and **HLS** (Hue, Lightness, Saturation). When **RGB** is current, adjust color intensity using the **Red**, **Green**, and **Blue** slider bars. When **HLS** is current, adjust color intensity using the **Hue**, **Lightness**, and **Saturation** slider bars. (Note: These slider bars replace the Red, Green, and Blue slider bars.)

    The **Select Color** button calls a standard Windows Select Color dialog box whereas the **Select Indexed** button presents AutoCAD's Select Color dialog box.

    The box in the lower-left corner of the frame will present the color that you've defined.
  - The next frame provides values AutoCAD uses to determine where to begin and end the fog. Values are percentages (from 0 to 1) of the distance from the camera to the back working plane.
  - The lat frame provides values AutoCAD uses to determine how much fog to place at the **Near** and **Far Distance** points.

Well, what do you think? I counted a least seven primary dialog boxes. I wouldn't dare count the number of variables! Do you begin to see why there are infinite possibilities to render a drawing? Do you begin to see why this is the most challenging lesson in our course? (Remember, we have just covered the Render dialog box. We haven't looked at materials or lighting … yet!)

But let's pause to acquire some hands-on experience before we look at adding materials to our rendering. (Oh, yeah. Before we start, I should tell you how to remove the rendering from a drawing – simply regenerate the drawing! Kind of anticlimactic, isn't it?)

> You can also access all of the commands in this lesson via the View pull-down menu. Follow this path:
>
> *View – Render – [command]*

Do This: 11.2.1	Discovering Rendering

I. Open the *rendering project.dwg* file in the C:\Steps3D\Lesson11 folder. The drawing looks like Figure 11.008.

II. Follow these steps.

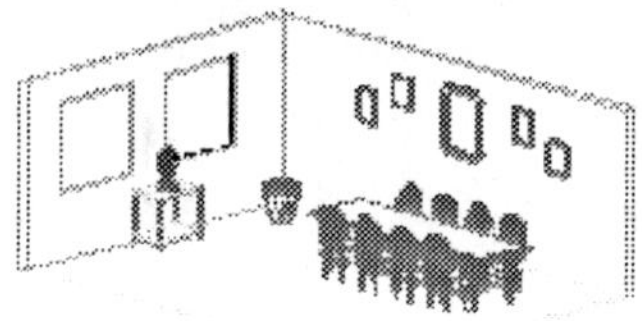

Figure 11.008

TOOLS	COMMAND SEQUENCE	STEPS
	**Command:** ***shademode***	1. First, we'll compare basic types of rendering. Set the **Shademode** system variable to **Gouraud**. Note the time required to complete the command. Your drawing looks like Figure 11.009.

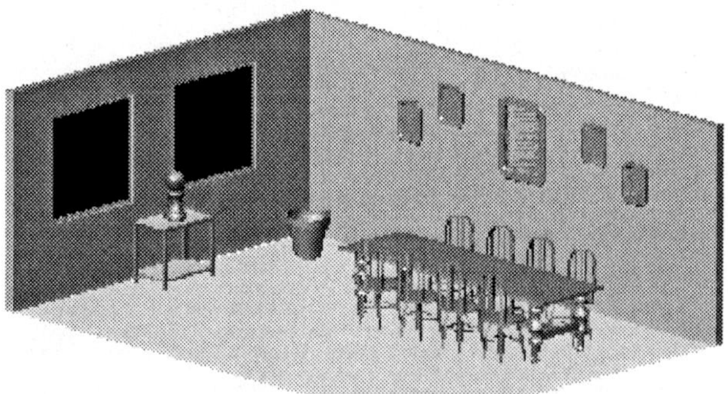

Figure 11.009

TOOLS	COMMAND SEQUENCE	STEPS
 Render Button	**Command:** ***rr***	2. Enter the ***Render*** command. Alternately, you can pick the **Render** button on the Render toolbar.

Tools	Command Sequence	Steps
		3. Using the **Render** type, as shown, pick the **Render** button to render the drawing. (Use defaults for all other settings.) Compare the differences between the basic **Render** type and the **Gouraud** Shademode. ***Render*** took just a little more time but produced a finer-quality image than Shademode. Unfortunately, however, *you can't work on a rendered drawing as you can on a shaded drawing.*
		4. Compare the **Render** type rendering to a **Photo Raytrace** rendering. Repeat Steps 2 and 3 using the **Photo Raytrace** setting. Your drawing looks like Figure 11.010. Notice the differences. The **Photo Raytrace** rendering took longer but produced a slightly better quality in the rendering.

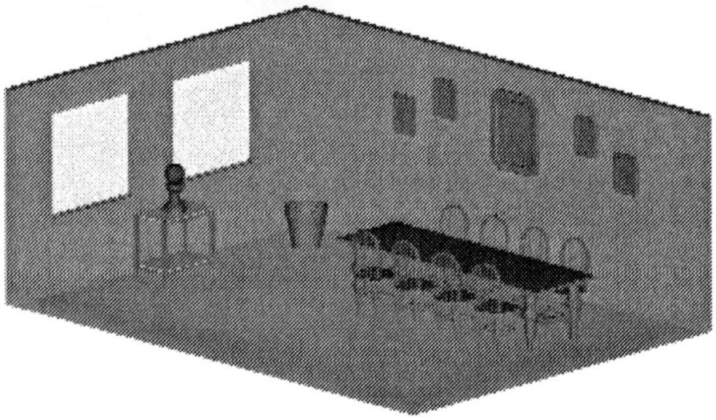

Figure 11.010

Tools	Command Sequence	Steps
	**Command:** ***[enter]***	5. Let's try some of the options. Repeat the ***Render*** command.
		6. (We'll continue with the **Photo Raytrace** type of rendering throughout the rest of the lesson.) This time, render only part of the drawing. (Refer to Figure 11.011.) Tell AutoCAD to **Query for Selections** (for specific objects to render). Send the rendering to the Render Window … … and speed the rendering by setting the **Sub Sampling** to 2:1. Pick the **Render** button to continue.

TOOLS	COMMAND SEQUENCE	STEPS

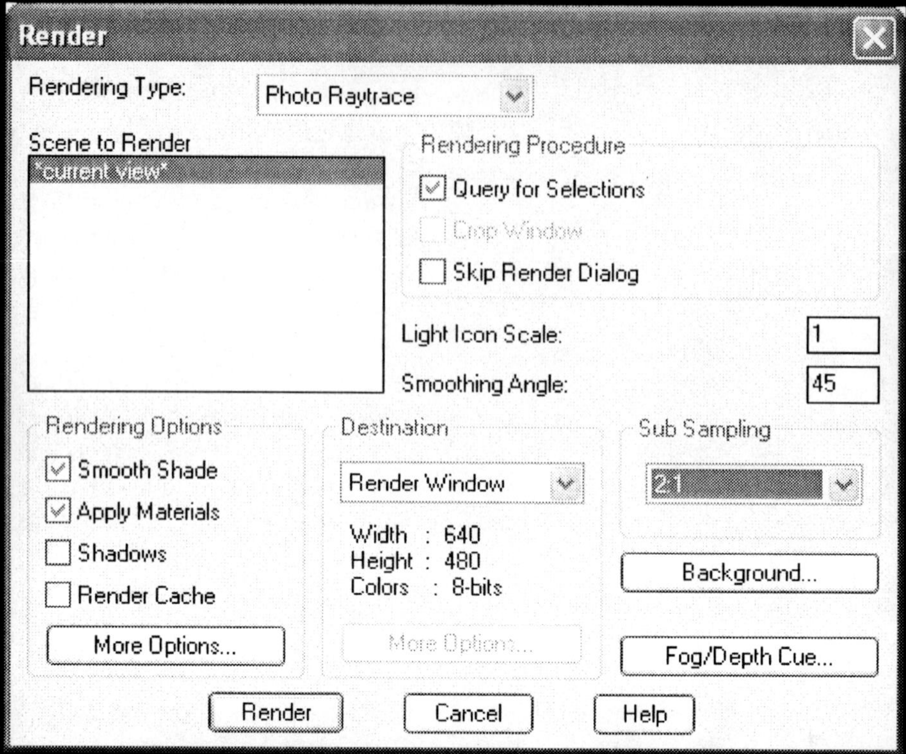

Figure 11.011

	Select objects: **Select objects: *[enter]***	7. AutoCAD prompts you to select the objects to render. Place a selection window around the table and chairs. The render window looks like Figure 11.012.

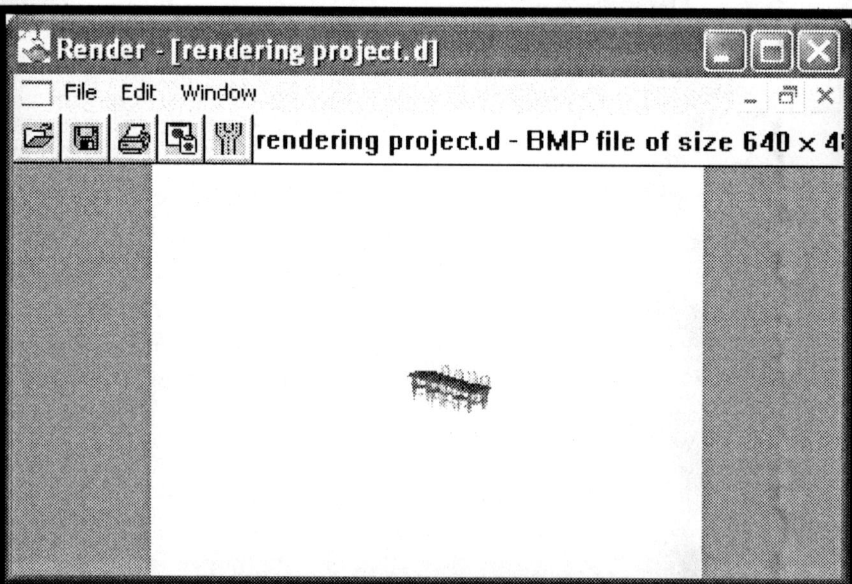

Figure 11.012

Tools	Command Sequence	Steps
	Consider the results: • The rendering moved fairly quickly because you only rendered part of the drawing and your **Sub Sampling** was halved. • The rendering was placed into a separate window. Take a moment to examine the buttons and pull-down menus for the Render window. Pay particular attention to the **Options** selection under the File pull-down menu. There you can adjust the quality of future images. • Notice the poor quality of the rendering. This is a result of the **Sub Sampling** changes you made. You can do this to render quickly – to make adjustments in the settings – but you'll want the final rendering to use the **1:1 Sub Sampling** settings. Pick the **X** in the upper right corner of the Render window to close it.	
	**Command:** ***rr***	8. Repeat the ***Render*** command and set the **Rendering Procedure**, **Sub Sampling**, and **Destination** frames to their defaults.
		9. Let's play with the background. Pick the **Background** button.
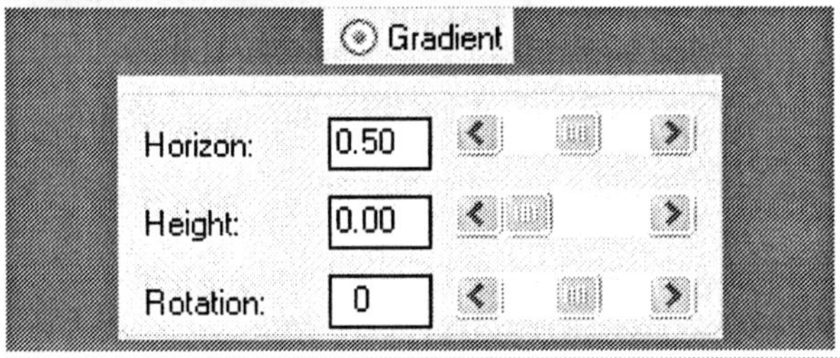		10. We'll use a two-color gradient background. Put a bullet next to **Gradient**, and set the **Height** to **0**.
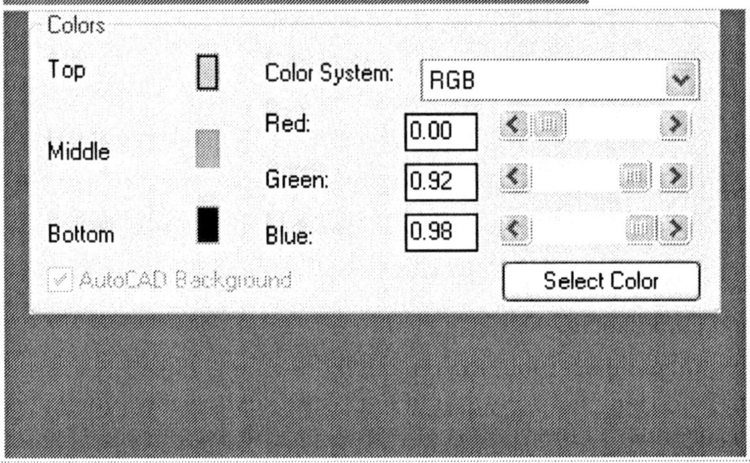		11. Now we'll change the background colors to mimic blue sky and green grass. Pick the color box next to the word **Top** in the **Colors** frame. Then slide the **Red** color bar all the way to the left, the **Green** color bar about three-fourths to the right, and the **Blue** color bar all the way to the right. The color box should show a light blue color (adjust the slider bars until you like the color).

TOOLS	COMMAND SEQUENCE	STEPS
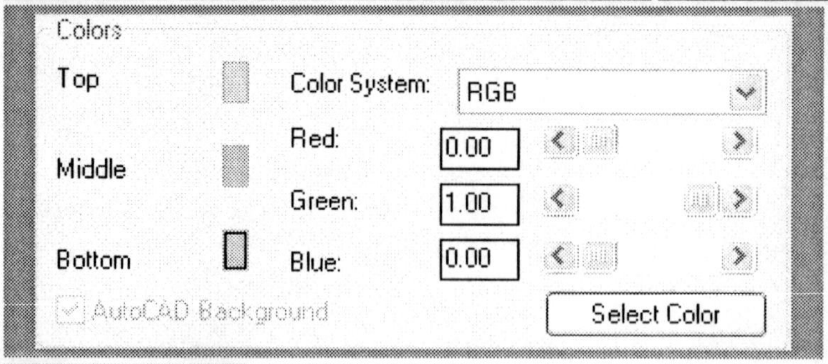		12. Repeat Step 11 for the bottom color. Set the color bars as indicated for green.
		13. Pick the **OK** button to return to the Render dialog box.
		14. Pick the **Render** button to complete the command. Notice the change in the background (Figure 11.013).

Figure 11.013

TOOLS	COMMAND SEQUENCE	STEPS
	Command: *re* **Command: *shademode***	15. Regen the drawing and reset the Shademode to **2D**.
	**Command: *qsave***	16. Save the drawing, but don't exit.

We could easily spend a hundred pages exploring the rest of the possibilities, but you should have the general idea. Take some time (once you complete the lesson) to continue exploring on your own.

You've seen the basics of the ***Render*** command, but so far, the rendered drawing is fairly unimpressive. It still looks like a cartoon – bright and colorful but not real. Next we'll begin to add reality to our image by assigning material values to the various objects.

Let's proceed.

11.3 Adding Materials to Make Your Solids Look Real

As you'll soon see, adding materials to an object can mean the difference between colorful cartoon images and images that come close to photographic realism. And luckily, you can accomplish it fairly easily.

To understand materials, think of them as paint (or wallpaper). The object doesn't actually become wood (or granite, etc.). Rather, it has the image of wood painted onto it. AutoCAD achieves this by attaching an image file (usually a bitmap) to the surfaces of the objects. The only trick involved for you, then, is to know which image file to use. AutoCAD provides a library full of possible images from which to choose. If these don't satisfy your needs, however, AutoCAD helps to create new images (or modify old ones)!

AutoCAD makes four categories of materials available: *Standard* (plain with no markings), *Granite*, *Marble*, or *Wood*. The last three contain markings that reflect properties of that type of material.

Begin the process of attaching materials in your drawing with the ***RMat*** command. (Alternately, you can pick the **Materials** button on the Render toolbar.) AutoCAD presents the Materials dialog box (Figure 11.014).

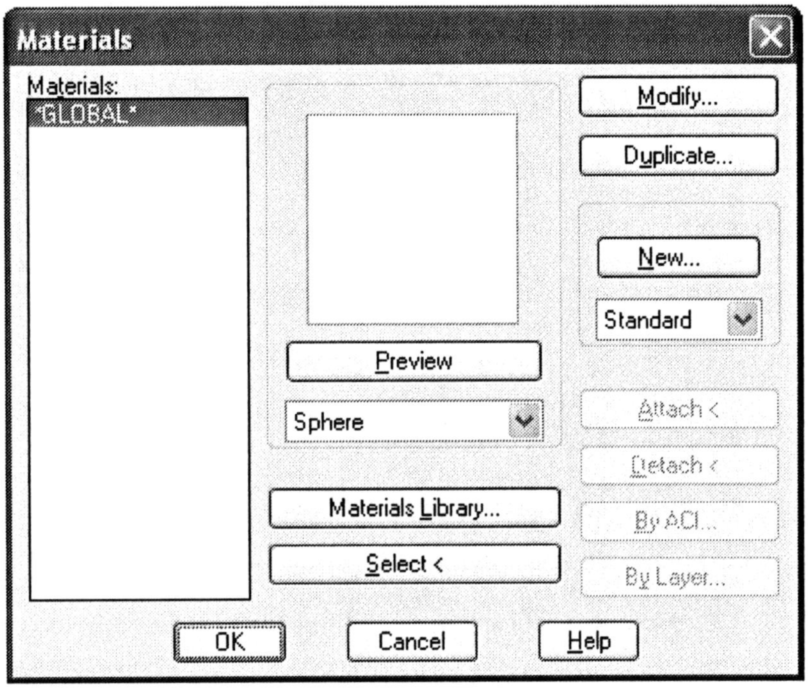

Figure 11.014

Let's take a look.

- The first thing you'll probably notice is the **Materials** list box on the left side of the dialog box. Here AutoCAD will list all the materials currently loaded and available for use. As with linetypes, AutoCAD has no materials automatically loaded to save drawing file size.
- The frame to the right of the **Materials** list box allows you to preview the material currently highlighted in the list box. Use the selection control box at the bottom of the frame to tell AutoCAD that you wish to preview the material against a **Sphere** or a **Cube**. Use the **Preview** button to create the preview at the top of the frame.
- Use the **Select** button (below the **Preview** frame) to identify materials already assigned to specific objects. It returns you to the graphics screen where you can select an object. AutoCAD then returns you to the Materials dialog box and highlights the material currently attached to the selected object.
- With attribute variations for the specific category of material, you'll use the **New**, **Modify**, and **Duplicate** buttons to adjust material attributes using the same dialog boxes we'll discuss (in a moment) as the New … Material dialog box.
- Use the **Attach** button to attach the material selected in the Materials list box to an object in the drawing. AutoCAD returns you to the graphics screen to make your selection(s). The **Attach** button will also highlight the objects to which the selected material is currently attached.

- Use the **Detach** button to remove (unattach) material that has been attached to an object. AutoCAD returns you to the graphics screen to select the object(s).
- Use the **By ACI** button to attach materials to objects according to the color of the objects.
- Use the **By Layer** button to attach materials to objects according to the layer on which the objects reside.
- The **Materials Library** button presents the Materials Library dialog box (Figure 11.015). This is where you'll load materials to use in your drawing.

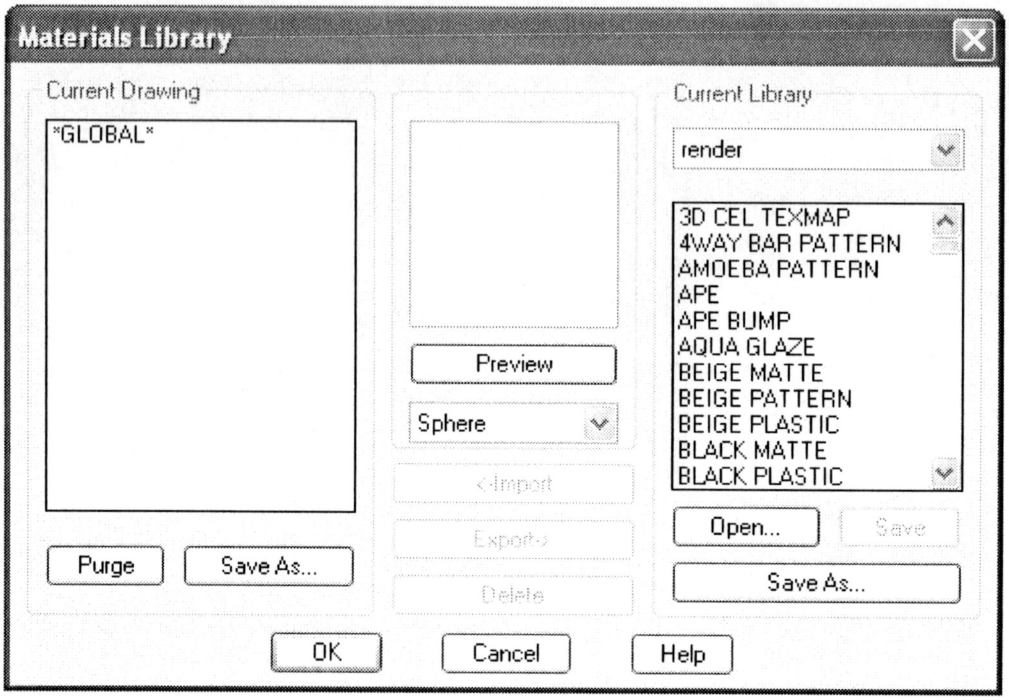

Figure 11.015

Let's take a look at our options.

- The **Current Drawing** frame presents a list box that shows all the materials loaded and available in the current drawing. Use the **Purge** button to remove unused materials (to reduce drawing size) and the **Save As** button to save the list (as an MLI file). You can use a saved list in another drawing.
- Use the **Preview** frame just as you did in the Materials dialog box.
- The **Current Library** frame presents all the materials available in the library shown in the selection control box (in the top of the frame). Use the **Open** button to open a different MLI file (*Materials LIbrary*). Use the **Save** or **Save As** buttons to save the library.
- Use the **Import** and **Export** buttons to load or unload materials highlighted in the **Current Drawing** or **Current Library** list box.
- The **Delete** button will remove selected materials from the current drawing *or the material library*. (Note: *It isn't a good idea to remove materials from the library as they then become unavailable for future use.*)
- The **OK** button will return you to the Materials dialog box.

You can also access the Materials Library using the ***MatLib*** command or by selecting the Materials Library button on the Render toolbar. If you access the Materials dialog box in one of these ways, the **OK** button will return you to the command prompt.

- Use the **New** button on the Materials dialog box to create user-defined materials. When you pick the **New** button, AutoCAD presents one of the four New … Material dialog boxes. The specific dialog box presented depends on the materials category displayed in the selection box below the **New** button.
  - The New Standard Material dialog box (Figure 11.016) allows you to create a solid-color material. This type of material is most useful for plastic or metal surfaces.
    - Begin by entering a unique name for your new material in the **Material Name** text box.
    - The **Attributes** frame presents several ways to adjust the appearance of the new material. Adjust the value of the attributes using the **Value** and **Color** frames.
      - **Color/Pattern**: This attribute allows you to adjust the main color of the material.
      - **Ambient**: Use this attribute to adjust the shadow color of the material.
      - **Reflection**: This attribute allows you to change the reflective (highlighted) color of the new material.

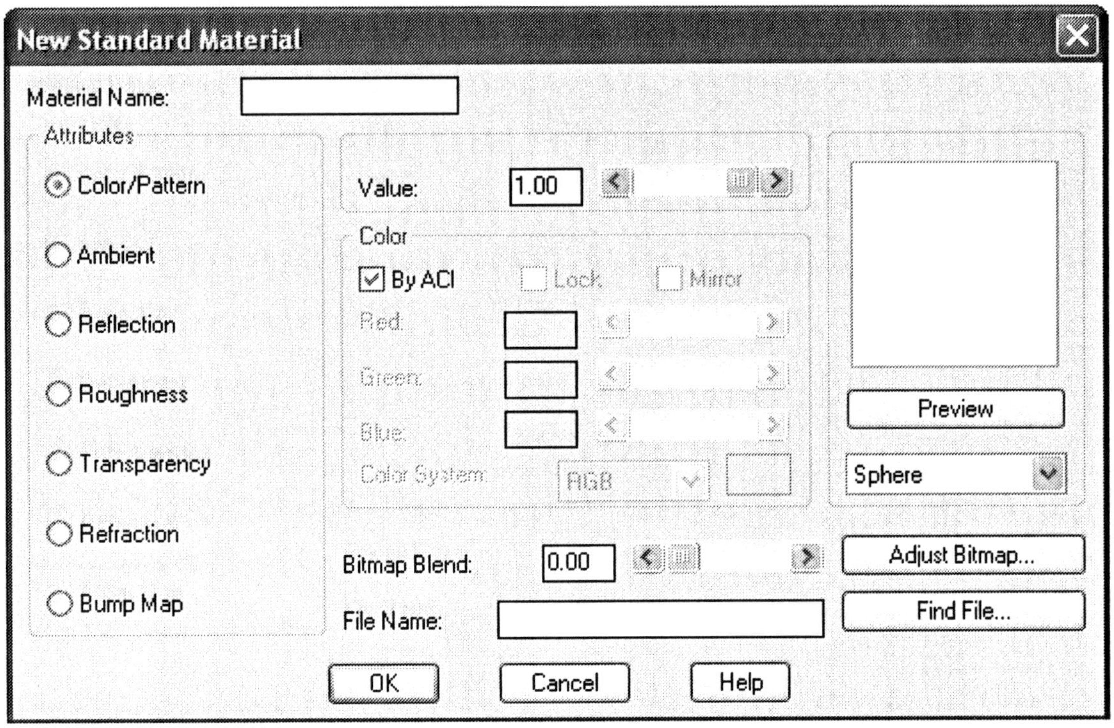

Figure 11.016

      - **Roughness**: Roughness refers to the size of the reflective area (the shininess of the object). A higher value means a larger reflective area.
      - **Transparency**: Controls the transparency of the material. A value of **1** is opaque. Use the **File Name** text box to assign an opacity map (a bitmap) to this material. When creating a **Photo Raytrace** or **Photo Real** rendering, transparency tends to diminish toward the edges of the object.

- **Refraction**: (Applies only to the **Photo Raytrace** render type.) Allows you to adjust how much the light will "bend" around the material. Use this in conjunction with **Transparency** to control what happens to light passing through an object.
- **Bump Map**: (According to AutoCAD's glossary, a bump map is "a map in which brightness values are translated into apparent changes in the height of the surface of an object.") A bullet next to the **Bump Map** attribute allows you to specify a bump map in the **File Name** text box.

- The **Preview** frame in the New … Materials dialog box works just as the other preview frames we've discussed.
- To assign a bitmap to your new material, use the **Find File** button in the lower-right corner of the dialog box. This is how you will create a new material using a scanned image.
- Access the Adjust Material Bitmap Placement dialog box (Figure 11.017) with the **Adjust Bitmap** button. Adjust the size and location of the bitmap here.

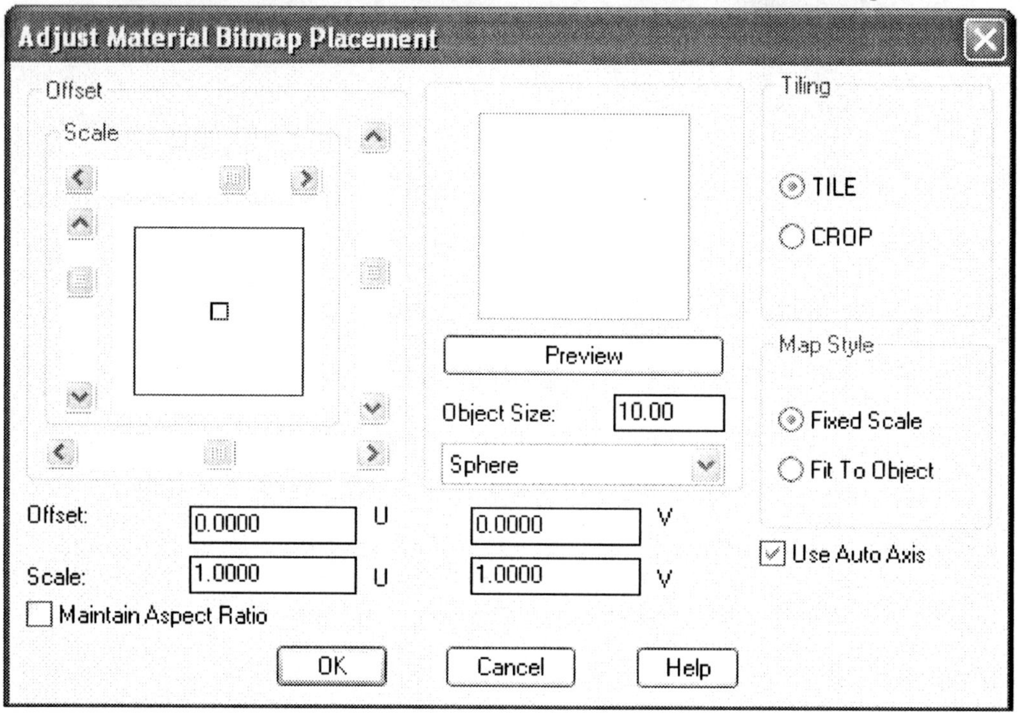

Figure 11.017

As stated previously, with attribute variations for the specific category of material, you'll use the **New**, **Modify**, and **Duplicate** buttons to adjust material attributes using the same dialog boxes.

This section has produced a lot of "material" to absorb, but it'll be easier to understand after you've completed an exercise.

Let's begin.

You can assign materials to any object or block in a drawing, but 3D faces and 3D solids produce the best results.

Do This: 11.3.1	Adding Materials for Rendering

I. Be sure you're still in the *Rendering Project.dwg* file in the C:\Steps3D\Lesson11 folder.

II. Follow these steps.

TOOLS	COMMAND SEQUENCE	STEPS
	**Command:** ***rmat***	1. Open the Materials dialog box.
		2. Access the Materials Library dialog box by picking the **Materials Library** button.
		3. In the **Current Library** frame's list box, select **Amoeba Pattern**. [Note: If the **Amoeba Pattern** (or any of the materials used in this exercise) isn't available, pick the down arrow next to **render** and select the **Rendering Project** library. If you don't see that library, use the **Open** button to access it in the C:\Steps3D\Lesson11 folder).
		4. Pick the **Import** button to copy this material to the current drawing.
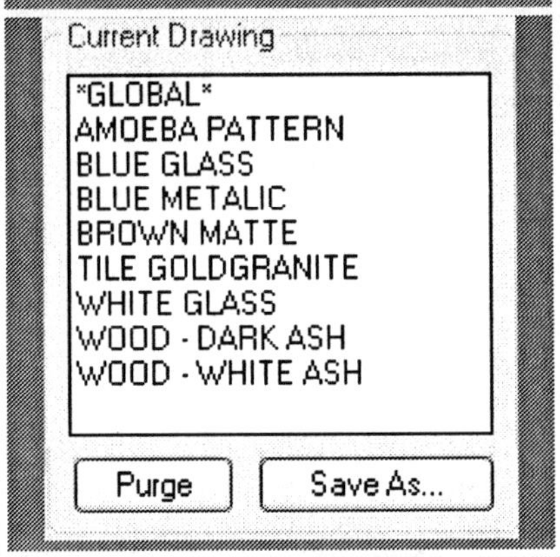		5. Repeat Steps 2 and 3 to import the materials shown.
		6. Pick the **OK** button to continue. AutoCAD returns you to the Materials dialog box.

Tools	Command Sequence	Steps
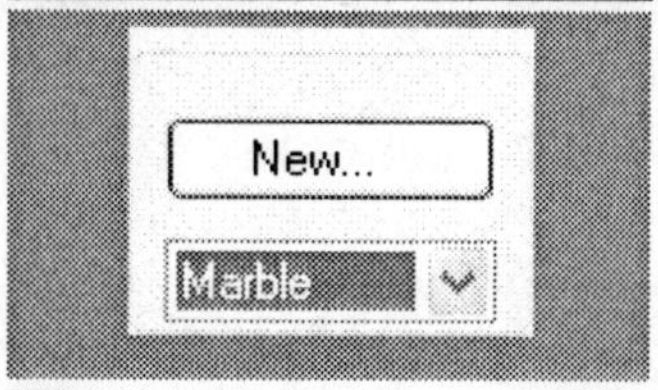		7. Let's create an additional material. Select **Marble** in the **New** frame control box, and then pick the **New** button. AutoCAD presents the New Marble Material dialog box.
		8. Call the material *MyMarble*.
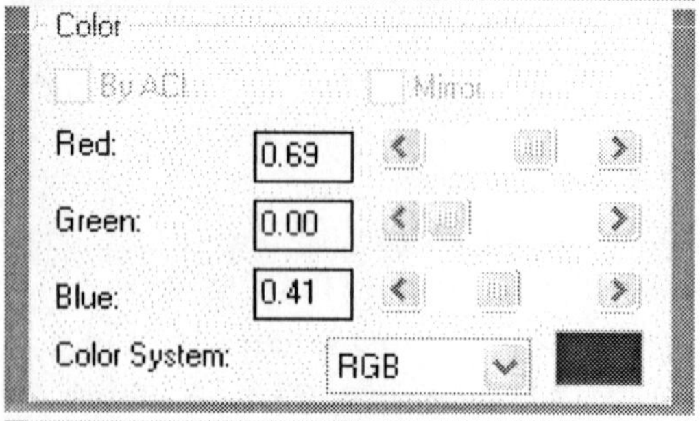		9. Put a bullet next to **Vein Color** in the **Attributes** frame and set the colors as indicated.
		10. Put a bullet next to **Scale** in the **Attributes** frame and set the value to **6.25**.
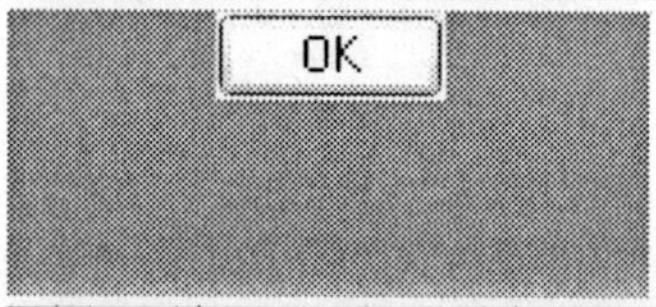		11. Pick the **OK** button to continue. Notice that **MYMARBLE** has been added to the list of available materials (Figure 11.018).

Figure 11.018

Tools	Command Sequence	Steps
		12. Now let's attach materials to objects in our drawing. Select the **Amoeba Pattern** material from the **Materials** list and then pick the **Attach** button.
	Select objects to attach "AMOEBA PATTERN" to: **Select objects:** ***[enter]***	13. AutoCAD returns to the graphics screen and asks you to select the objects to which you want to attach the material. Pick the floor.
		14. Complete the command.

Tools	Command Sequence	Steps
	Command: ***rr***	15. We'll use the **Amoeba Pattern** to simulate carpet. Render just the floor (use the **Query for Selections** procedure and select the floor.)
		16. Repeat the ***RMat*** command and Steps 12 through 14 to assign the materials indicated in Figure 11.019.

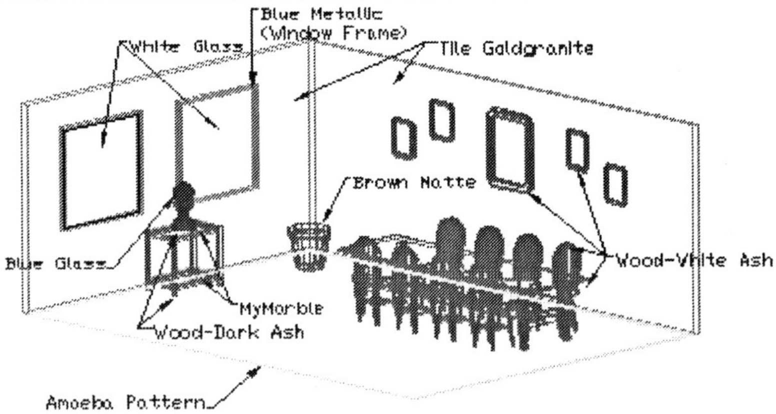

Figure 11.019

Tools	Command Sequence	Steps
	Command: ***rr***	17. Create a **Photo Raytrace** rendering of the entire drawing. Notice the time it takes to render. Your drawing looks like Figure 11.020. (Compare it with Figure 11.013 in our last exercise.)

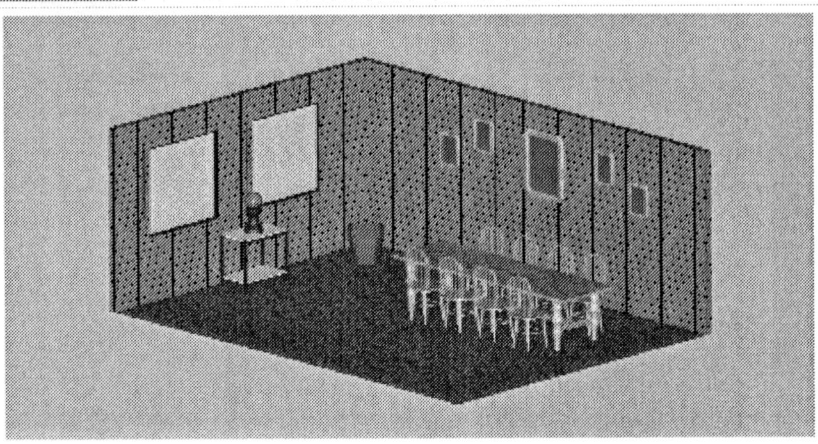

Figure 11.020

Tools	Command Sequence	Steps
	Command: ***qsave***	18. Save the drawing, but don't exit.

You can say, "Wow!" if you want, but we're not quite there yet. We need to add some more touches to make our image look as real as possible.

Our next step will be to add some graphics – pictures in the frames, a plant in the pot, and a tree outside. Then we'll add some lighting as a final touch.

11.4 Special Effects – Adding Other Graphic Images to Your Drawing

When I was in my late twenties – some centuries back – I'd had my own apartment for many years. I was comfortable. Then one day my mother asked if I'd give my baby sister a place to live for a while. She was still young and recently out on her own. My first reaction, as you might imagine, was to be aghast at the idea (I mean, who wants their baby sister living with them?). Still, being of good Irish stock, I couldn't very well say no to Mama.

When she moved in, my sister completely redecorated my bachelor pad. This really didn't bother me much (it needed a good cleaning anyway). But when she brought in a truckload of plants … I was amazed by how much life they brought into my apartment.

We need a plant (and some photographs) to make our dining room as homey as Sue made my apartment. (She left some time later, taking her plants, and I felt completely abandoned! Gosh, I missed those plants!)

Adding additional graphics – the AutoCAD term is *Landscaping* – to your drawing is very much like adding materials. You'll find a Landscaping Library – *Render.lli* – in the /Support folder. This library holds a few graphics you can use, but AutoCAD also allows you to import graphics into this file. You can even create your own library.

First, let's consider the few differences between attaching materials and landscaping.

- You don't have to attach landscaping to an object. It can stand freely in your drawing.
- You can't access the landscape library from the Landscape New dialog box. AutoCAD provides separate commands – ***LSNew*** to access the Landscape New dialog box, and ***LSLib*** to access the Landscape Library dialog box.
- AutoCAD also provides a separate command for modifying the size and position of landscaping – ***LSEdit***. But luckily, it uses the same dialog box as the ***LSNew*** command.

Let's begin with the Landscape New dialog box (Figure 11.021).

- Select the image you wish to use from the list box in the upper-left corner of the dialog box.
- Preview the image by picking the **Preview** button in the frame next to the list box.
- The **Geometry** frame has two options.
  - **Single Face** (a single occurrence of the image) or **Crossing Faces** (two images crossing at 90° through the center – like a cardboard Christmas tree). I recommend the **Single Face** option. **Crossing Faces** don't look real enough for me.
  - A check in the **View Aligned** box will ensure that the image appears

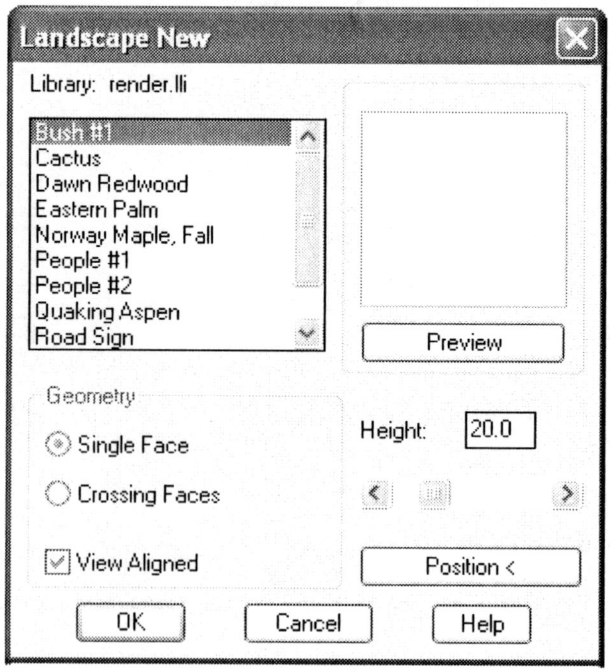

Figure 11.021

to be three-dimensional (flat against the screen) from any viewing angle. Remove the check to view the image from one direction only (you'll see these in the exercises).

- Use the **Height** text box to control the size of the image.
- Pick the **Position** button to place the image in the drawing. AutoCAD will prompt

  Choose the location of the base of the landscape object

Let's add a plant inside our dining room and a tree outside.

Do This: 11.4.1	Adding Landscaping for Rendering

I. Be sure you're still in the *Rendering Project.dwg* file in the C:\Steps3D\Lesson11 folder.

II. Follow these steps.

TOOLS	COMMAND SEQUENCE	STEPS
	**Command: *lsnew***	1. Enter the ***LSNew*** command. Alternately, you can pick the **Landscape New** button on the Render toolbar.
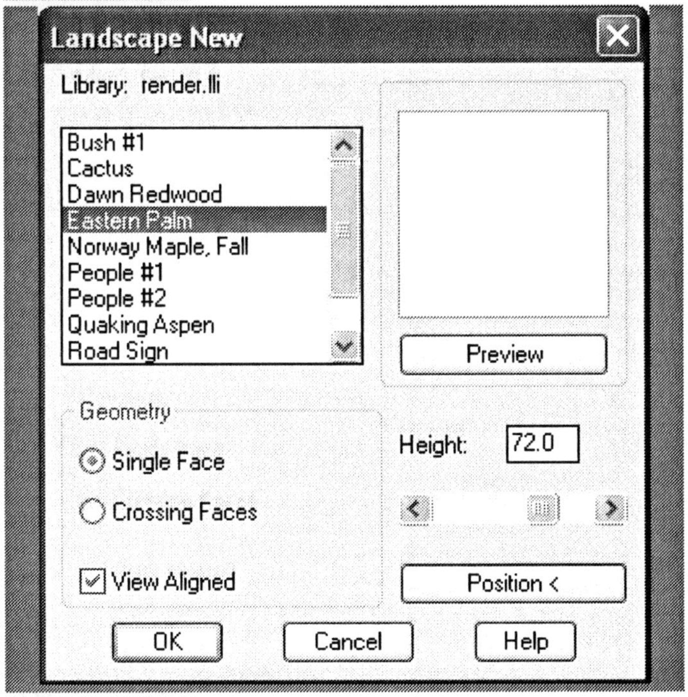		2. Select the **Eastern Palm** image and set the parameters indicated. Pick the **Position** button to continue.
	Choose the location of the base of the landscape object *12'1,14'7,1'3*	3. AutoCAD asks where to place the image. We'll put it in the planter; enter the coordinates shown. AutoCAD returns to the Landscape New dialog box.
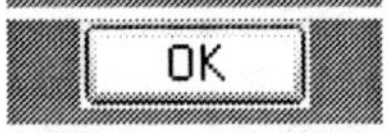		4. Complete the command.
		5. Repeat Steps 1 through 4 to add the **Dawn Redwood** outside the windows. Make the image 7' high, single faced,

Tools	Command Sequence	Steps
		and view aligned. Position it at coordinates **6', 14', 0**. Your drawing looks like Figure 11.022. (Notice that markers appear to locate the images. Like materials, the actual image won't appear until the scene is rendered.)

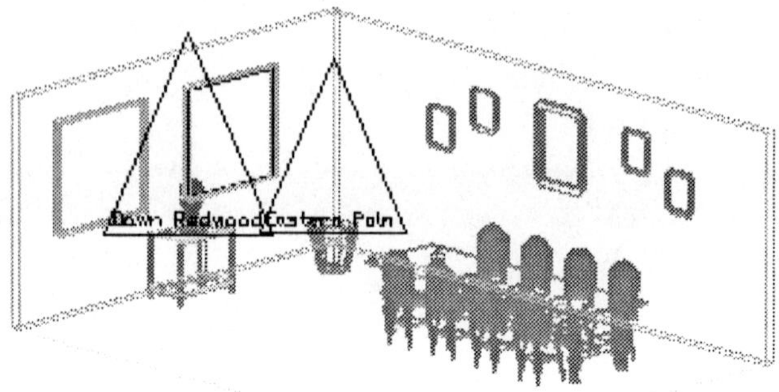

Figure 11.022

Tools	Command Sequence	Steps
	Command: ***qsave***	6. Save the drawing, but don't exit.

We'll render the scene after our next exercise. First, however, let's take a look at the landscape library.

Call the Landscape Library dialog box (Figure 11.023) with the ***LSLib*** command. For such a tiny box, this one provides some very useful opportunities. Let's look at these.

Figure 11.023

- The list box, of course, shows the images available in the library file indicated above the box.
- Use the **Delete** button to remove an image from the library.
- Use the **Save** button to save changes to the current library or to create a new library.
- Use the **Open** button to open another library file.
- The **Modify** and **New** buttons call the same dialog box (Figure 11.024), although it has two different names. Use this box to add (or modify) an image to the library.
  - The **Default Geometry** frame looks like the **Geometry** frame of the Landscape New dialog box. Here you can set the defaults for your new image that will appear in the Landscape New dialog box.
  - Enter a unique name for your new image in the **Name** text box.
  - Enter the image file (and location) you wish to use in the **Image File** text box, or use the **Find File** button to locate it.

- Enter the **Opacity Map File** name (and location) in the appropriate text box, or use the **Find File** button to locate it. (AutoCAD's glossary defines an opacity map as the "[p]rojection of opaque and transparent areas onto objects, creating the effect of a solid surface with holes or gaps.") As AutoCAD requires the entry of both an image file and an opacity map file, I generally specify the same file in both text boxes. This doesn't, however, allow for shadows. But AutoCAD apparently doesn't provide opacity maps for its images.

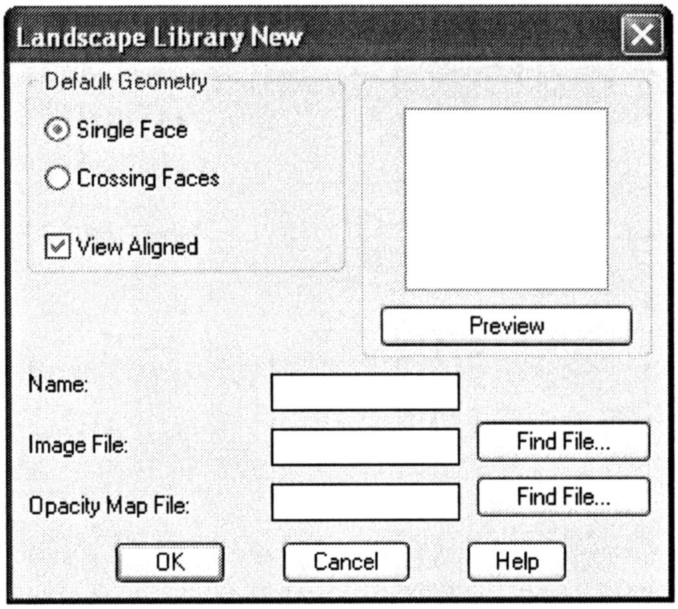

Figure 11.024

We'll add some graphics to our landscape library and place them in the picture frames on our wall.

Do This: 11.4.2	Adding Images to the Landscaping Library

I. Be sure you're still in the *Rendering Project.dwg* file in the C:\Steps3D\Lesson11 folder.

II. Thaw the **Marker** layer.

III. Restore the **frames** view and rotate the UCS 180° about the Z-axis.

IV. Follow these steps.

TOOLS	COMMAND SEQUENCE	STEPS
	**Command:** ***lslib***	1. Enter the ***LSLib*** command. Alternately, you can pick the **Landscape Library** button on the Render toolbar.
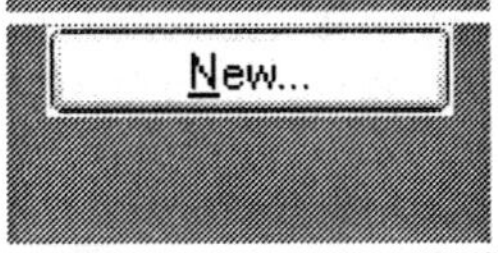		2. Pick the **New** button to access the Landscape Library New dialog box (Figure 11.024).
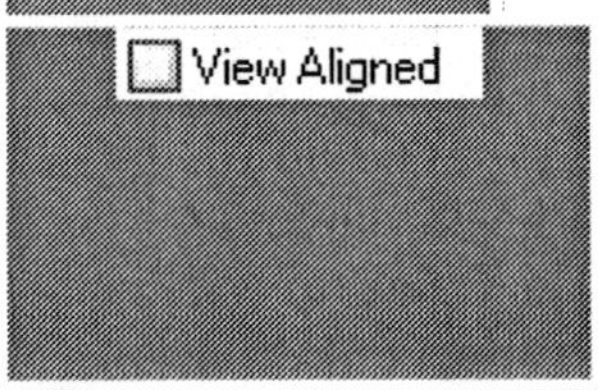		3. We'll want to view the images within the frames regardless of our viewing angle, so remove the check from the **View Aligned** box in the **Default Geometry** frame.
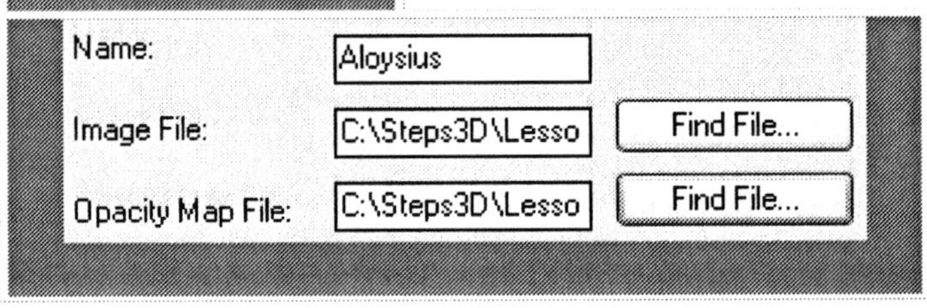		4. Call the new image *Aloysius*, and use the **Find File** button to locate the *Aloysius.gif* file in the C:\Steps3D\Lesson11 folder. (Be sure to look for a gif file). Do this for both the **Image File** and the **Opacity Map File**.

TOOLS	COMMAND SEQUENCE	STEPS
OK		5. Pick the **OK** button to continue. AutoCAD returns to the Landscape Library dialog box. Notice that **Aloysius** has become an available image.
		6. Repeat Steps 2 through 5 for the following images: *Barbara, Kevin, Starbuck*, and *Boys*. (These are all gif files located in the C:\Steps3D\Lesson11 folder.)
OK		7. Complete the command.
		8. AutoCAD presents the Landscape Library Modification message box (Figure 11.025) giving you the opportunity to discard your changes. Pick the **Save Changes** button.

Figure 11.025

TOOLS	COMMAND SEQUENCE	STEPS
Open		9. AutoCAD asks for the location of the library file to save. It defaults to the *Render.lli* file. Pick the **Open** button to complete the command.
	Command: *lsnew*	10. Now we'll insert the new images. Enter the ***LSNew*** command.
		11. Notice that the new images are available in the list box. Select **Kevin**. Give the image a height of **14"** and position it at the node of the lower-left frame. (Pick the **OK** button to complete the command.)
		12. Repeat Steps 9 and 10 for the remaining images. From left to right, they are Kevin, Starbuck, Barbara (large frame – 28"), Aloysius, and Boys. Your drawing looks like Figure 11.026.

TOOLS	COMMAND SEQUENCE	STEPS

Figure 11.026

TOOLS	COMMAND SEQUENCE	STEPS
	**Command:** ***v***	13. Restore the **Base** view and the UCS. Freeze the **Marker** layer.
	**Command:** ***rr***	14. Render the drawing. It looks like Figure 11.027.

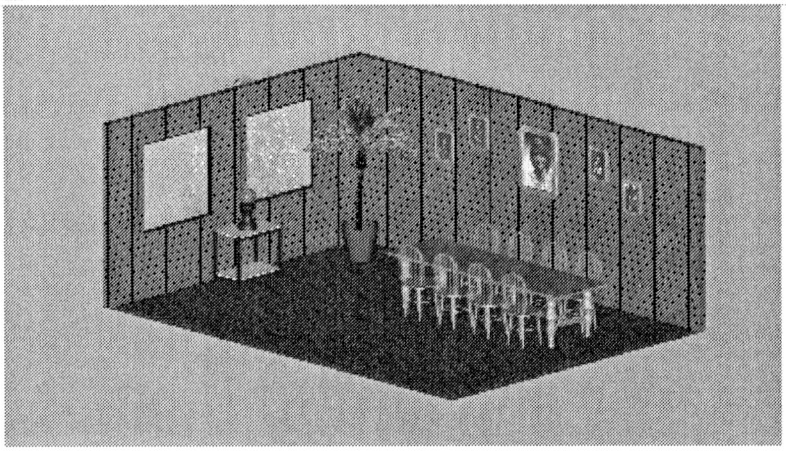

Figure 11.027

TOOLS	COMMAND SEQUENCE	STEPS
	**Command:** ***qsave***	15. Save the drawing, but don't exit.

Compare our latest rendering with previous ones. Our dining room looks better and better! But it still doesn't quite look real. We have one more thing to add!

11.5 Lights

Then God said, "Let there be light," and there was light. God saw how good the light was. … - the first day.

Genesis 1:3

Perhaps we can better understand the importance of light when we consider that it was the first thing He created.

AutoCAD provides four types of lighting. These include one that AutoCAD defines and you control – **Ambient Light** – and three that you define and control – **Point Light**, **Spotlight**, and **Distant Light**.

- **Ambient Light** lights all surfaces with equal intensity. It has no actual source, but you can control its intensity.

- A **Point Light** works like a light bulb. It spreads rays in all directions from a single source. It dissipates as it moves away from the source, and it casts shadows.
- A **Spotlight** works like a **Point Light**, except that it can be pointed in a single direction.
- A **Distant Light** mimics the sun. In fact, you can define the location of **Distant Light** in terms of planetary location and time of day.

You can use any combination of one, two, three, or all types of light in your rendering.

Call the Lights dialog box (Figure 11.028) with the ***Light*** command.

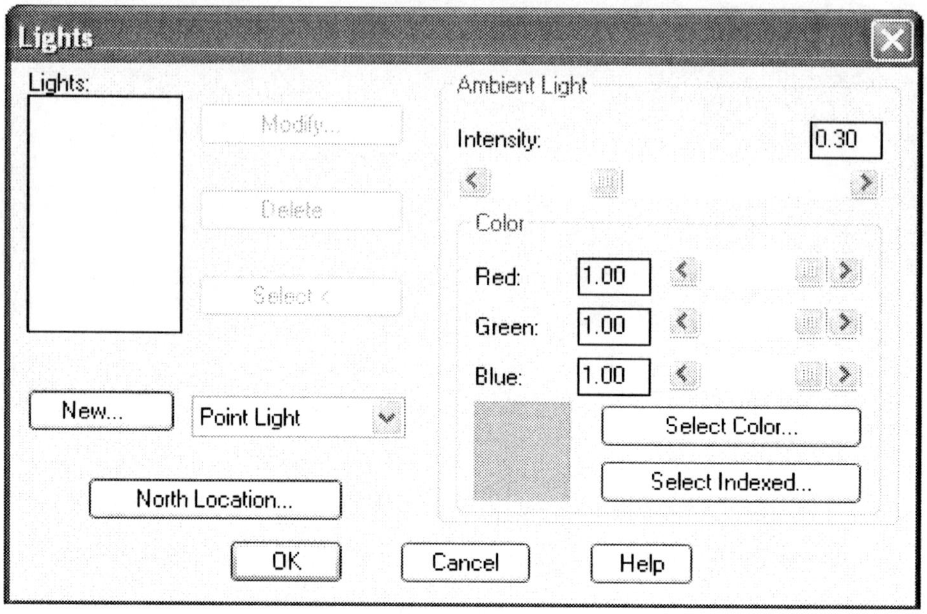

Figure 11.028

- Let's begin with the **Ambient Light** frame.
  - The **Color** subframe allows you to determine the color of the ambient light. Full **Red**, **Green**, and **Blue** make white light. You can change the color by degrees using the slider bars or select a color using the **Select Custom Color** (or **Select Indexed**) button.
  - Control the intensity (brightness) of the **Ambient Light** using the **Intensity** slider bar or by entering a number from 0 (no light) to 1 (brightest setting) in the text box.
- The **Lights** list box shows all of the lights currently in the drawing.
- The **Modify** button calls a specific dialog box for **Point Light**, **Spotlight**, or **Distant Light**, depending on which has been selected in the list box. These dialog boxes are identical to their New … Light dialog box counterparts discussed later.
- The **Delete** button, of course, removes a light from the drawing's database. A deleted light is no longer available for use.
- The **Select** button returns you to the graphics screen where you can select a light icon (see Figures 11.029, 11.030, and 11.031). AutoCAD then returns you to the Lights dialog box and identifies the light you've selected by highlighting it in the list box.

Figure 11.029: Point Light Icon

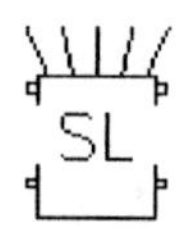

Figure 11.030: Spotlight Icon

Figure 11.031: Distant Light Icon

- The **North Location** button presents the North Location dialog box (Figure 11.032). Here you can set north according to a specific UCS, the WCS, or by angle in the XY-plane.

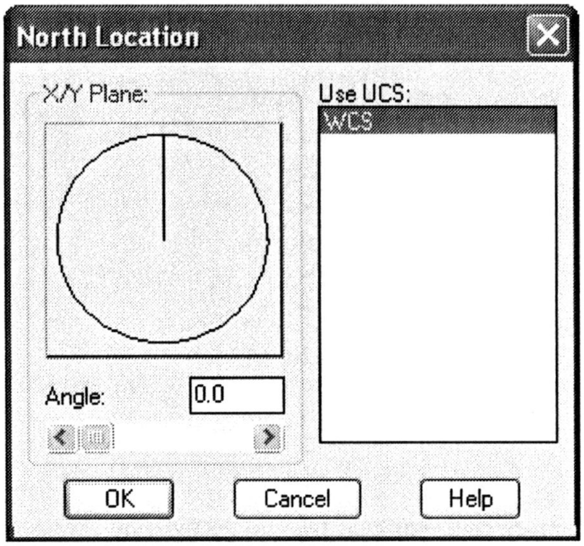

Figure 11.032

To add a new light to the drawing, select the type of light you want from the control box next to the **New** button, and then pick the **New** button. The dialog box presented depends on the type of light you want to add. Let's look at each.

- AutoCAD has divided the New Point Light dialog box (Figure 11.033) into four frames and one open area.
  - Put a unique name for your new point light in the **Light Name** text box in the upper-left corner of the dialog box. Limit the name to eight characters or less.
  - Use either the slider bar or the text box to enter the **Intensity** of the light. It might be easier for you to think in terms of wattage – how powerful a light bulb would you use?
  - The **Modify** button in the **Position** frame returns you to the graphics screen and prompts you to

    Enter light location <current>:

    You can pick a point on the screen or enter coordinates for your point light. (Remember that if you pick an arbitrary point, AutoCAD assumes a Z-axis value of 0.)
  - The **Show** button presents the Show Point Light Position dialog box (Figure 11.034). This box identifies the X, Y, and Z coordinates of the light.

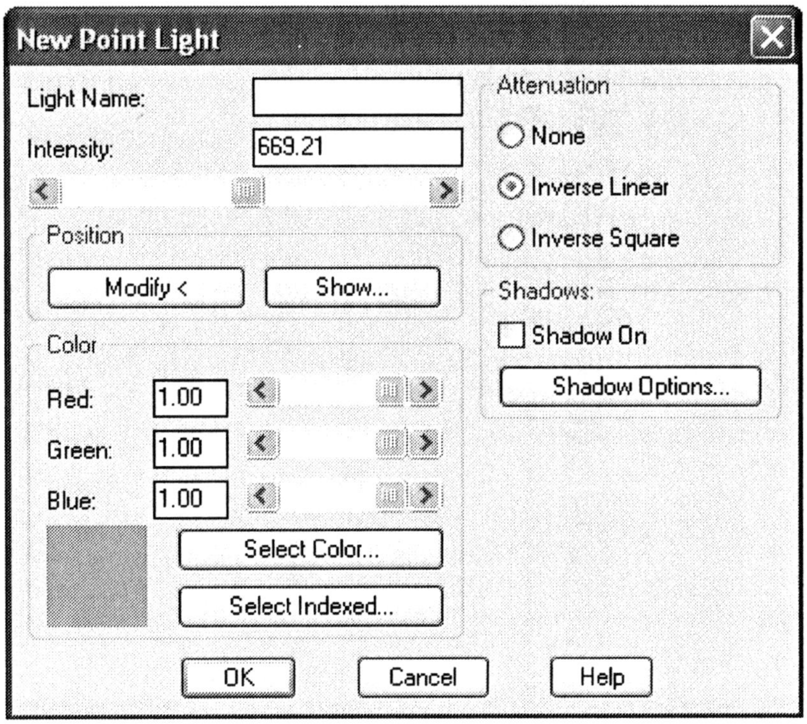

Figure 11.033

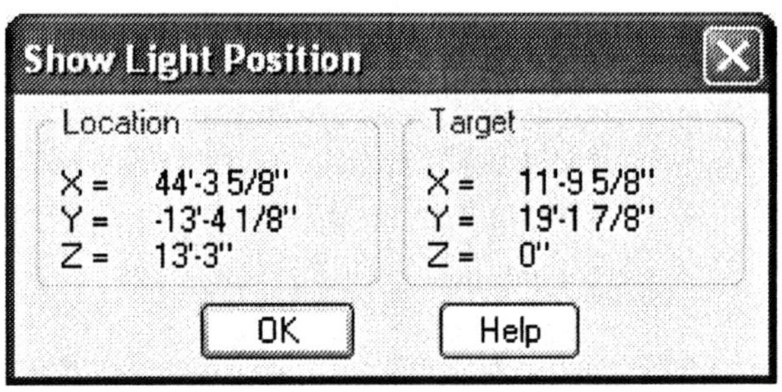

Figure 11.034

- The **Color** frame works as it has in other Rendering dialog boxes. Use this one to set the color of the point light.
- **Attenuation** refers to how light dissipates over distance.
    - **None** means that distant objects will reflect as much light from this point light as closer objects reflect.
    - **Inverse Linear** is normal reflectivity. In other words, an object twice as far away as another object will reflect half the light the closer one reflects. This is the default setting and appropriate for most situations.
    - **Inverse Square** is similar to **Inverse Linear** but works with squared numbers. That is, in our previous example, the distant object will reflect one-quarter the light that the closer one reflected.
- For a more realistic rendering, place a check in the **Shadow On** box in the **Shadow** frame. You can control the type of shadow with settings in the Shadow Options dialog box (Figure 11.035) accessed by picking the **Shadow Options** button.
    - The default option (a check in the **Shadow Volumes/Ray Traced Shadows** box) is the "normal" shadow display.

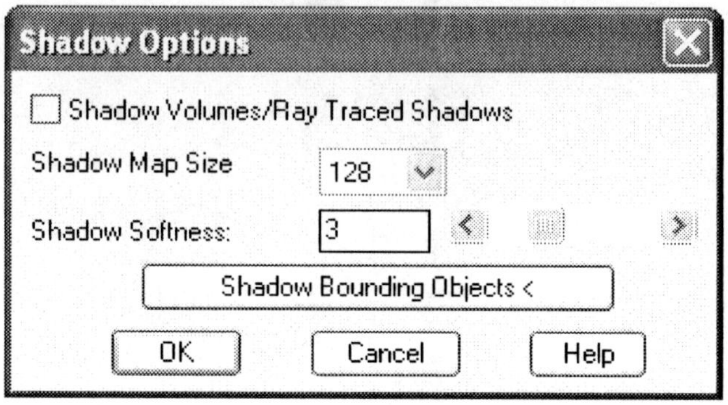

Figure 11.035

    - Control the accuracy of the shadow with the value in the **Shadow Map Size** control box. The larger the number (from 64 to 4096), the more accurate the shadow. However, larger numbers mean more rendering time. I've never found the need to increase the number beyond the default 128.
    - **Shadow Softness** refers to sharpness of the shadow. Again, I've never found the need to alter the default.
    - The **Shadow Bounding Objects** button allows you to modify a selected set of objects whose shadows are clipped by the bounding box.

- The New Spotlight dialog box (Figure 11.036) contains the same prompts as the New Point Light dialog box with two additions.
    - The **Hotspot** value determines the angle (from 0° to 160°) of the main beam of light (the center of the light). Use this to assign a direction for your spotlight to point.
    - The **Falloff** value determines the size of the full cone of light (the field angle). Use this

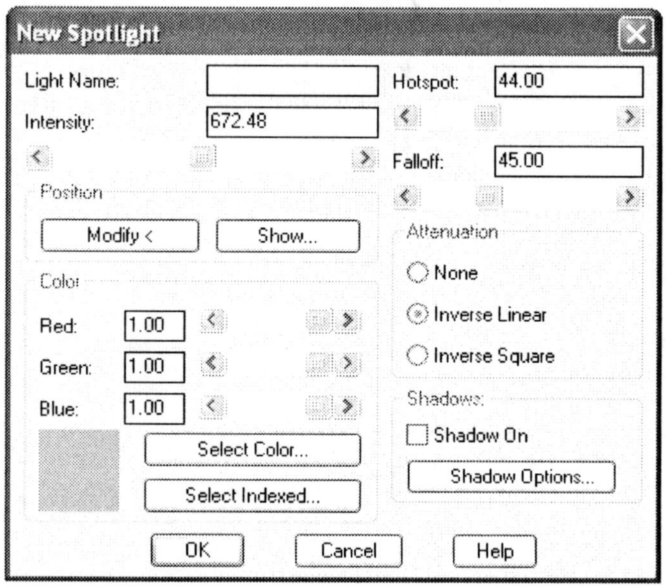

Figure 11.036

to focus or widen the spotlight.

- The New Distant Light dialog box (Figure 11.037) contains the same **Name**, **Intensity**, **Color**, and **Shadow** options as the first two. But here we have some useful tools for locating the light source (the sun).

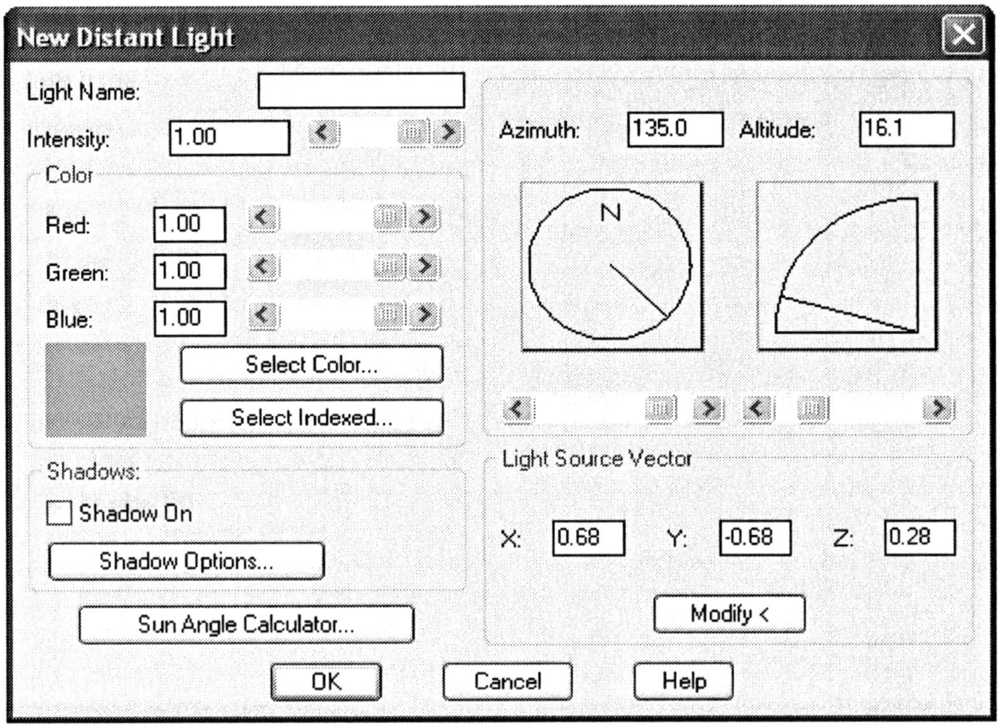

Figure 11.037

  - Use the upper-right frame to locate the sun by **Azimuth** (N-S-E-W direction) and **Altitude** (how high in the sky).
  - Use the **Light Source Vector** frame to locate the sun by coordinates. The **Modify** button allows you to locate the light by location and direction.
  - The **Sun Angle Calculator** button presents the preferred method for locating the sun –

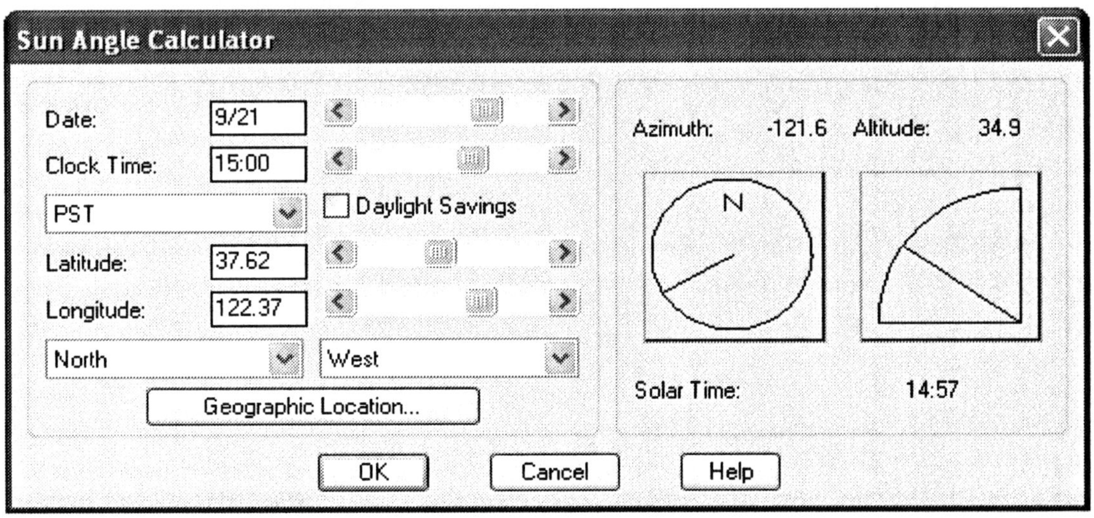

Figure 11.038

the Sun Angle Calculator dialog box (Figure 11.038). Here you can allow AutoCAD to determine the location of the sun by entering a date, time and location (latitude and longitude).

Sound difficult? Pick the **Geographic Location** button. AutoCAD present the Geographic Location dialog box (Figure 11.039). Here you can select a city anywhere in the world using the **City** control box. Alternately, you can simply pick a point on the map! (Use the **Continental** control box above the map to change the map shown.)

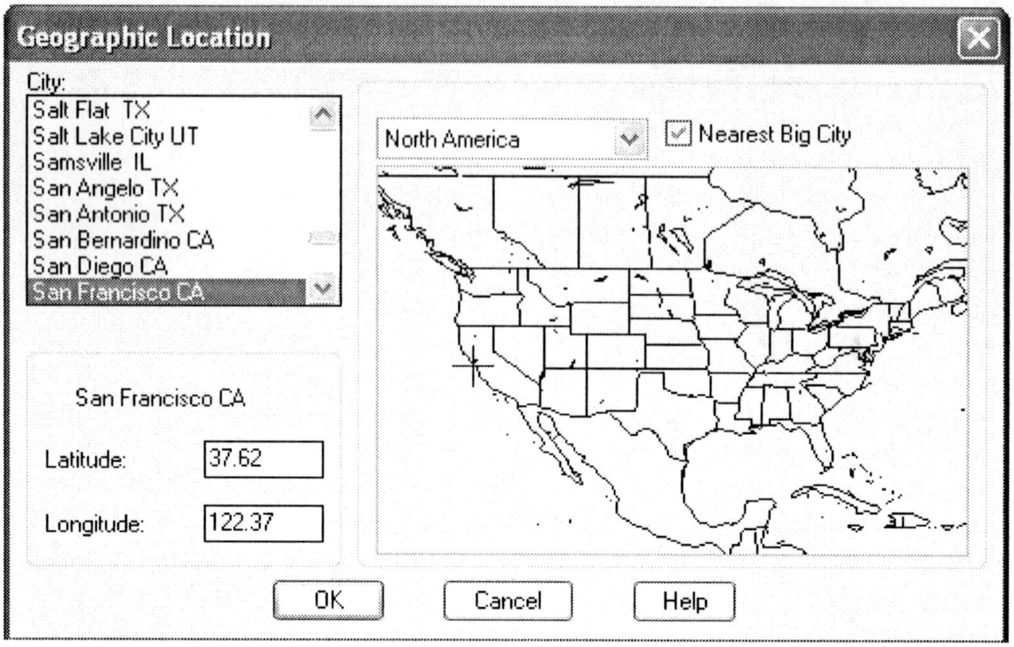

Figure 11.039

It just can't get any easier than that!

With so much information to manage, does your appreciate for dialog boxes continue to grow?

We'll add three lights to our dining room – two point lights and a distant light.

Do This: 11.5.1	Let There Be Light!

I. Be sure you're still in the *Rendering Project.dwg* file in the C:\Steps3D\Lesson11 folder.

II. Thaw the **Marker** layer and regenerate the drawing.

III. Follow these steps.

Tools	**Command Sequence**	**Steps**
	**Command:** ***light***	1. We'll begin by placing a light bulb in our lamp. Enter the ***Light*** command or pick the **Lights** button on the Render toolbar.
		2. Set the type of light to **Point Light** and pick the **New** button. AutoCAD presents the New Point Light dialog box (Figure 11.033).

Tools	Command Sequence	Steps
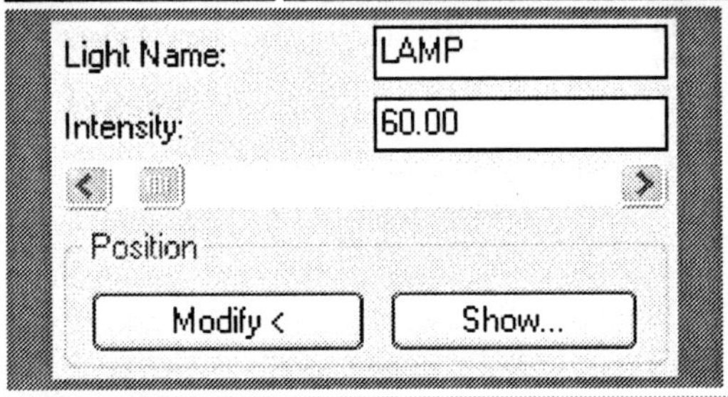		3. Call the new light *Lamp*, set the intensity to that of a 60-watt bulb, and then pick the **Modify** button in the **Position** frame.
	Enter light location <current>:	4. Use OSNAPs to locate the light in the center of the lamp (between the windows). (A node on the **Marker** layer marks the spot.)
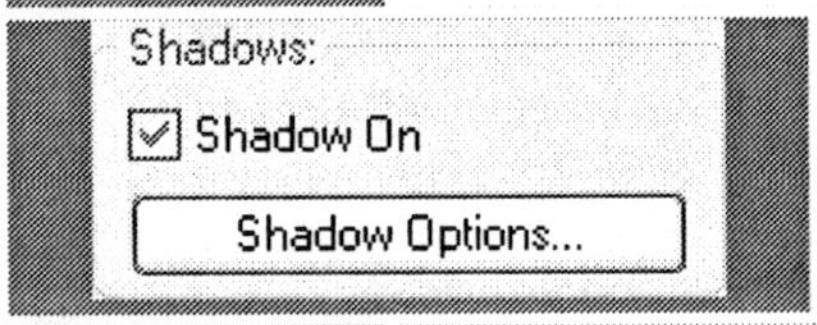		5. Place a check in the **Shadow On** check box to be sure our lamp casts shadows.
		6. Pick the **OK** button to return to the Lights dialog box and again to complete the command. AutoCAD places an icon showing you where the light will be located (Figure 11.040). (If you can't see the icon, use the ***RPref*** command and set the **Light Icon Scale** to **24**.)

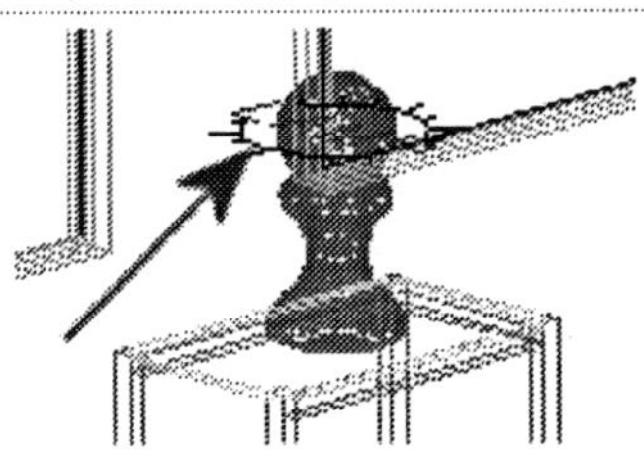

Figure 11.040

Tools	Command Sequence	Steps
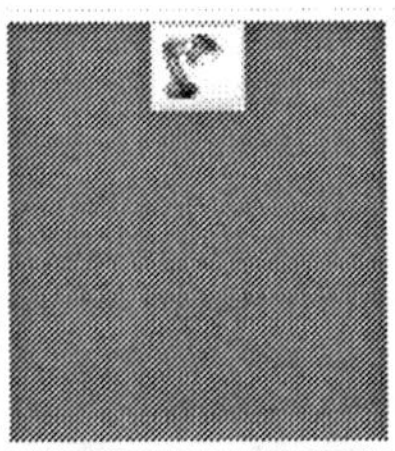	**Command:** ***light***	7. Repeat Steps 1 through 6 to add another point light (call it *Ceiling*). The light should have the intensity of a 90-watt bulb and be located at coordinates **27'6, 4', 8'**. Remember to turn on the shadows.
	**Command:** ***[enter]***	8. Now let's locate the sun. Repeat the ***Light*** command.
		9. Set the type of light to **Distant Light** and pick the **New** button. AutoCAD presents the New Distant Light dialog box.
		10. Call the new light *Sun*.

Tools	Command Sequence	Steps
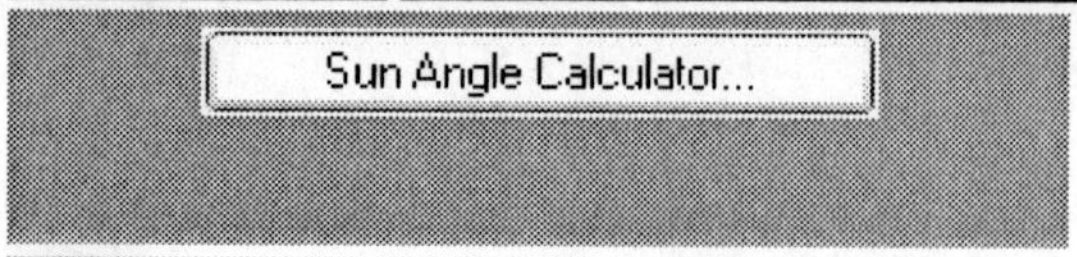		11. Pick the **Sun Angle Calculator** button to call the sun Angle Calculator dialog box.
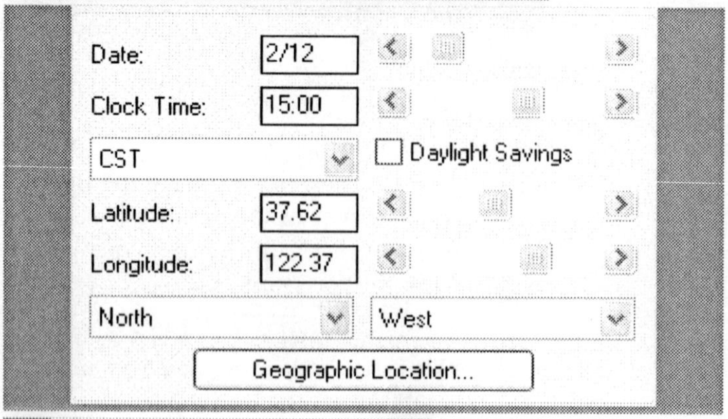		12. Set the date/time to March 12 at 3:00pm CST. Then pick the **Geographic Location** button to continue.
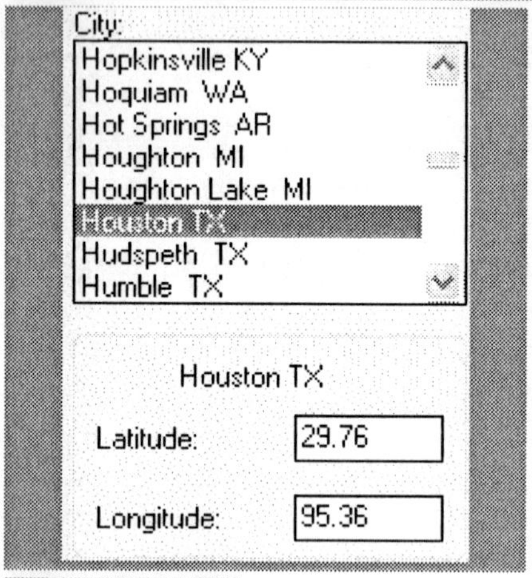		13. Pick a point near **Houston, Texas**, as indicated.
		14. Pick the **OK** button twice to return to the New Distant Light dialog box.
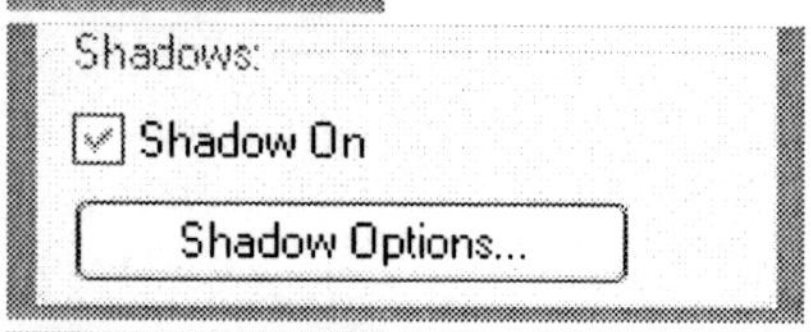		15. Be sure shadows are activated.
		16. Pick the **OK** button twice to complete the command.
	**Command:** ***qsave***	17. Save the drawing, but don't exit.

We've added the lights, but we need to create a scene to use the lights. We'll do that next and then render our drawing.

11.6 Creating a Scene

Perhaps the easiest part of rendering – creating a scene – involves nothing more than telling AutoCAD which lights to use and which view to render.

The ***Scene*** command calls the Scenes dialog box (Figure 11.041). AutoCAD lists available scenes in the list box and offers three choices (buttons) of things you can do.

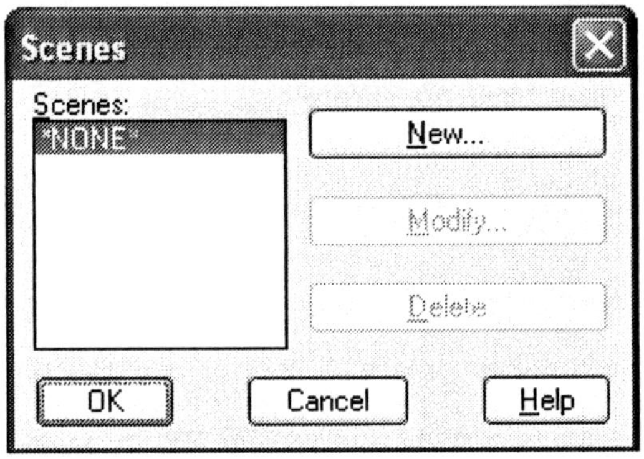

Figure 11.041

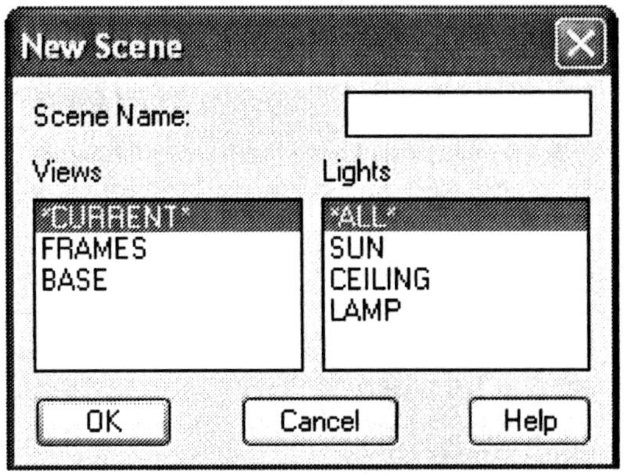

Figure 11.042

- The **New** and **Modify** buttons call the dialog box shown in Figure 11.042. Here you designate which (existing) view to use in this scene along with which light(s) to use.

  The only real difference between the New Scene and Modify Scene dialog boxes is that the New Scene dialog box requires that you give your new scene a unique name. The Modify Scene dialog box allows you change the name, view, or lights but doesn't require anything.
- The **Delete** button, of course, allows you to remove a scene from the drawing's database.

Let's make a scene!

Do This: 11.6.1	Make a Scene!

I. Be sure you're still in the *Rendering Project.dwg* file in the C:\Steps3D\Lesson11 folder.

II. Follow these steps.

TOOLS	COMMAND SEQUENCE	STEPS
	**Command: *scene***	1. Enter the ***Scene*** command. Alternately, you can pick the **Scene** button on the Render toolbar.
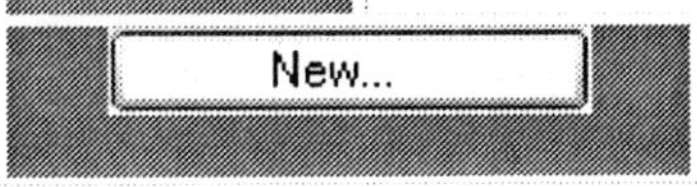		2. Pick the **New** button in the Scene dialog box.

TOOLS	COMMAND SEQUENCE	STEPS

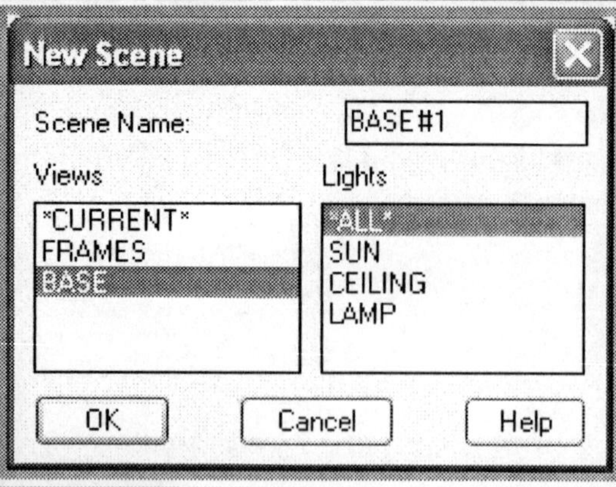

3. Call the scene *BASE#1*. Select the **Base** view and use **All** the lights in the drawing. Pick the **OK** button to continue.

4. Notice that **BASE#1** is listed in the available Scenes list box (Figure 11.043).

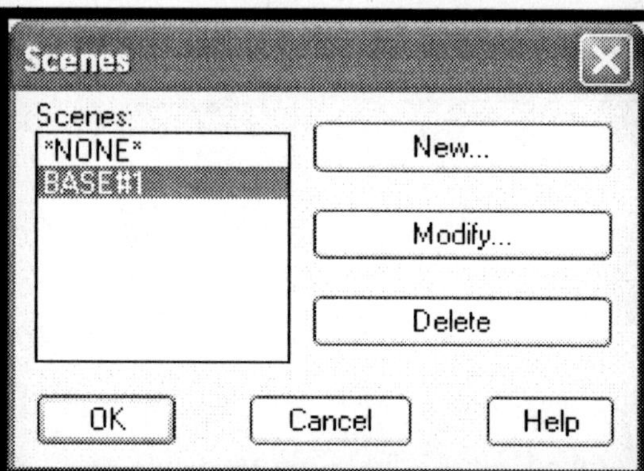

Figure 11.043

5. Repeat Steps 1 through 3 to create the scenes shown in the following chart.

SCENE	VIEW	LIGHT(S)
BASE#2	Base	Ceiling
BASE#3	Base	Ceiling and Sun
BASE#4	Base	Ceiling and Lamp
FRAME#1	Frames	Ceiling

6. The Scenes dialog box looks like Figure 11.044. Complete the command.

TOOLS	COMMAND SEQUENCE	STEPS
	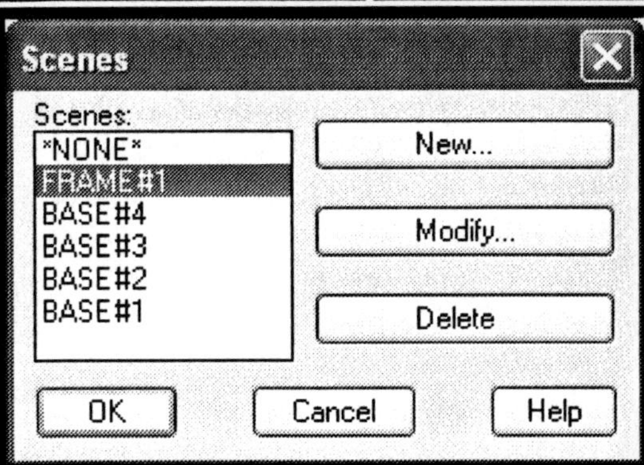 Figure 11.044	
	**Command:** ***scene***	7. Now let's modify our scenes. Repeat the ***Scene*** command.
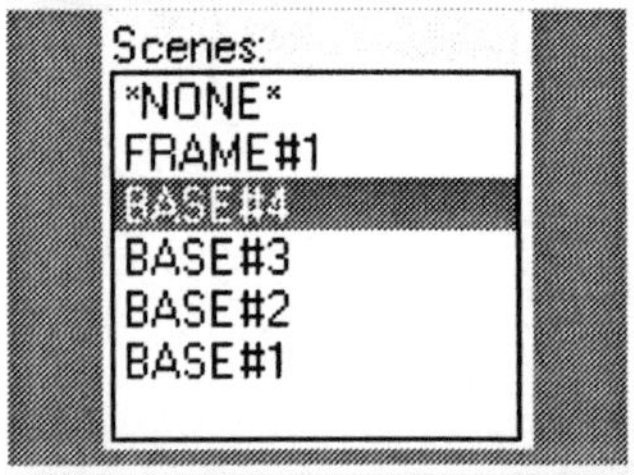 		8. Select the **BASE#4** scene in the **Scenes** list box and pick the **Modify** button.
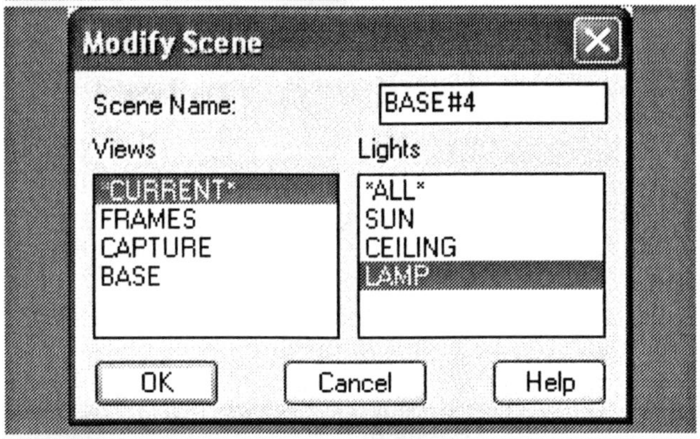 		9. Remove the **CEILING** light. Pick the **OK** button to continue.
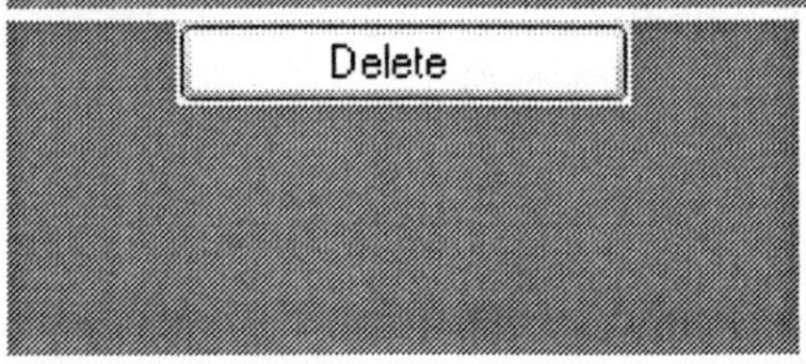 		10. Remove the **BASE#3** scene by selecting it in the list box and then picking the **Delete** button. AutoCAD prompts with an "are you sure" dialog box. (Pick the **OK** button to confirm the deletion.)
 		11. Pick the **OK** button again to complete the command.
	**Command:** ***rr***	12. Enter the ***Render*** command.
		13. Notice the scenes you just created are now shown in the **Scene to Render** list box. Render each for comparison. (The **BASE#1** rendering is shown in Figure 11.045.)

Tools	Command Sequence	Steps
	 Figure 11.045	
	Command: ***qsave***	14. Save the drawing.

Did you notice the lights and shadows? This image is truly photographic quality.

And now the bad news: AutoCAD isn't designed to plot a rendered drawing. (Do you really want *more* plotting parameters?!) You can, however, render the drawing to a file and print the rendered file through a graphics program (such as MS Paint).

11.7 Extra Steps

You should experiment with a couple other (less frequently needed) rendering tools.

- The ***RPref*** command (Rendering Preferences) presents essentially the same dialog box used by the ***Render*** command. The biggest difference is the replacement of the Render button with an **OK** button. Use this box to set rendering defaults.
- The ***Stats*** command (for render Statistics) presents the Statistics dialog box, which provides detailed information about the last rendering. Details include type of rendering, scene rendered, time to completion, the size of the rendering, and much, much more.

Redo the renderings on each of the scenes you created in the last exercise and compare the statistics on each. Then set up the rendering preferences you'd like to use for defaults in the future.

11.8 What Have We Learned?

Items covered in this lesson include:

- *Rendering commands and techniques*
  - *Assigning materials*
  - *Adding graphics*
  - *Adding lights*
  - *Creating scenes*
- *Commands*
  - ***Render***
  - ***RMat***
  - ***MatLib***
  - ***LSNew***
  - ***LSLib***
  - ***LSEdit***
  - ***Light***
  - ***Scene***
  - ***RPref***
  - ***Stats***

You've accomplished a great deal with the completion of Lesson 11. In this lesson, you've conquered the bridge between CAD operation and art. This is as far as AutoCAD goes toward the creation of design imagery. (Beyond this, you'll have to get some tools and build the objects!)

Let's tackle some final exercises and answer some questions.

11.9 Exercises

1. Open the *emerald11.dwg* file in the C:\Steps3D\Lesson11 folder. My rendering appears in Figure 11.046.

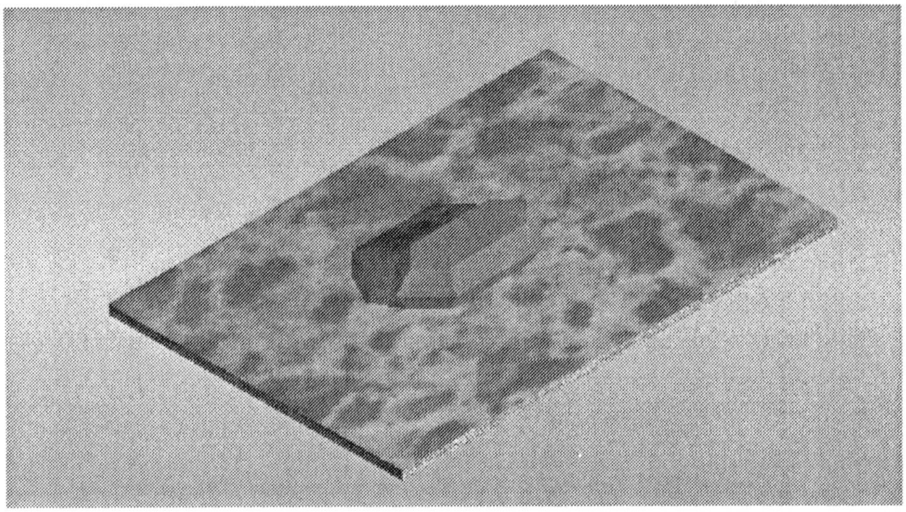

Figure 11046

   1.1. Assign materials to the emerald and tabletop. (I used **Green Glass** and **Mottled Marble**. I changed the transparency of the glass to a value of **0.45**.)

   1.2. Assign a background. (I used a three-color gradient.)

   1.3. Place al light(s). (I used a single-point light with an intensity of **10.7** located at coordinates **18,4,9**.)

   1.4. Render the drawing.

   1.5. Repeat Steps 1.1 to 1.4 using a different set of assignments.

2. Open the *caster11.dwg* file in the C:\Steps3D\Lesson11 folder. My rendering appears in Figure 11.047.

   2.1. Assign materials to the objects. I used

      2.1.1. **Bright Olive** (bearings)

      2.1.2. **Bumpy Metal** (flooring)

      2.1.3. **Chrome Gifmap** (wheel)

      2.1.4. **Dark Olive Matte** (axle, spindle, bearing case)

      2.1.5. **Sand Texture** (frame)

   2.2. Place a light(s). (I used a single-point light with an intensity of **27.45** located at coordinates **0,-10,4**.)

Figure 11.047

   2.3. Render the drawing.

   2.4. Repeat Steps 2.1 through 2.3 using a different set of assignments.

3. Open the *pipe11.dwg* file in the C:\Steps3D\Lesson11 folder. My rendering appears in Figure 11.048.
   3.1. Assign materials to the objects. I used
      3.1.1. **Beige Plastic** (tank and nozzle)
      3.1.2. **Blue Matte** (large pipe and fittings)
      3.1.3. **Brown Matte** (model platform)
      3.1.4. **Olive Metal** (pipe rack)
      3.1.5. **Red Matte** (small pipe)
   3.2. Assign a background. (I used the **HLS** Color System for a two-color gradient. I made both colors a light blue/cyan color.)
   3.3. Place a light(s). (I used a distant light with an ambient light intensity of **0.45**. The unit is located in a plant in New Orleans. The graphic was created at 3:00 P.M. in late September.)
   3.4. Render the drawing.

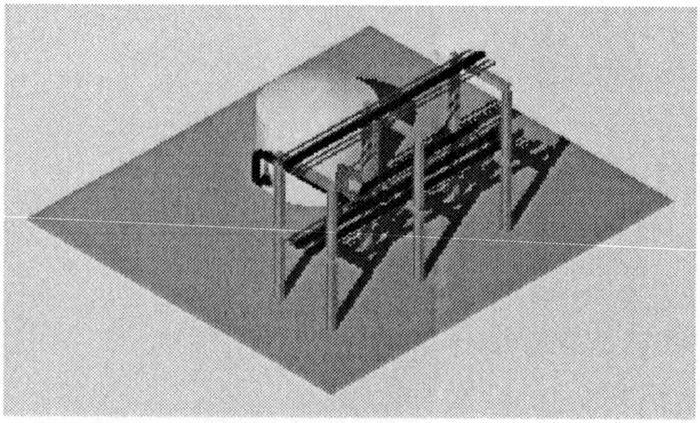

Figure 11.048

4. Open the *train11.dwg* file in the C:\Steps3D\Lesson11 folder. My rendering appears in Figure 11.49.
   4.1. Assign materials to objects. I used:
      4.1.1. **Ape** (front of train)
      4.1.2. **Blue Plastic** (cab, base, top of cattle guard)
      4.1.3. **Chrome Gifmap** (wheels, smokestack, bell)
      4.1.4. **Cream Plastic** (backdrop)
      4.1.5. **Red Plastic** (water tank, undercarriage, base of cattle guard, flag)
      4.1.6. **Wood White Ash** (bell frame, box, flagpole)
   4.2. Attach a landscape person in the cab. (I used **People #2**.)
   4.3. Assign a background. (I used a solid green background.)
   4.4. Place a light(s). (I used a single spotlight located at 0,-60,0, and a target location of **4,1,4.5**. Ambient light intensity is **0.5**.)
   4.5. Render the drawing.

Figure 11.049

You have quite a variety of exercises from which to choose in this lesson. Your next assignment is to return to any of the exercises you've completed in the text, assign materials, graphics, and lights, as you deem necessary and then render the drawings. Following are several exercises you can do. (If you haven't completed these drawings, they're available in the C:\Steps3D\Lesson11 folder.)

Figure 11.050

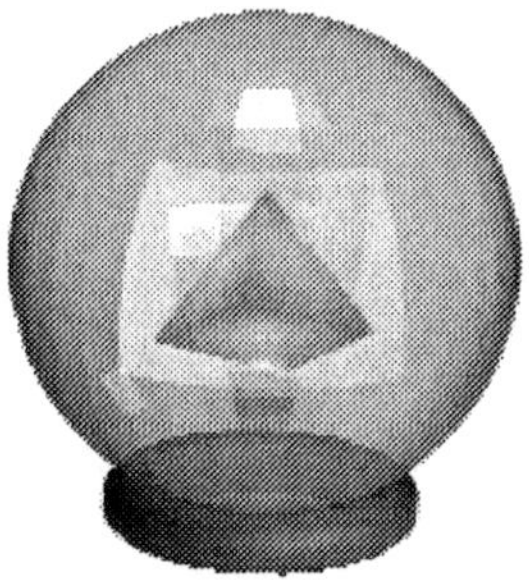

Figure 11.051

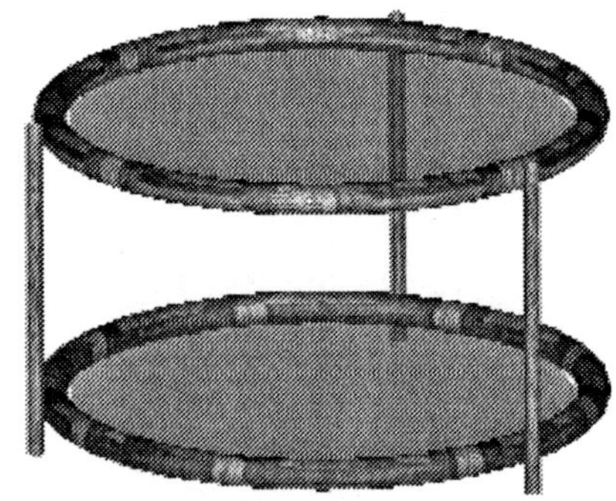

Figure 11.052

Figure 11.053

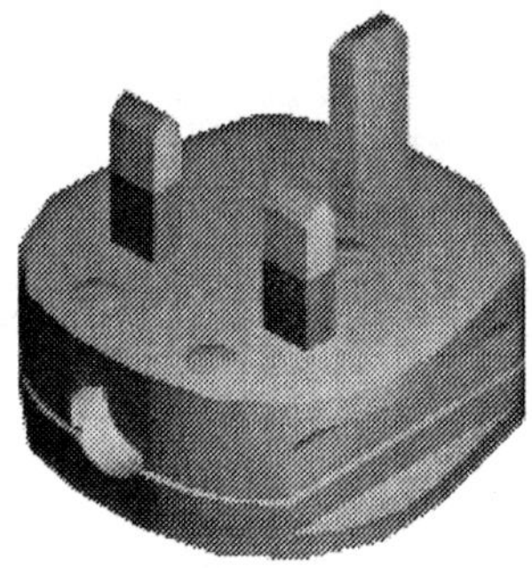

Figure 11.054

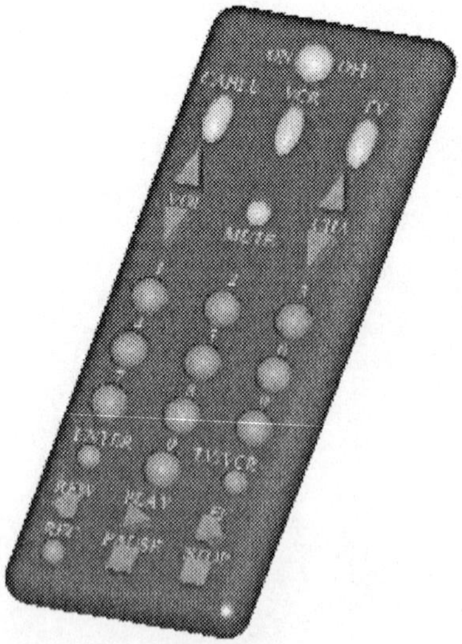

Figure 11.055

Figure 11.056

Figure 11.057

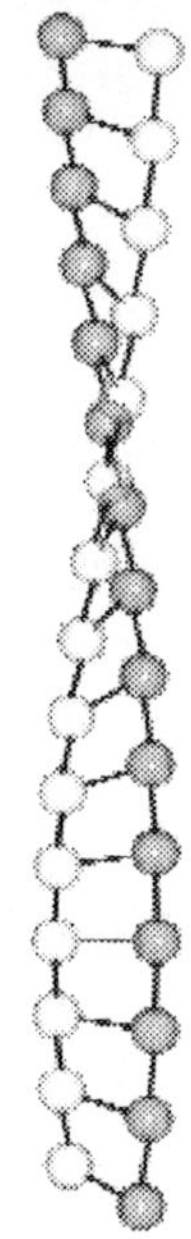

Figure 11.058

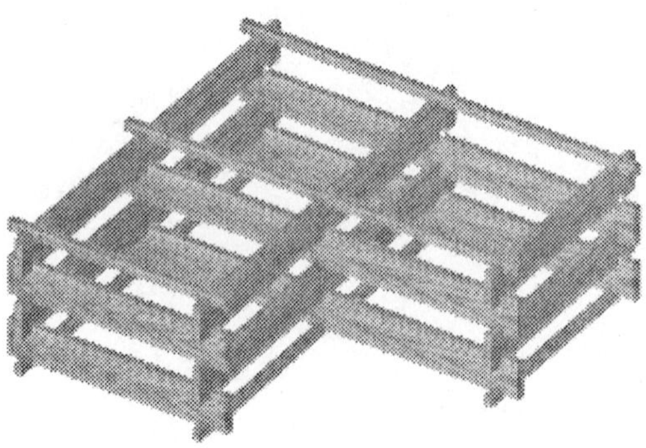

Figure 11.059

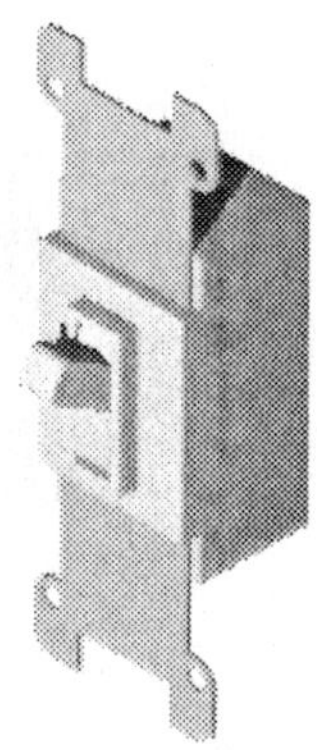

Figure 11.060

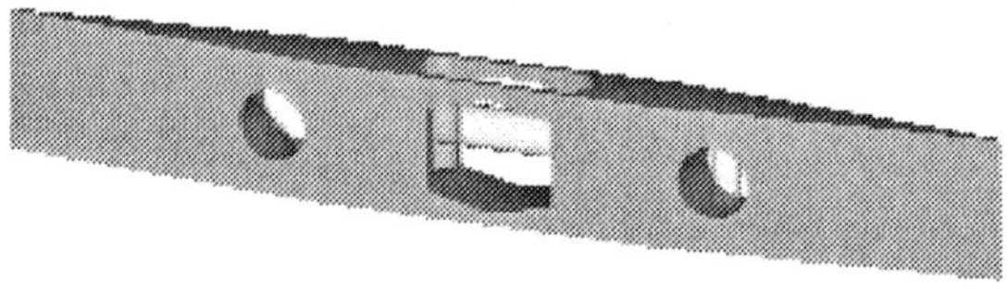

Figure 11.061

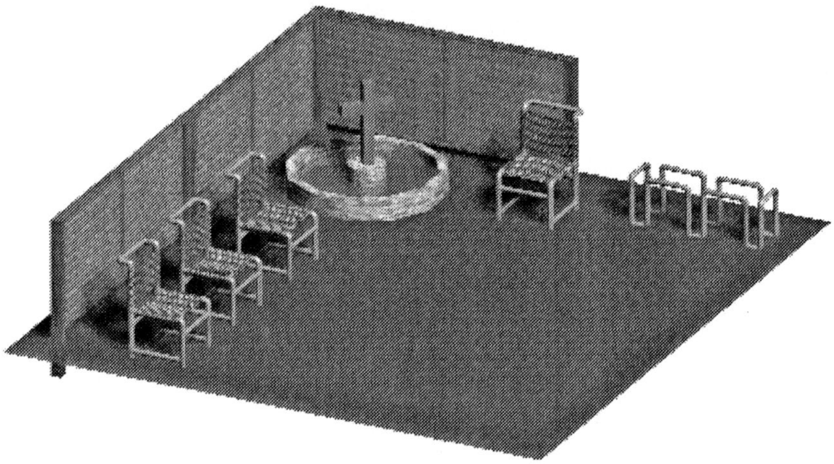

Figure 11.062

18. Using other objects you've created in Lessons 4 through 11, redesign the patio scene. Some suggestions follow:
    18.1. Replace the fountain with the planter box.
    18.2. Place planters in the planter box and add plants.
    18.3. Add a sidewalk or some decking.
19. Using other objects you've created, redesign the room we created in our rendering project. Some suggestions follow:
    19.1. Replace the plant with the standing lamp (Lesson 4).
    19.2. Put the 3D chess set on the table.
    19.3. Change the pictures on the wall.

11.10 For Web-Based Review Questions, visit: http://www.uneedcad.com/2006/Files/06R11-3D.pdf

Appendix – A: Drawing Scales

SCALE (=1')	SCALE FACTOR	DIMENSIONS OF DRAWING WHEN FINAL PLOT SIZE IS: 8½"x11"	11"x17"	17"x22"	22"x34"	24"x36"
1/16"	192	136'x176'	176'x272'	272'x352'	352'x544'	384'x576'
3/32"	128	90'8x117'4	117'4x181'4	181'4x234'8	234'8x362'8	256'x384'
1/8"	96	68'x88'	88'x136'	136'x176'	176'x272'	192'x288'
3/16"	64	45'4x58'8	58'8x90'8	90'8x117'4	117'4x181'4	128'x192'
¼"	48	34'x44'	44'x68'	68'x88'	88'x136'	96'x144'
3/8"	32	22'8x29'4	29'4x45'4	45'4x58'8	58'8x90'8	64'x96'
½"	24	17'x22'	22'x34'	34'x44'	44'x68'	48'x72'
¾"	16	11'4x14'8	14'8x22'8	22'8x29'4	29'4x45'4	32'x48'
1"	12	8'x6'11	11'x17'	17'x22'	22'x34'	24'x36'
1½"	8	5'8x7'4	7'4x11'4	11'4x14'8	14'8x22'8	16'x24'
3"	4	34"x44"	3'8x5'8	8'x6'11	7'4x11'4	8'x12'
(1" =)						
10'	120	85'x110'	110'x170'	170'x220'	220'x340'	240'x360'
20'	240	170'x220'	220'x340'	340'x440'	440'x680'	480'x720'
25'	300	212'6x275'	275'x425'	425'x550'	550'x850'	600'x900'
30'	360	255'x330'	330'x510'	510'x660'	660'x1020'	720'x1080'
40'	480	340'x440'	440'x680'	680'x880'	880'x1360'	960'x1440'
50'	600	425'x550'	550'x850'	850'x1100'	1100'x1700'	1200'x1800'
60'	720	510'x660'	660'x1020'	1020'x1320'	1320'x2040'	1440'x2160'
80'	960	680'x880'	880'x1360'	1360'x1760'	1760'x2720'	1920'x2880'
100'	1200	850'x1100'	1100'x1700'	1700'x2200'	2200'x3400'	2400'x3600'
200'	2400	1700'x2200'	2200'x3400'	3400'x4400'	4400'x6800'	4800'x7200'

Appendix – B: Project Drawings

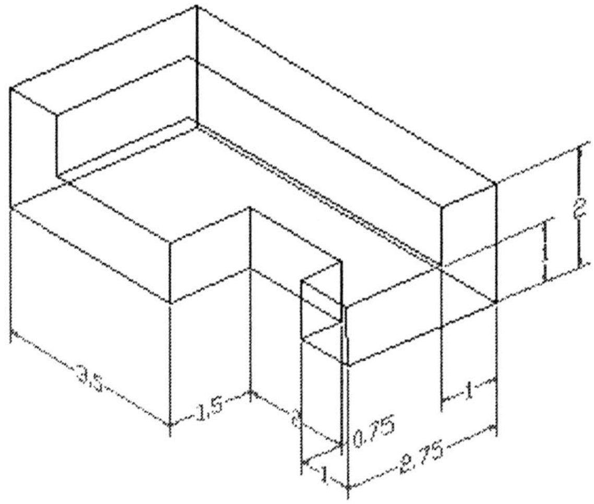

Figure B-1w

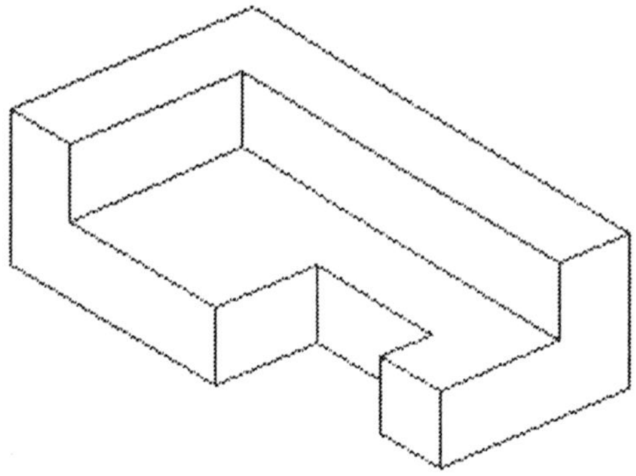

Figure B-1su

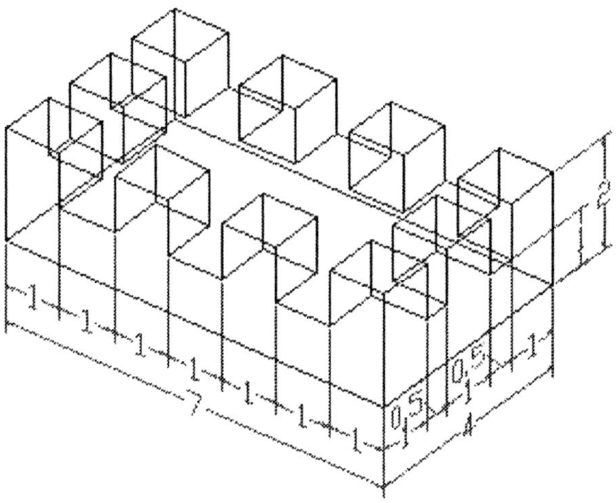

Figure B-2w

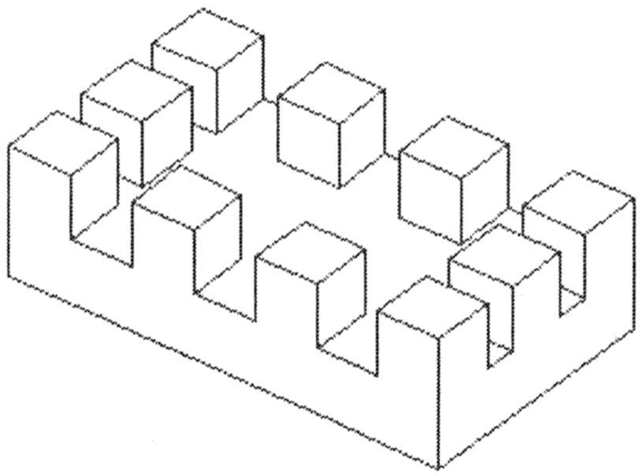

Figure B-2su

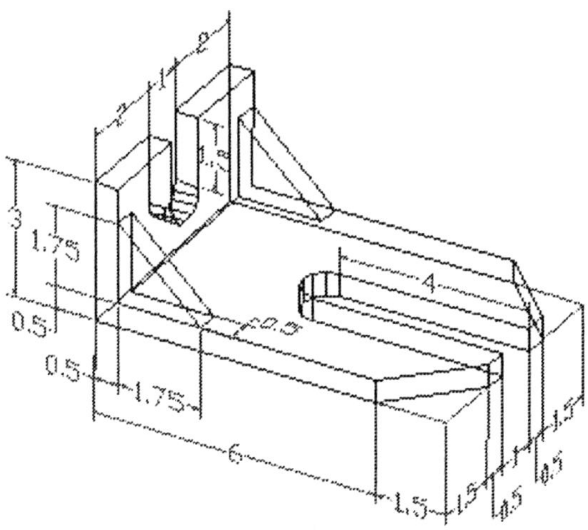

Figure B-3w

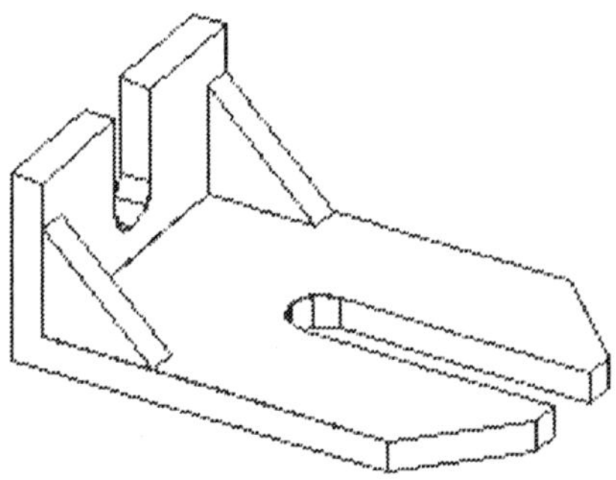

Figure B-3su

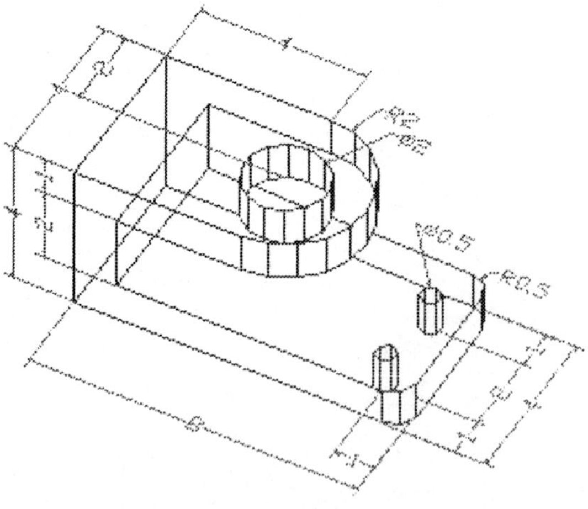

Figure B-4w

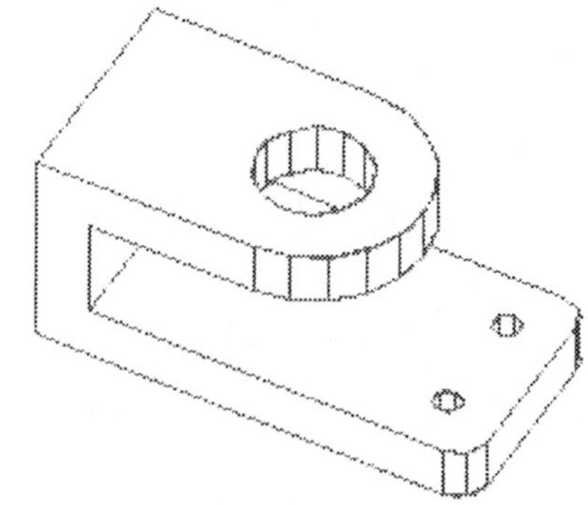

Figure B-4su

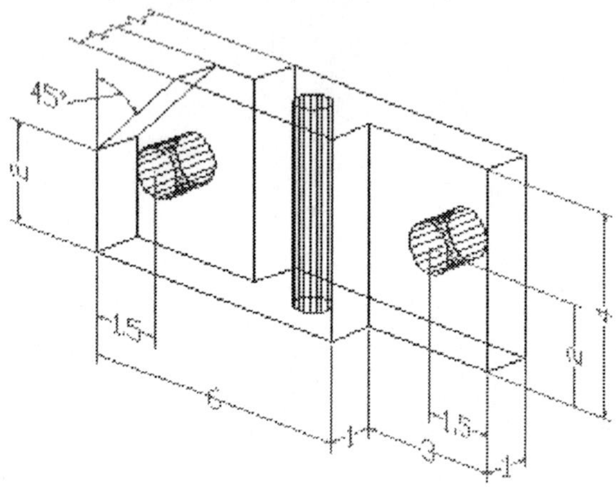

Figure B-5w

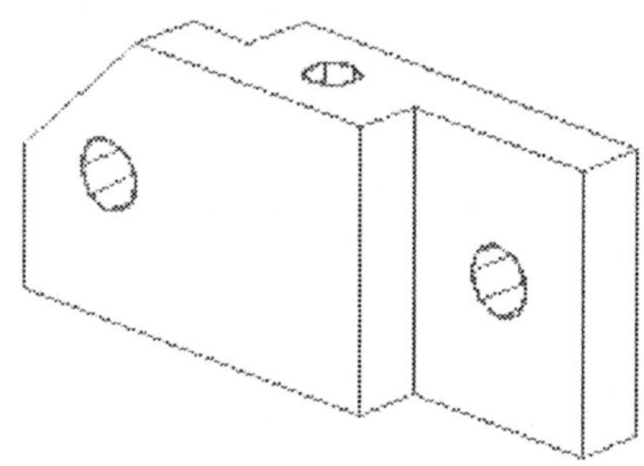

Figure B-5su

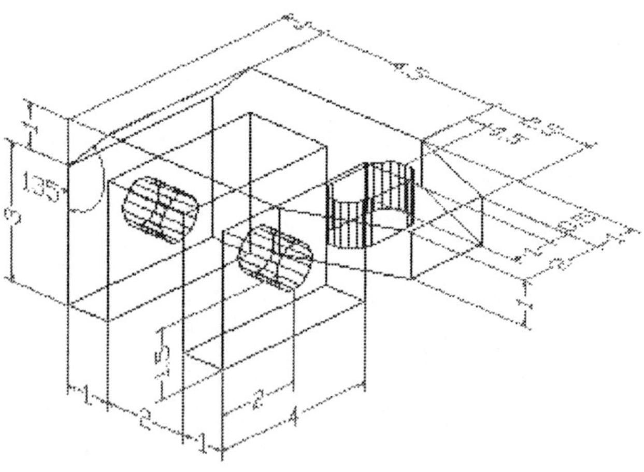

Figure B-6w

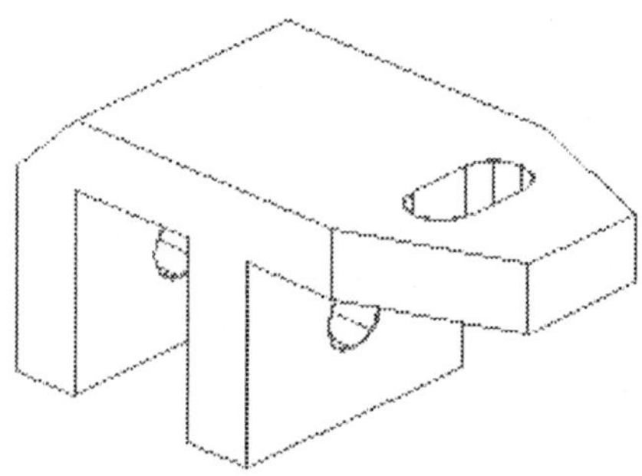

Figure B-6su

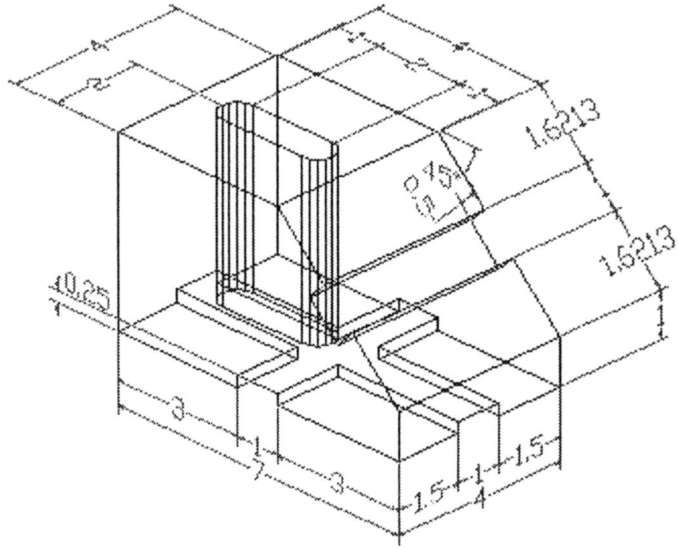

Figure B-7w

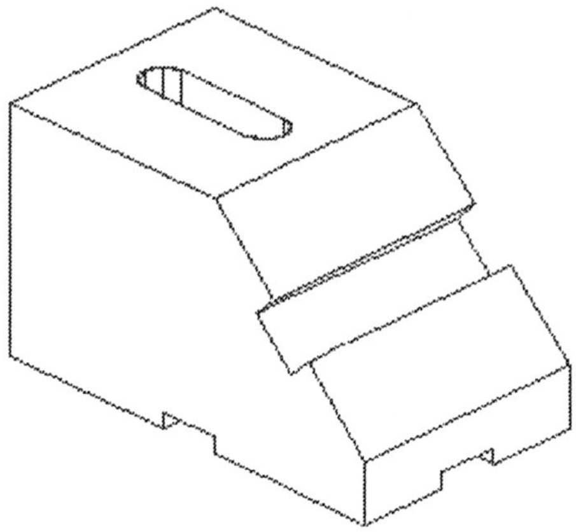

Figure B-7su

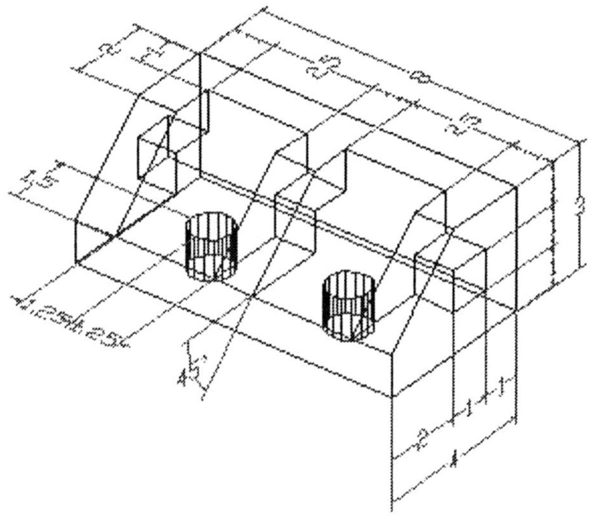

Figure B-8w

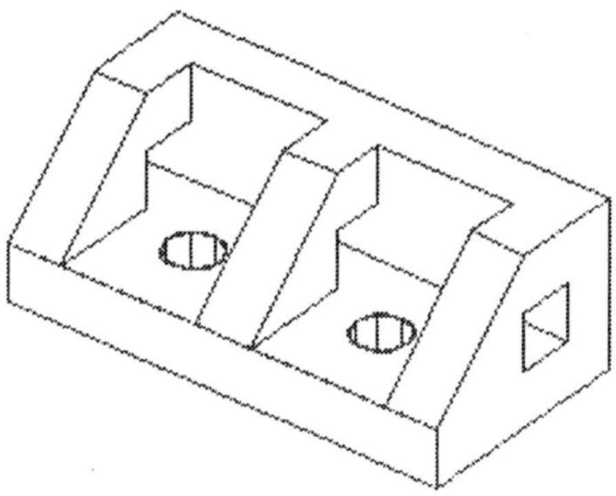

Figure B-8su

[Note: Appendix C has been replaced with web addresses for chapter review questions and their answers.]

Appendix – D: Additional Projects

Here are two drawings that might challenge you.

1. The pieces for Aloysius' Cabin can be found at our web site – http://www.prenhall.com/sykes. They include:

    1.1. *longlog.dwg*

    1.2. *medlog.dwg*

    1.3. *shortlog.dwg*

    1.4. *longhalflog.dwg*

    1.5. *medhalflog.dwg*

    1.6. *roof.dwg*

    1.7. *gable.dwg*

    1.8. *HS Logo.bmp*

2. The bookshelves are 7'-6" square x 11½" deep, with a ½" plywood back inserted into rabbets on the sides. Use your own spacing for the internal shelves.

3. Search the web for additional projects. Try to draw some of them. I used keywords like "3D render" and "solid model" and found these:

    3.1. http://www.blackline.com/drafting/drawings/buildings.htm [This is a good site to see some dwf files.]

    3.2. http://www.pleione.com/pithouse/ [This site has some cool renderings.]

    3.3. http://www.powerup.com.au/~bkhcg/ [More really good rendering.]

    3.4. http://www.gsmmedia.com/cad/renderings.html [Fair renderings.]

    3.5. http://www.imperial-design.com/cad.html [This is an excellent dwf site for experimenting with different views in a dwf file. Try to draw the bicycle assembly in dwg2.]

    3.6. http://www.archvisual.co.nz/~vismedia/portfilio.htm

    3.7. http://ww3.foxinternet.net/kkcassady/KGCRes/Academic/ME491/slide1.html [This is the excellent surface model of a sailboat I mentioned earlier.]

    3.8. http://www.academix.com/3d_fant.htm [Here is an excellent series of rendered models.]

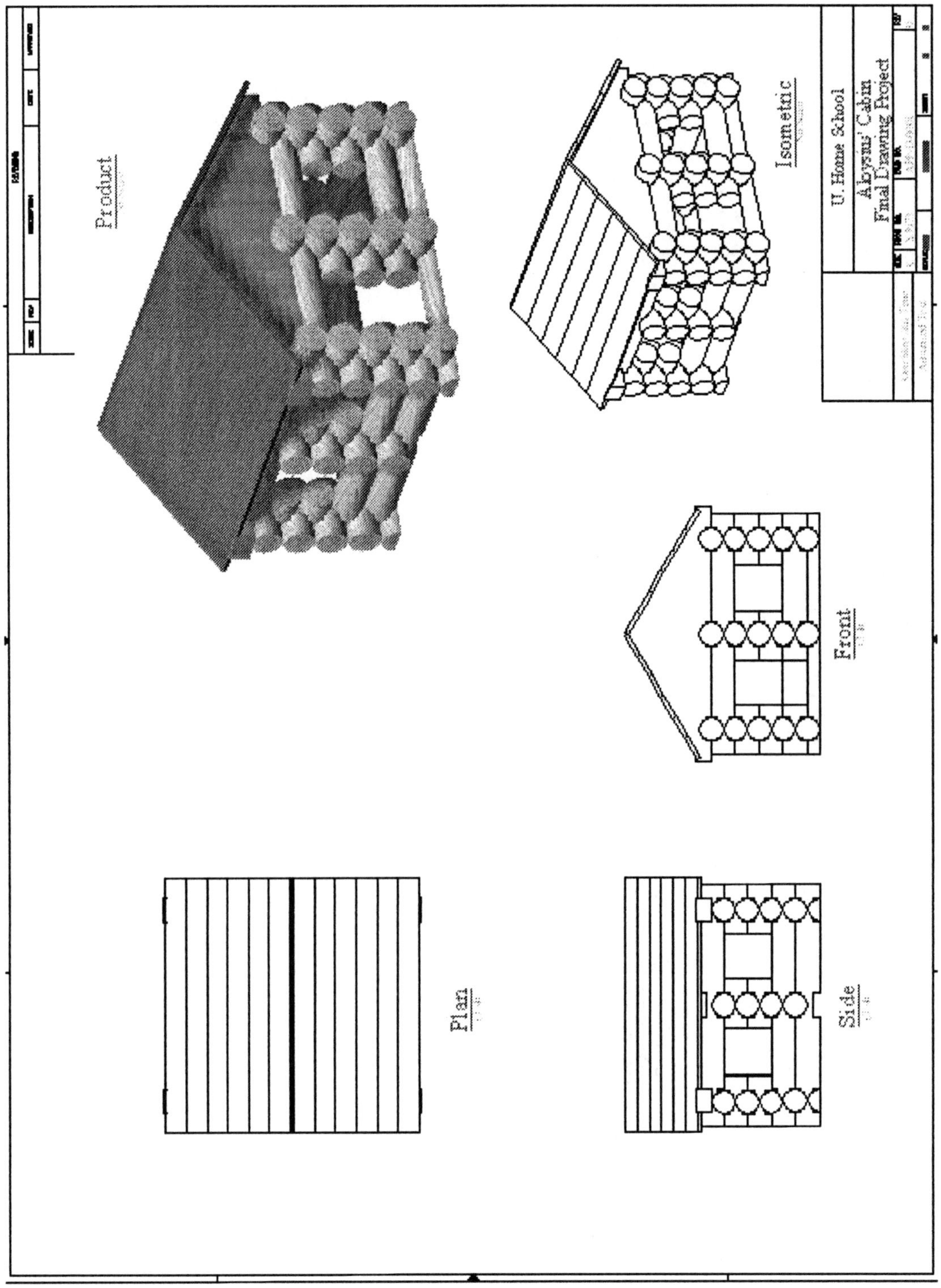
Product
Isometric
Front
Plan
Side
U. Home School
Final Drawing Project

Figure D1

Figure D2

Printed in the United States
58558LVS00005B/39-42